Challenges and Advances in Computational Chemistry and Physics

Volume 35

Series Editor

Jerzy Leszczynski, Department of Chemistry and Biochemistry, Jackson State University, Jackson, MS, USA

This book series provides reviews on the most recent developments in computational chemistry and physics. It covers both the method developments and their applications. Each volume consists of chapters devoted to the one research area. The series highlights the most notable advances in applications of the computational methods. The volumes include nanotechnology, material sciences, molecular biology, structures and bonding in molecular complexes, and atmospheric chemistry. The authors are recruited from among the most prominent researchers in their research areas. As computational chemistry and physics is one of the most rapidly advancing scientific areas such timely overviews are desired by chemists, physicists, molecular biologists and material scientists. The books are intended for graduate students and researchers.

All contributions to edited volumes should undergo standard peer review to ensure high scientific quality, while monographs should be reviewed by at least two experts in the field. Submitted manuscripts will be reviewed and decided by the series editor, Prof. Jerzy Leszczynski.

Supratik Kar · Jerzy Leszczynski
Editors

Current Trends in Computational Modeling for Drug Discovery

Springer

Editors
Supratik Kar
Chemometrics and Molecular Modeling Laboratory
Department of Chemistry
Kean University
Union, NJ, USA

Jerzy Leszczynski
Department of Chemistry, Physics and Atmospheric Science
Jackson State University
Jackson, MS, USA

ISSN 2542-4491 ISSN 2542-4483 (electronic)
Challenges and Advances in Computational Chemistry and Physics
ISBN 978-3-031-33870-0 ISBN 978-3-031-33871-7 (eBook)
https://doi.org/10.1007/978-3-031-33871-7

This Springer imprint is published by the registered company Springer Nature Switzerland AG
The registered company address is: Gewerbestrasse 11, 6330 Cham, Switzerland

For COVID-19 HEROES [Frontline workers, Health Care Professionals, First responders, Researchers and Scientists worked for Vaccines and Small drug molecules]

Preface

Computer-aided drug design (CADD) approaches are one of the rapidly growing research areas to minimize the experimental efforts to support and seed up the drug design and discovery [1, 2]. The discovery time of small drug molecules not only decreased over the years but also the late-stage drug failure is also reduced due to the early prediction of absorption, distribution, metabolism, excretion toxicity (ADMET) profile. Along with the explanation of molecular basis for the expected biological response, CADD offers precise prediction of possible derivatives and structural scaffolds that would improve therapeutic activity [3, 4]. CADD involves a series of computational techniques or in silico approaches which comprise combinatorial chemistry, quantitative structure–active relationships (QSARs), pharmacophore modeling, rigid, flexible, covalent, and quantum polarized docking, molecular dynamics simulation followed by molecular mechanics with generalized Born and surface area solvation (MM/GBSA) and Poisson–Boltzmann surface area (PBSA) to perform virtual screening, lead optimization, de novo design, and so forth. Nowadays, among the major in silico approaches, drug repurposing through computational chemistry is one of the popular one where scientists discover new uses of already approved drugs by regulatory agencies for another disease to provide the quickest possible transition from bench to bedside. CADD through in silico techniques helps in the identification and optimization of new drugs employing leverage of chemical and biological evidence about targets and ligands using computational power. Not only that, QSAR and machine learning (ML) models offer removal of unwanted molecules with undesirable ADMET profile to ease the selection of the most hopeful candidates [1, 2, 5].

Over the years, CADD has made major contributions to speed up the drug discovery process through the amalgamation of in silico approaches with experimental efforts. Indeed, several marketed drugs such as indinavir, captopril, dorzolamide, ritonavir, oseltamivir, boceprevir, nolatrexed, tirofiban, imatinib, zanamivir, and nelfinavir have been identified or optimized with the aid of molecular modeling techniques [6]. We are extremely hopeful that the number will be increased to manifold, and without any doubt, we can say that we are living in the era of CADD and artificial intelligence (AI)-based drug design and discovery!

The book includes ten chapters encompassing current and advanced computational modeling techniques and their real-world application for drug design and discovery for different diseases covering different therapeutic classes.

Chapter 1 by Chakraborti and S talks about multiple components of structure-based drug discovery (SBDD), its workflow, and associated challenges. Authors also provided the possible limitations and how these limitations can be overcome which is extremely important in drug design and discovery.

Chapter 2 prepared by Khatun et al. deals with the structural biology of class IIb histone deacetylases (HDACs) and talks about how in silico techniques including the virtual screening approaches have been implemented to design HDAC6 and HDAC10 inhibitors. Furthermore, the interactions of class IIb HDACs with their inhibitors are also emphasized comprehensively to offer a detail insight. This chapter presents knowledge for designing newer class IIb HDAC inhibitors in future.

Chapter 3 by Yadav et al. offers important findings involving computational modeling of small compounds as multitarget-directed ligands (MTDLs) with potential anti-AD activity which could afford vital leads for discovering new molecules as novel AD therapeutics to be used for the management of Alzheimer's disease (AD).

Chapter 4 by Purohit et al. emphasizes the fundamentals of computer modeling and discusses the relationship between in silico experiments and viral infections followed by role of computational model in the development of antiviral agents.

Chapter 5 by Gautam and Kumar recapitulates the experimentally tested antivirals as well as the in silico approaches to identify inhibitors for Nipah virus which will be helpful for the researchers in antiviral drug discovery against NiV.

Chapter 6 by Gomatam et al. illustrates an overview of the various computational strategies that have been reported in the discovery of drugs for HIV. A comprehensive overview of several structure-based and ligand-based computational methods is presented followed by some notable applications of these methods in the discovery of novel anti-HIV compounds. Authors also discussed the emergence of powerful machine learning algorithms which have proved to be useful in the design of new lead molecules and in the development of theoretical models that can predict resistance to antiretroviral therapy.

Chapter 7 prepared by Chatterjee et al. discusses severe fever with thrombocytopenia syndrome virus (SFTSV) disease and its causative agent, epidemiology, pathogenesis, diagnosis, and recent development in the treatment in form of identification of potential lead using computational modeling.

Chapter 8 by Benfenati et al. thoroughly discusses computational toxicological aspects in drug design and discovery, screening adverse effects along with existing in silico tools or future perspectives.

Chapter 9 by Banerjee and Roy demonstrates the read-across and RASAR tools and different quality and evaluation metrics associated with this research developed in the Drug Theoretics and Cheminformatics (DTC) Laboratory and their applications in prediction of different activity/toxicity endpoints.

Chapter 10 by Kar and Leszczynski summarizes major drug databases covering drug molecules, chemicals, therapeutic targets, metabolomics, and peptides which are major resources for drug discovery employing drug repurposing, high throughput, and virtual screening.

The editors convey their gratefulness to all the authors for their knowledge informative contributions. Furthermore, we thank the reviewers for their time, expertise, and fruitful comments to improve the book's quality. We firmly believe that this edited book will be helpful to all the early days researcher as well as seasoned ones in the field of CADD irrespective of discipline the budding researcher and experts in this specific field.

Union, NJ, USA — Supratik Kar
Jackson, MS, USA — Jerzy Leszczynski

References

1. Roy K, Kar S, Das RN (2015) Understanding the basics of QSAR for applications in pharmaceutical sciences and risk assessment. Academic press
2. Roy K, Kar S, Das RN (2015) A primer on QSAR/QSPR modeling: fundamental concepts. Springer
3. Kar S, Leszczynski L (2020) Open access in silico tools to predict the ADMET profiling of drug candidates. Expert Opin Drug Discov 15:1473–1487
4. Kar S, Roy K, Leszczynski L (2020) In silico tools and software to predict ADMET of new drug candidates. In: Silico methods for predicting drug toxicity, Benfenati E (ed). Humana, New York, NY, pp 85–115
5. Kar S, Sanderson H, Roy K, Benfenati E, Leszczynski J (2022) Green chemistry in the synthesis of pharmaceuticals. Chem Rev 122:3637–3710
6. Baig H, Ahmad K, Roy S, et al (2016) Computer aided drug design: success and limitations. Curr Pharm Des 22:572–581

Contents

Contributors

Sk. Abdul Amin Natural Science Laboratory, Division of Medicinal and Pharmaceutical Chemistry, Department of Pharmaceutical Technology, Jadavpur University, Kolkata, India

Arkaprava Banerjee Department of Pharmaceutical Technology, Drug Theoretics and Cheminformatics (DTC) Laboratory, Jadavpur University, Kolkata, India

Rahul Barot Faculty of Pharmacy, The Maharaja Sayajirao University of Baroda, Vadodara, Gujarat, India

Emilio Benfenati Istituto di Ricerche Farmacologiche Mario Negri IRCCS, Milano, Italy

Pobitra Borah School of Pharmacy, Graphic Era Hill University, Dehradun, Uttarakhand, India

Sohini Chakraborti Centre for Targeted Protein Degradation, Division of Biological Chemistry and Drug Discovery, School of Life Sciences, University of Dundee, Dundee, UK

Shilpa Chatterjee Department of Biomedical Science, College of Medicine, Chosun University, Gwangju, Republic of Korea

Monica Chauhan Faculty of Pharmacy, The Maharaja Sayajirao University of Baroda, Vadodara, Gujarat, India

Evans Coutinho Department of Pharmaceutical Chemistry, Bombay College of Pharmacy, Mumbai, India

Pran Kishore Deb Department of Pharmaceutical Sciences, Faculty of Pharmacy, Philadelphia University, Amman, Jordan

Merwyn D'costa Department of Pharmaceutical Chemistry, Bombay College of Pharmacy, Mumbai, India

Sakshi Gautam Virology Unit and Bioinformatics Centre, Institute of Microbial Technology, Council of Scientific and Industrial Research (CSIR), Chandigarh, India; Academy of Scientific and Innovative Research (AcSIR), Ghaziabad, India

Shovanlal Gayen Laboratory of Drug Design and Discovery, Department of Pharmaceutical Technology, Jadavpur University, Kolkata, India

Anish Gomatam Department of Pharmaceutical Chemistry, Bombay College of Pharmacy, Mumbai, India

Sangeeta Hazarika School of Pharmacy, Graphic Era Hill University, Dehradun, Uttarakhand, India;
Department of Pharmaceutical Engineering and Technology, Indian Institute of Technology (Banaras Hindu University), Varanasi, Uttar Pradesh, India

Tarun Jha Natural Science Laboratory, Division of Medicinal and Pharmaceutical Chemistry, Department of Pharmaceutical Technology, Jadavpur University, Kolkata, India

Gaurav Joshi School of Pharmacy, Graphic Era Hill University, Dehradun, Uttarakhand, India

Karan Joshi Faculty of Pharmacy, The Maharaja Sayajirao University of Baroda, Vadodara, Gujarat, India

Supratik Kar Department of Chemistry, Chemometrics and Molecular Modeling Laboratory, Kean University, Union, NJ, USA

Afreen Khan Department of Pharmaceutical Chemistry, Bombay College of Pharmacy, Mumbai, India

Samima Khatun Laboratory of Drug Design and Discovery, Department of Pharmaceutical Technology, Jadavpur University, Kolkata, India

Manoj Kumar Virology Unit and Bioinformatics Centre, Institute of Microbial Technology, Council of Scientific and Industrial Research (CSIR), Chandigarh, India; Academy of Scientific and Innovative Research (AcSIR), Ghaziabad, India

Jerzy Leszczynski Department of Chemistry, Physics and Atmospheric Sciences, Interdisciplinary Center for Nanotoxicity, Jackson State University, Jackson, MS, USA

Anna Lombardo Istituto di Ricerche Farmacologiche Mario Negri IRCCS, Milano, Italy

Davide Luciani Istituto di Ricerche Farmacologiche Mario Negri IRCCS, Milano, Italy

Arindam Maity Department of Pharmaceutical Technology, JIS University, Kolkata, India

Prashant R. Murumkar Faculty of Pharmacy, The Maharaja Sayajirao University of Baroda, Vadodara, Gujarat, India

Priyank Purohit School of Pharmacy, Graphic Era Hill University, Dehradun, Uttarakhand, India

Kavita Raikuvar Department of Pharmaceutical Chemistry, Bombay College of Pharmacy, Mumbai, India

Kunal Roy Department of Pharmaceutical Technology, Drug Theoretics and Cheminformatics (DTC) Laboratory, Jadavpur University, Kolkata, India

S. Sachchidanand Department of Bioinformatics, Zydus Research Centre, Ahmedabad, India

Gianluca Selvestrel Istituto di Ricerche Farmacologiche Mario Negri IRCCS, Milano, Italy

Debanjan Sen Department of Pharmaceutical Technology, BCDA College of Pharmacy & Technology, Hridaypur, Kolkata, India

Mange Ram Yadav Centre of Research for Development, Parul University, Vadodara, Gujarat, India

Rasana Yadav Faculty of Pharmacy, The Maharaja Sayajirao University of Baroda, Vadodara, Gujarat, India

Chapter 1
SBDD and Its Challenges

Sohini Chakraborti and S. Sachchidanand

Abstract Proteins are the important biological macromolecules that are targeted by most of the existing drugs. SBDD play a critical role in design of drug-like, novel, potent, and safe modulators. It is a joint effort from structural biologists and computational scientists, which considers various limitations of the techniques and suitably guides drug designers. Identifying a novel, potent, and safe drug-like molecule is a long challenging path, and throughout this discovery journey, SBDD provides crucial guiding light at different stages. SBDD involves the use of structural data of target proteins to identify suitable ligand candidates that might bind the protein of interest and modulate its functions, resulting in therapeutic benefit. In this chapter, we provide an overview of computational SBDD workflow, and the various challenges associated with it. We also discuss strategies that could be adopted to tackle the challenges by making the best use of available information.

Keywords Structure-based drug discovery (SBDD) · Structure selection · Ligand screening · Binding affinity prediction · Protein flexibility

1.1 Introduction

The biological functions carried out by a living cell are governed by complex molecular recognition defined mostly by various non-covalent interactions among biological macromolecules and small molecule—macromolecule complexes that are

---

Dedicated to the memory of Late Professor N. Srinivasan

S. Chakraborti (✉)
Centre for Targeted Protein Degradation, Division of Biological Chemistry and Drug Discovery, School of Life Sciences, University of Dundee, 1 James Lindsay Place, Dundee DD1 5JJ, UK
e-mail: schakraborti001@dundee.ac.uk

S. Sachchidanand
Department of Bioinformatics, Zydus Research Centre, Ahmedabad 380058, India

S. Kar and J. Leszczynski (eds.), *Current Trends in Computational Modeling for Drug Discovery*, Challenges and Advances in Computational Chemistry and Physics 35,
https://doi.org/10.1007/978-3-031-33871-7_1

present in a cellular milieu [1]. Though the origin of molecular recognition is microscopic in nature, its consequences are macroscopic. The fundamental basis for molecular recognition is inherent in the potential energy surface represented by the interaction energy of two or more molecules as a function of their mutual separation and orientation. The feasibility and strength of any molecular recognition event are dictated by the extent to which the three-dimensional (3D) structures of the interacting partners complement each other in their shapes and electrostatic features, given the coherence in their spatiotemporal existence [2]. It is important to emphasize here that molecular recognition in aqueous biological system is complex with respect to its predictabilities towards binding while mystery like the role of water in binding still to be resolved. Proteins, carbohydrates, and nucleic acids are the important biological macromolecules that maintain the life processes by interacting with diverse binding partners called ligands. This chapter would focus on the interactions between proteins and small molecules (drug-like ligands).

Proteins play a versatile role in maintenance of life and their 3D structures influence their interactions and functions. Factors influencing changes in the 3D structure of proteins could eventually alter cellular functions due to changes in their interaction profiles with their interactomes, which ultimately might lead to disease phenotypes. Therefore, understanding the 3D structure of proteins (static as well as dynamic) is of great importance to investigate the cause of the disease at a molecular level [3]. Such understanding guides the rational design of therapeutic agents (drugs) that can be targeted against the protein of interest to modulate its function and obtain the desired pharmacological response—the approach being termed as Structure-Based Drug Design (SBDD). SBDD approaches employ a collection of computational techniques which provide insights from the 3D static structure of a protein and its complexes, study their dynamic behaviour at atomic level to guide the design of modulators targeted against the protein of interest. While SBDD approaches are routinely used in any drug discovery program, the ability to interpret and rationally use the multiple layers of information that these approaches provide would ultimately lead to designing of better modulators. SBDD played an important role in the approval of the HIV-1 protease inhibitors in the 1990s [4], and gave major boost to the approach. Since then, structure-based approaches contributed to the approval of several new drugs in different therapeutic areas [5].

The structural information of the target protein helps in optimizing its interactions with the potential drug candidates and thus guides in improving its potency [6, 7] (Fig. 1.1). Not only the structure of the target protein but also the 3D structure of related proteins (homologs) also play an important role in SBDD by guiding to achieve specificity in interactions [8] (Fig. 1.1), for example, design of JAKs inhibitors [9, 10]. Such specificity in interactions contributes to designing of a safer drug candidate which is a crucial objective in drug discovery. Though serendipity and high throughput screening (HTS) play an important role in drug discovery and design, SBDD approach despite its limitations provides a rational foundation towards discovery of new drugs. Structure-based virtual screening has evolved with the power of improved and sophisticated computational resources to explore larger chemical space and provide initial hits that are validated by screening of much smaller set of

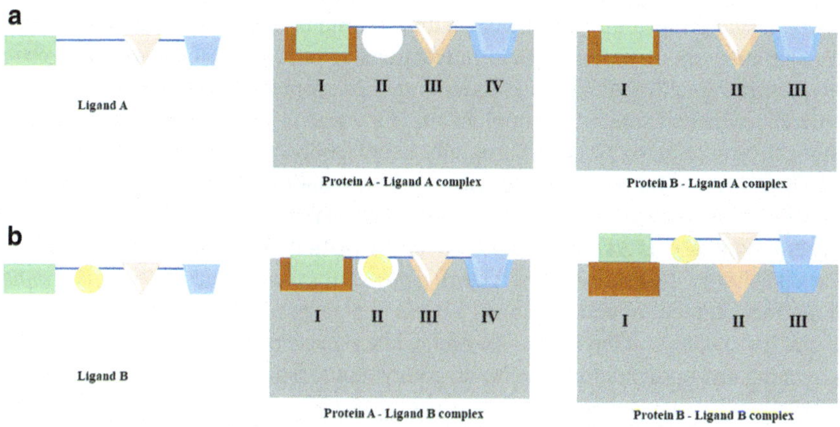

Fig. 1.1 **a** The Ligand A fits well in Protein A (target) and Protein B (off target) due to similar binding sites. **b** The modified Ligand B fits well in Protein A but not in Protein B. The modification of the ligand helped it to fit better in sub-pocket II of Protein A that is absent in Protein B

test molecules through wet lab experiments, therefore, resulting in better hit rates than HTS [11–13].

In this chapter, we present an overview of the general workflow of computational SBDD pipeline, and the common challenges associated with it. Learning from our experiences, we have further discussed various strategies that could be adopted to tackle such challenges and making the best use of available information.

1.2 Overview on Structure-Based Drug Design (SBDD)

SBDD is the outcome of collaboration between structural biologists and computational chemists, who integrate their respective techniques to derive understanding from structural models of protein–ligand complexes. Together with medicinal chemists and other drug discovery scientists, the understanding translates to designing novel and potent ligands against the therapeutic target protein of interest. The main objective of the SBDD approach is to understand the binding pose (conformation and orientation) of a ligand molecule in the protein binding site (an active or allosteric site). The bioactive conformation of the ligand in the protein binding pocket (determined either by experimental techniques or predicted by computational techniques) provides molecular insights to design and optimize potent New Chemical Entities (NCEs). The insights derived from the structural data primarily involve probing the shape and the electronic complementarity of protein–ligand complexes. The complementary stereoelectronic features of protein and ligand are governed by favourable thermodynamic parameters through various types of intra- and intermolecular interactions, such as hydrogen bonding, π-π stacking, hydrophobic contacts [2, 14]. Therefore, binding of a ligand to a protein is the result of energetic gain from the establishment of multiple interactions. SBDD strategies allow the optimization of protein–ligand interactions to improve potency and selectivity by preserving and optimizing selected drug-like properties. Validating predictions with experimental studies and implementing the learning to improve computational models are crucial to any SBDD program. Using feedback from various experiments, SBDD is also useful in designing multi-parameter optimized molecules that helps in understanding the structure activity relationships (SAR). This understanding can aid in resolving various ADMET (absorption, distribution, metablolism, excretion, and toxicity) issues. Optimization of ADMET properties using structural data is another important segment of computational SBDD as elaborated elsewhere [8, 15].

In this chapter, we have focussed on the key aspects of SBDD that involve ligand screening and optimization. The basic requirements to initiate a SBDD program are: (i) availability of a 'suitable' protein structure, (ii) 3D structure of ligand molecules, and (iii) a tool to predict the ligand binding pose in the protein pocket of interest. Each of these essential requirements for SBDD is discussed below.

1.2.1 Protein Structure

A SBDD approach would require a 3D structure of the apo-/holo-protein that can be obtained either by experimental or computational means. The availability of 'good quality' (as discussed in Sect. 1.3.1) structural data of the target protein, preferably bound to a known binder (e.g. endogenous ligand, cofactor, known inhibitor), is important for SBDD. Such information helps to decipher the location, shape, and composition of the binding sites (specific regions) on a protein structure that could mediate its interactions with potential drug molecules and modulate the desired biological functions. The size, volume, degree of preorganization (conformational flexibility/rigidity), polarity, and polarizability analysis of the binding sites are very useful. The features of the binding site of interest are therefore the guiding cues to identify agents that engages in interactions with critical residues of the target protein and are a good fit in terms of shape and electrostatic features [15]. The residues critical for obtaining desired biological response could be identified from various sources such as mutational data [16], pre-existing SAR of modulators [17], or could be anticipated from bioinformatic analyses of protein sequence conservation [18].

The publicly available structural repository, the Protein Data Bank (PDB) [19], holds more than 190,900 (as of September 2022, https://www.rcsb.org/stats/summary) structures of biological macromolecules that contain at least one protein molecule. In this section, we briefly discuss various techniques for obtaining 3D structural data of proteins that are essential pre-requisites for a SBDD pipeline. Readers are encouraged to refer to the cited literatures for details of each technique that is beyond the scope of discussion in this chapter. Three experimental techniques primarily contribute to providing atomic coordinates of protein and protein–ligand complexes. These are macromolecular X-ray crystallography (MX) [20], nuclear magnetic resonance (NMR) [21], and cryogenic electron microscopy (Cryo-EM) [22] with each having its own advantages and limitations. Computationally, the structure can either be generated by homology modelling/comparative modelling [23] (if possible) or can be extracted from the AlphaFold database [24] (if a 'suitable' structure is available). The quality of structural information derived from protein models predicted by classical threading-based methods (TBMs) [25] is not suitable for SBDD. However, an improved TBM combined with structural approaches, FINDSITE^{comb}, has been demonstrated to identify potential binders of a target protein, and its performance to a great extent is shown to be insensitive to structural quality [26]. In the absence of any suitable structure of the target protein, the FINDSITE^{comb} technique could be useful during early drug discovery stage for identification of potential in silico hits from large chemical space. Advanced ligand design and lead optimization stages demand high-quality of structural data, and hence, the current TBMs are unlikely to provide reliable information.

1.2.1.1 Experimental Protein Models

MX involves crystallization of the protein molecules that leads to locking the protein in a single conformation within a crystal lattice. Under physiological conditions, proteins '*jiggle*' and '*wiggle*' to perform their functions [27]. Therefore, in its native state, a protein exists as an ensemble of structural conformations. It could be possible that the conformation of the protein or its specific regions as trapped during crystallization is not relevant to native biological conditions. This could be, especially, for the regions of the protein that have higher flexibility or when the crystallization conditions are far away from the physiological environment of the target protein [28, 29]. Further, determining high-resolution protein structures bound to small molecules, that is a desirable pre-requisite for SBDD, is a challenging task and errors in structural data are not uncommon [30, 31]. Proteins that lack stable secondary structures due to higher flexibility are disordered and are generally not amenable to MX—a problem that can be overcome with NMR as the latter allows capturing the structural information of a protein in solution state (wherever size of the protein is not a limitation to NMR). NMR spectroscopy is a very useful tool that not only provides the structural information of the protein in solution phase but can also help in understanding the dynamics of a wide range of biological macromolecules and hence can provide better functional insights than MX. However, NMR is currently restricted to small and medium-sized proteins [32]. The technique has limitations with respect to the speed, and size of molecules that can be tackled when compared to X-ray diffraction methods (for well-diffracting crystals). Wherever it is possible, NMR spectroscopy facilitate atomic resolution studies of sparsely populated, transiently formed biomolecular conformations that exchange with the native state [33]. The dynamics of the macromolecules and their complexes can also be studied when only X-ray structure is available, using computational techniques like molecular dynamics simulations (as discussed in Sect. 1.3.5). Cryo-EM has undergone a '*resolution revolution*' in the recent years and is rapidly emerging as an important tool for SBDD [34]. This technique has the potential to capture structures of large macromolecular assemblies in near-native conformations. Currently, the application of Cryo-EM is mostly restricted to large proteins/protein assemblies with less success for smaller ones. By achieving atomic resolution using Cryo-EM, the structural assembly of proteins and their complexes can be understood that cannot be easily examined by X-ray crystallography [35].

1.2.1.2 Computational Protein Models

In the absence of an experimental protein structure, computational techniques that help to predict 3D structures of proteins, such as homology modelling, can guide SBDD approaches [36]. Such modelling techniques use the known structural information (template) of related proteins (homologs) to predict 3D structures of the target protein. The artificial intelligence (AI)-based methods like AlphaFold [37] and RosTTAFold [38] have shown remarkable success in the recent times. Albeit these methods do not require structural information of related proteins for predicting structures of target proteins, but the prediction algorithms are trained on the existing structural data available in the PDB. Hence, it is likely that success of these methods would be higher for protein classes and their conformations that are well represented in the PDB [39]. Studies have demonstrated that refinement of computational protein models by molecular dynamic simulations to generate suitable conformational states improves the accuracy of the models and hence their predictability efficiencies [40, 41].

Once the protein structure is in place, the second input for SBDD studies is the availability of drug-like libraries and collection of potential analogues/design ideas for screening. SBDD can help in screening of the physical ligand libraries based on experimental 3D coordinates of protein–ligand complex [42, 43], and/or it can facilitate virtual screening of in silico library of compounds to identify hits [44]. The former is time taking as it involves experimental methods, whereas the latter is comparatively faster and require lesser investment of resources. The validated hits identified from screening would then move through various phases of SBDD to obtain lead compounds that are then optimized.

1.2.2 Ligands

Besides protein structure, the 3D structures of ligand molecules are the other important inputs for SBDD. The ligand library intended for screening could be a small number of analogous compounds or a large set of drug-like molecules that preferably follow rule of five (Ro5) [45] and passed through pan-assay interference compounds (PAINS) [46] and rapid elimination of swill (REOS) [47] structural filters. Screening large and diverse set of molecules against the target ensures maximum coverage of the chemical space and helps to identify multiple starting points with diverse scaffolds for hit to lead generation stage. The objective of the SBDD program influences the design of the input ligand library. The main objectives that guide library design are: (i) hit generation, (ii) fragment (MW < 300 Da) identification for fragment-based drug design (FBDD), (iii) hit to lead generation, and (iv) lead optimization. The discovery libraries are designed to address the first objective, i.e. hit generation. Fragment library is designed to identify fragments to be linked using FBDD approach to design and identify hits. Fragment-based design requires screening fragments (MW < 300 Da), which are not intrinsically drug-like, but become fragments of drug-like compounds upon combining [48]. The next goal of library design is to identify a lead against a target. Such libraries are known as focussed library, which are built around certain structural motif (known to be active against the target of interest) or against identified pharmacophoric features known to be important for binding [49].

1.2.2.1 Library Design

A library is a collection of already synthesized/synthesizable compounds or fragments that could be screened against the therapeutic target. The chemical space and its diversity accounted for within the library are inversely proportional to the amount of information available for the target's binding site. Selection or design of library is dependent on its intended use. For example, if library is required for limited target classes, a focussed library would serve the purpose. The clustering density (degree of structural similarity of library members) must be high where repetitive screen for similar targets is desired while for diverse targets, a library with lower density would ensure maximum degree of diversity in their collection. Chemoinformatic tools play significant role in designing compound library for an identified therapeutic target [50]. Figure 1.2 summarizes several stages involved in library design.

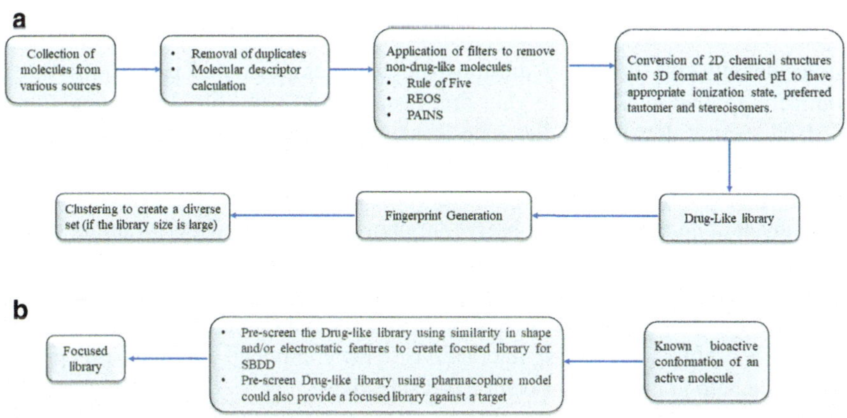

Fig. 1.2 Examples of the steps involved in designing compound library for virtual screening, **a** by building a diverse set drug-like library, **b** additional steps involved in building a focussed drug-like library

1.2.2.2 Ligand Screening/Optimization

Upon obtaining satisfactory quality of 3D structural information of target protein and drug-like molecules, the next step in SBDD is in silico screening/optimization. Screening of physical stock of millions of compounds is not only practically challenging but also poorly rewarding. To accelerate this process, computational methods play a crucial role in predicting potential binders and their binding modes. Predicted potential binders are then evaluated using various biochemical/biophysical/cell-based assay methods to identify and rank order the validated hits. A good binder(s) is/are selected for 3D structure determination in complex with the target protein which helps in validating the binding pose of the compound/s obtained by computational studies and understanding SAR for further optimization of the initial binders.

In drug discovery, multiple cycles of optimization of a lead are carried out (without significant compromise on affinity towards the target) before declaring a candidate. For example, the designing of ligands can be done to complement the binding site features (viz., shape and electrostatics) of target and optimization would further fine tune intermolecular interactions and steric complementarity with the binding site for achieving optimum thermodynamic parameters. Every cycle of optimization, be it improving metabolic stability [51] or ADME properties [52] or Cyp liabilities [53], requires a new set of SAR, involving new designs of molecules around the lead.

It is important to emphasize here that there is a rule of 'no rule' to design drugs. Various tips and tricks from past examples and experiences might sometime work rationally and sometime serendipitously. Therefore, reinforcing methods with other techniques and integrating knowledge from various reliable sources are advantageous in any drug discovery program.

1.2.3 Molecular Docking Simulations

Molecular docking is one of the most popular computational techniques that is routinely applied in SBDD to gain first insights into plausible design hypotheses [54]. Molecular docking studies help in identifying compounds against the target protein through predicted docking pose at the binding site of interest. The quality of pose is assessed using docking score. Molecular docking analysis also provides enrichment of huge compound library (enrichment from docking is its ability to rank large proportion of the active compounds at the top of the proposed list for experimental evaluation) for screening only a few hundreds of compounds by a process called structure-based virtual screening. Suggested binding mode of ligand through docking is useful in optimizing interactions of the compound with the target protein and hence help in improving potency and selectivity.

Any docking method would require the 3D structural information of protein target and ligands. As discussed later in Sect. 1.3, the docking method has its own limitations, and success of this technique depends on several factors, for example, quality of protein and ligand structures, identifying correct ionization profile of binding site residues, nature of binding site (rigid vs. flexible), selection of force field and scoring functions. Over the past few years, numerous artificial intelligence/machine learning (AI/ML)-based techniques have also been reported that predict compound binding affinity and binding modes with greater accuracies than existing methods [55–58]. However, one of the major challenges of AI/ML-based techniques is its dependence on availability of large datasets that are required to train most of these algorithms [5, 59]. Therefore, in unique and novel cases with limited data, AI/ML methods are unlikely to make meaningful predictions.

1.3 Crucial Components and Challenges in Computational SBDD

The success of any computational SBDD program is largely dependent on the extent to which the drug-target interactions (as it happens under physiological conditions) could be mimicked in silico. Realistic representation of the biophysical events within the computational algorithm is likely to result in accurate predictions. Unfortunately, the improvement in accuracy of predictions comes with the compromise in speed of the calculations. To strike a balance between speed and accuracy, it is necessary to incorporate approximations in the algorithms. It is important that depending on the question to be addressed and availability of resources, one should adopt the appropriate strategy at each stage in the computational drug discovery pipeline. In the following paragraphs, we discuss about the crucial components in computational SBDD workflows, associated challenges, and ways to handle such challenges (Fig. 1.3).

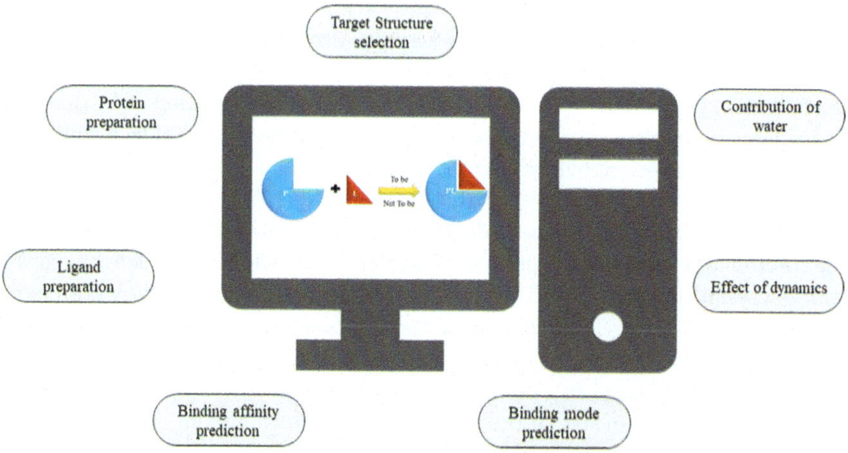

Fig. 1.3 Various components in computational SBDD. *P* Protein, *L* Ligand, *PL* Protein–Ligand complex

1.3.1 Target Structure Selection

The foremost step in computational SBDD is to select the appropriate structure/s of the target protein that will be subsequently used for compound screening and optimization. The success of ligand screening/optimization is greatly dependent on the quality of the input protein structure. High-quality crystal structures of protein–ligand complex provide the platform to generate a sound design hypothesis. There are many literatures discussing various components of SBDD that we have highlighted in Sects. 1.3.2–1.3.5. However, to the best of our knowledge, there are hardly any published literature that provides a comprehensive guide to aid decision-making for selecting the suitable starting structure/s. This is mostly because structure selection for computational SBDD is dependent on several factors that are beyond generalization. Here, we have aimed to discuss various factors that could help the beginners in the field to form an idea about the rational thought process that generally influences input structure selection. In an ideal situation, a SBDD project demands the prior availability of multiple high-quality structures of the apo- and holo-protein of interest. However, it is difficult to satisfy all desired criteria for structure selection and rational judgement needs to be applied to select the best suitable inputs. Table 1.1 presents a few hypothetical scenarios that may aid decision-making to choose a suitable structure in certain practical scenarios. We emphasize that these are only guidelines and case-specific decisions could be influenced by several practical limitations.

1.3.1.1 Quality of Structure

While 'resolution' of X-ray structure is one of the common indicators of its quality, it is not always directly related to the accuracy of data. Resolution is a measure of the quantity of data collected and not its quality. While experienced researchers might be aware, it might not be obvious to novice users that high-resolution structural data need not always guarantee the reliability of local structural data such as protein–ligand binding sites [60]. It is recommended to use the combination of R_{free} and the diffraction component precision index (DPI) of a structure instead of its resolution to get a better impression about the overall model quality and hence the reliability of the atomic positions within that model [61, 62]. There are several reports that emphasize errors in high-resolution structural data is not uncommon, and it is advised to verify the co-ordinates against the experimental evidence, such as electron density maps in case of crystal structures [31, 63]. Quality parameters such as real space correlation coefficient (RSCC) [64] and electron density scores for individual atoms (EDIA) [65] quantify the electron density fit of the structural entity and are useful indicators of quality of coordinates in local regions of a structure. RSCC $\geq$ 0.9 and EDIA $\geq$ 0.8 indicate good fit of the co-ordinates with experimental data. The RCSB PDB has recently introduced the ligand quality slider feature and included assessment on experimental data fitting and geometry of bound ligands of interest (https://www.rcsb.org/docs/general-help/ligand-structure-quality-in-pdb-structures). This is one of the

Table 1.1 Guide for target structure selection for virtual screening using SBDD approach to identify potential hit compounds against target Protein X that has more than one structure available as a potential starting point. Note that mutation of an amino acid residue in Protein X causes Disease Y and this mutation locks Protein X in an inactive conformation

Parameters	Structure 1	Structure 2	Remark
Quality: (a) Resolution (b) Ligand RSCC	1.8 Å 0.45	2.5 Å 0.98	Though Structure 1 has better resolution, the electron density fit of the ligand as indicated by RSCC is poor (See Section 1.3.1.1). So, Structure 2 should be preferred. However, visual inspection of the ligand pose in the crystal structure against the electron density map is recommended
Sequence	Wild type	Disease relevant mutation	Structure 2 should be preferred as it has the relevant mutation in its sequence
Ligand bound/ unbound	Bound (activator)	Bound (inhibitor)	As the study would aim to design an inhibitor, so an inhibitor bound structure, i.e. Structure 2 should be preferred
Conformational state	Active	Inactive	The aim of the project is to target the inactive state of the protein of the interest. Hence, Structure 2 should be preferred
Crystallization condition pH	3.5	6.8	Selection of the target protein structure should be based on the pH of its environment under physiological condition

easiest ways to quickly verify the quality of the ligand bound to protein of interest that is deposited in the PDB. In the absence of target protein structure with satisfactory quality in the PDB, the PDB-REDO [66] database could be searched to check for the availability of a better quality model.

1.3.1.2 Sequence Information

Examining the amino acid sequence of the target protein is another important component of structure selection. Depending upon whether targeting the wild type or a mutated protein (such as in many cancers [67]) is of interest, care should be taken to choose the appropriate starting structures. A single change of amino acid can appreciably alter the structure of a protein and hence affect its binding with its partners (especially when such change is in the binding site). If a suitable structure of the target protein is not available, the common practice is to use the structure of a close homolog. While choosing a structural homolog, it should be ensured that the binding site features are largely conserved between the target protein and the chosen homolog so that there are minimal chances of interference with prediction outcomes. Computational tools like ProBis [68] are helpful to compare the structural similarity of protein–ligand binding sites.

1.3.1.3 Apo/Holo Conformation

The ligand bound state (holo) of the protein is convenient to use for SBDD when the screened ligands are intended to bind to the known site. The apo state structure of a protein does not provide the information on ligand binding site, unless otherwise any other experimental evidence exists. Computational tools such as Fpocket [69], SiteMap [70] are useful to predict potential druggable sites (regions amenable to functional modulation upon binding of drug molecules) on a protein structure. The ligand bound (holo) or unbound (apo) state of the protein structure may influence the screening outcomes. It has been shown earlier that holo state structure gives better enrichment compared to apo state structures as the protein binding site in the former is already preformed to accommodate similar ligands [71]. It is known that ligands with different size and/or belonging to different chemical classes may trigger varying conformational changes in the protein binding site—a phenomenon termed as induced fit [72]. If multiple experimental structures of the target protein bound to different chemical classes of ligands show significant conformational changes in the binding site, it is worth considering an ensemble of structures that are representatives of each chemical class of ligands. Such an approach would avoid bias and minimize the chances of missing promising hits that might prefer one conformation over the other. Also, a ligand that is predicted to bind to majority of the conformers in the structural ensemble would have higher likelihood of binding to the target protein. However, sometimes the screening program might be intended to identify new ligands that have pharmacophores similar to a particular known binder [73]. In such scenarios, it is justified to use only the protein structure that is bound to a ligand with desired pharmacophoric features rather than using an ensemble approach. In the absence of appropriate holo state structures, induced fit docking [74] and molecular dynamics approaches [75], as discussed later, could be helpful to obtain suitable conformation of the target protein that can then be used as starting structure/s.

1.3.1.4 Effect of Neighbouring Residues

It could be possible that the ligand binding site of interest lies at or proximal to protein–protein interfaces. Under biological conditions, these interfaces could be formed by homomers or heteromers (e.g. in a multi-protein complex). In a slightly different scenario, one may encounter a situation where the structure of only a single domain from a multi-domain protein target of interest is available. If the ligand binding site is close to domain–domain interface, it could be possible that binding of a ligand to one domain is affected by the contribution of residues from a neighbouring domain. Such scenarios require careful assessment and if possible, multiple chains/

domains that form the ligand binding site should be preferred to account for the contribution from all binding site residues during the ligand binding event.

1.3.1.5 Flexibility Signatures

Dealing with flexibility signatures of ligand binding site is crucial to structure selection in computational studies. Higher flexibility of a protein residue could be manifested as multiple conformations of its side chain, higher B-factors, and missing regions in the electron density maps of crystal structure [27]. In case of multiple conformations, the common practice is to consider the one that has the highest occupancy value. In our experience, it is worthy to consider each of the multiple conformations of the side chains as separate input structural models for ligand screening/optimization in order to be closer to realistic conditions. This also enhances the chances of identifying promising hits that prefer to bind to one of the conformers more strongly than other as mentioned earlier in Sect. 1.3.1.3. A structure that has missing residues in its ligand binding site could not be used for the prediction of ligand binding mode and its binding affinity towards the target using SBDD approaches. If no other suitable structure of the target protein is available, it is necessary to build the missing residues using appropriate tools like Modeller [76], Prime [77].

1.3.1.6 Protein Conformational State

Proteins can sample a multitude of conformational states in the energy landscape. These different conformational states are often associated with distinct biological functions mediated by the protein [27, 78]. Kinases, the popular drug targets, are known to exist in multiple conformations [79]. The binding site geometry of the active state of kinases is known to be more conserved than the inactive state conformations among all kinases [80]. Thus, designing conformation-specific kinase targeted drugs could help in avoiding undesired effects [81–83]. Again, post-translational modification of a protein under biological conditions may affect its conformation and thus may alter drug binding capacity [84]. It is, therefore, essential to ensure that the input structure selected for SBDD represents the conformation of the protein that is intended to be targeted. If such a structure of the target protein or its homolog is unavailable, molecular dynamics and other conformational sampling techniques could be employed to predict the conformations that are likely to be closer to the physiological conditions [85].

1.3.1.7 Unavailability of Target Structure

The structure selection criteria discussed so far assume that experimental structures of the target proteins are available. However, determining structures of many targets, such as membrane proteins and intrinsically disordered proteins, are highly challenging. Computational models could be used in the SBDD programs intended to target proteins for which experimental structures are unavailable (as mentioned in Sect. 1.2). Scenarios where an experimental structure of the target protein or its close homologue is known but is unsatisfactory for SBDD would also require computational models as the starting point. It is important that any predicted structural model of protein used in SBDD should satisfy the quality evaluation [86] and represent the conformational state intended to be targeted by the designed modulator. Detailed discussion of computational structure prediction methods is beyond the scope of this chapter but can be found elsewhere [87].

1.3.2 Target and Ligand 3D Structure Preparation

The structure of the target protein obtained from the PDB, and the ligand structures obtained from chemical libraries are generally not suitable for computational studies as it is. These structures require pre-treatment, as discussed below, to fix certain issues and obtaining reliable predictions from the models.

1.3.2.1 Protein Structure Preparation

A typical PDB file of protein and/or protein–ligand complexes might not contain all the information that are required for initiating any modelling studies. The co-ordinates of hydrogen atoms are generally absent in the macromolecular crystal structures, unless the structure is of ultra-high resolution [88]. It is important that the hydrogen atoms are added to the protein structure before using them for any SBDD applications. It should also be ensured that the added hydrogen atoms have the right geometry to optimize the local hydrogen bonding network and the final structure used as input should be free of steric clashes. This might sometimes require flipping the side chain of certain residues like histidine, asparagine, and glutamine. Assignment of protonation and tautomerization states of the binding site residues (especially His, which can be neutral with a proton either on Nδ or Nε or have a positive or negative charge) play a critical role. Incorrect ionization would interfere with docking scores and hence affect the rank order of the screened compounds. The issues with missing atoms as mentioned in Sect. 1.3.1.5 need to be fixed, and assignment of right charges to the protein residues at the desired pH should be ensured to mimic the biological conditions. Removal of co-ordinates of water molecules from the binding site of a protein crystal structure before docking simulation is generally recommended unless there is enough experimental evidence to believe that such water molecules play an important role in protein–ligand interaction. Freely available tools through the WHAT IF web interface [89] or paid tools like Protein Preparation Wizard [88] available through Schrodinger suite are few of the many computational tools that could be used for preparing the input protein structures.

1.3.2.2 Ligand Structure Preparation

The 2D or 3D structures of ligands could be obtained from publicly available chemical libraries such as PubChem [90], ChEMBL [91], BindingDB [92]. The 2D ligand structure either downloaded from chemical databases or drawn using chemical sketchers (like ChemDraw [93]) requires conversion into its 3D form for any SBDD studies. The 3D structures that are available from these libraries may not be suitable for direct use as it would require geometry optimization in a manner that is compatible with the force field [94] to be used in the subsequent steps. Further, assignment of bond orders, and charges, generation of right tautomer and ionization states of the ligands are essential prior to usage to represent the biological conditions. The library molecules must be passed through different filters (as discussed in Sect. 1.2.2) which are implemented to build drug-like/lead-like/fragment library. Apart from these filters, library molecules must also be filtered to exclude those that have toxic or metabolically unstable groups or molecules known to be chemically reactive which can interfere with assays (refer Sect. 1.2.2) [46, 47]. One of the important aspects of ligand preparation when it contains a chiral centre is to consider the stereoisomer that is relevant for biological activity. The information on bioactivity of different stereoisomers of a given ligand against the target of interest could be obtained from experimental studies. In the absence of such experimental data, it is wiser to consider all possible stereoisomers for at least the early stages of computational studies. OpenBabel (freely available) [95] and LigPrep (commercial tool available with Schrodinger Suite) [88] are two of the many available computational tools that help in ligand preparation.

1.3.3 Binding Affinity and Mode Prediction

The central focus of computational SBDD programs is: (a) virtual screening and (b) to predict the binding poses of the ligands in the target protein pocket [96]. The scoring algorithms differentiate between 'good' and 'bad' binding poses for individual ligand molecules and provide decent enrichment factors from virtual screening. Docking score is often wrongly interpreted as a measure of affinity of a ligand towards the target. Notably, docking scores are generally known to have poor correlation with experimental binding affinities [97, 98]. The accurate calculation of binding free energy (ΔG_{bind}) would require exhaustive sampling of the molecular system in explicit solvent environment both in bound and unbound state. Since it is time consuming and accurate in silico representation of the physical laws that govern the event is difficult and complex, several approximations are used to reduce the complexity of the system [99]. Most docking programs involve sampling the ligand conformations (with limited degrees of freedom) within the rigid pocket of the target protein and subsequently ranking the poses based on the goodness of fit using a scoring function. The scoring functions in most cases are just the approximate representation of the binding energy that completely or partly neglect the entropic contributions and only considers the enthalpic component. The enthalpy component is also a simplistic representation of the protein–ligand interactions. It disregards the solvent effects and so-called non-classical (for example, CH-π, halogen bonds, S···O) interactions that are important component of binding energy [100, 101]. We are probably yet to discover many such 'non-classical' interactions and thus far from incorporating those in scoring functions. Due to these limitations, it is quite common to encounter cases when the best ranked docked pose as suggested by any docking algorithm is not the biologically meaningful pose. However, even with the numerous limitations, docking simulations are undoubtedly one of the most useful computational tools to distinguish potential binders and non-binders from the vast chemical libraries in a reasonably less time [102]. With the remarkable advancement of computing resources in the past few years, employing flexibilities to the receptor and including solvent contribution to certain degrees even in early stage computational studies could now be routinely done within reasonable time [103]. Induced fit docking approaches that allow sampling the side chain conformations of the binding site residues in the presence of the ligand candidates are shown to improve prediction efficiencies [74]. Other advanced techniques to predict binding affinity and binding pose such as free energy perturbation (FEP), quantum mechanics/molecular mechanics (QM/MM), though computationally expensive, when rationally combined with the traditional rigid docking protocols could be helpful [104–106]. Analysing the results from any of these prediction approaches in the light of the biological understanding would ultimately dictate the success of these methods. Ensuring if the predicted pose of the ligand is engaged in interactions with the functionally important protein binding site residues and retain similar interaction fingerprints as that of known binders are a few strategies to select the meaningful poses from the pool of suggested docking solutions.

1.3.4 Contribution of Water

Under physiological conditions, the protein and the ligand molecules are solvated. The binding event between the target protein and the ligand necessitates desolvation of both the molecules. Desolvating the polar atoms of protein residues or ligand leads to unfavourable change in enthalpy but facilitates movement of thermodynamically stable water molecules into the bulk resulting in gaining of entropy [107, 108]. The degree of disorder of such water molecules prior to their replacement governs the thermodynamic signature of the water release (hydrophobic effect) and the effect can range from entropic to enthalpic [109]. It contributes to a favourable enthalpy term for the hydrophobic contacts in tight pockets and to a favourable entropy term for the polar interactions. Thus, water molecules play a crucial role in the drug-target binding event. The thermodynamic signatures of the water molecules in the binding site can guide in understanding if displacement, replacement, or retention of a particular water molecule could be helpful for improving potency and/ or achieving specificity [110,

111]. Explicit modelling of water in protein–ligand binding event is computationally expensive. Therefore, most docking algorithms either neglect the contribution of water or use a crude representation to account for it. Such a reductionist approach introduces inaccuracies in predicted binding affinity. Computational techniques that use implicit water model, like Prime-MMGBSA [112], to predict binding affinities are shown to better correlate with experimental binding affinity for congeneric series of ligands than traditional docking approaches. Albeit computationally expensive, exploiting thermodynamic signatures of binding site water molecules during advanced lead optimization phase could prove helpful particularly in scenarios where water thermodynamics play a dominant role.

1.3.5 Effect of Dynamics

One of the main reasons for the poor correlation between predicted binding affinities obtained from docking studies and experimental binding affinities is due to lack of consideration of the inherent plasticity of the target protein in the presence of the ligand [113]. From local rearrangement of side chain and/or backbone atoms of binding site residues to large-scale movement of loops or domains could be triggered during a protein–ligand binding event [114]. Such movements may dictate the stability of the interactions between the protein and ligand atoms and hence the binding affinity. The success rate of computational ligand screening programs that employ only rigid docking approach for proteins with flexible ligand binding sites is likely to be low [115]. Molecular dynamics (MD) simulations are useful tools to predict atomic movements of protein–ligand complexes in explicit solvent environment [116]. Thus, MD simulation can be used to test and validate the stability of protein–ligand complexes predicted from docking studies. The information derived from these simulations could help in incorporating suitable functional groups on the ligand to maximize its interactions with the protein. Further, MD snapshots could be used to generate ensemble of structures representing different conformations of the target protein [117]. The ensemble of structures could then be used to dock the libraries of ligands and this approach has shown to improve the efficiency of docking studies [118]. Undoubtedly, MD simulations are computationally expensive but with advent of graphical processing units (GPUs) and advancement of other computational resources, it is now feasible to simulate the dynamics of biomolecular systems at millisecond scales [119, 120]. Depending upon the question that is intended to be addressed, the right level of flexibility could be introduced into simulation, striking a balance between accuracy and speed [115].

1.4 Conclusion

In this chapter, we have discussed various components of computational SBDD and associated challenges. These challenges arise majorly due to limitations with respect to: (i) computing speed and (ii) accuracy of prediction. The former could be addressed by advancement in hardware as we are witnessing in the recent times with the advent of the GPUs [121] and high-performance clusters (HPCs) [122] era. Strategic investment in computing resources is, thus, essential in any modern-day drug discovery programs. The latter challenge requires better understanding of biology and the physical and chemical laws that drive the biological process. Understanding in detail how molecules interact with each other in physiological and pathological states is central to modern drug discovery. Accurate representation of such understanding in computer algorithms by minimizing the approximations is likely to improve prediction efficiencies.

We would like to emphasize that the possible ways to tackle some of the challenges as we have discussed in this chapter are based on common scenarios in computational SBDD. Case-specific issues might demand specialized strategies. In our opinion, the fundamental strategy to improve the confidence of any prediction is to rationally combine the use of multiple algorithms that work on varying principles and verify if there is a consensus in the outcomes. Selection of the best predicted solutions should be guided by understanding the governing physical and chemical laws and their biological relevance. Experimental validations are essential to test the correlation between theoretical and experimental studies. The learning from the experimental studies should be integrated to the computational pipelines to improve the predictive power of the algorithms. In other words, judicious integration of computational methods with experimental techniques by considering the limitations of both is one of the key drivers of any successful SBDD program.

References

1. Alberts B, Johnson A, Lewis J et al (2015) Molecular biology of the cell
2. Patrick G (2018) An introduction to medicinal chemistry (6th edn). Oxford University Press, Oxford
3. Anderson AC (2003) The process of structure-based drug design. Chem Biol 10:787–797. https://doi.org/10.1016/j.chembiol.2003.09.002
4. Wlodawer A, Vondrasek J (1998) INHIBITORS OF HIV-1 PROTEASE: a major success of structure-assisted drug design. Annu Rev Biophys Biomol Struct 27:249–284. https://doi.org/10.1146/annurev.biophys.27.1.249
5. Batool M, Ahmad B, Choi S (2019) A structure-based drug discovery paradigm. Int J Mol Sci 20:2783
6. Náray-Szabó G (1993) Analysis of molecular recognition: steric electrostatic and hydrophobic complementarity. J Mol Recognit 6:205–210. https://doi.org/10.1002/jmr.300060409
7. Yazhini A, Chakraborti S, Srinivasan N (2021) Protein structure, dynamics and assembly: implications for drug discovery—innovations and implementations of computer aided drug discovery strategies in rational drug design. In: Singh SK (ed) Springer, Singapore, pp 91–122

8. Stoll F, Göller AH, Hillisch A (2011) Utility of protein structures in overcoming ADMET-related issues of drug-like compounds. Drug Discov Today 16:530–538. https://doi.org/10.1016/j.drudis.2011.04.008
9. Schwartz DM, Kanno Y, Villarino A et al (2018) Erratum: JAK inhibition as a therapeutic strategy for immune and inflammatory diseases. Nat Rev Drug Discov 17:78. https://doi.org/10.1038/nrd.2017.267
10. Chen C, Yin Y, Shi G et al (2022) A highly selective JAK3 inhibitor is developed for treating rheumatoid arthritis by suppressing γc cytokine–related JAK-STAT signal. Sci Adv 8:eabo4363. https://doi.org/10.1126/sciadv.abo4363
11. Sadybekov AA, Sadybekov AV, Liu Y et al (2022) Synthon-based ligand discovery in virtual libraries of over 11 billion compounds. Nature 601:452–459. https://doi.org/10.1038/s41586-021-04220-9
12. Lionta E, Spyrou G, Vassilatis KD, Cournia Z (2014) Structure-based virtual screening for drug discovery: principles, applications and recent advances. Curr Top Med Chem 14:1923–1938
13. Kar S, Roy K (2013) How far can virtual screening take us in drug discovery? Expert Opin Drug Discov 8:245–261. https://doi.org/10.1517/17460441.2013.761204
14. Bissantz C, Kuhn B, Stahl M (2010) A medicinal chemist's guide to molecular interactions. J Med Chem 53:5061–5084. https://doi.org/10.1021/jm100112j
15. Burley SK (2021) Impact of structural biologists and the Protein Data Bank on small-molecule drug discovery and development. J Biol Chem. https://doi.org/10.1016/j.jbc.2021.100559
16. Chakraborti S, Chakraborty M, Bose A et al (2021) Identification of potential binders of MTB Universal Stress Protein (Rv1636) through an in silico approach and insights into compound selection for experimental validation. Front Mol Biosci 8:599221. https://doi.org/10.3389/fmolb.2021.599221
17. Verma H, Khatri B, Chakraborti S, Chatterjee J (2018) Increasing the bioactive space of peptide macrocycles by thioamide substitution. Chem Sci 9:2443–2451. https://doi.org/10.1039/C7SC04671E
18. Capra JA, Singh M (2007) Predicting functionally important residues from sequence conservation. Bioinformatics 23:1875–1882. https://doi.org/10.1093/bioinformatics/btm270
19. Berman HM, Westbrook J, Feng Z et al (2000) The Protein Data Bank. Nucleic Acids Res 28:235–242. https://doi.org/10.1093/nar/28.1.235
20. Rhodes G (2006) An overview of protein crystallography. In: Rhodes GBT-CMCC (ed) Complementary science. Academic Press, Burlington, pp 7–30
21. Howard MJ (1998) Protein NMR spectroscopy. Curr Biol 8:R331–R333. https://doi.org/10.1016/S0960-9822(98)70214-3
22. Savva C (2019) A beginner's guide to cryogenic electron microscopy. Biochem (Lond) 41:46–52. https://doi.org/10.1042/BIO04102046
23. Webb B, Eswar N, Fan H, Khuri N, Pieper U, Dong GQ, Sali A (2014) Comparative modeling of drug target proteins. In: Reedijk J (ed) Elsevier reference module in chemistry, molecular sciences and chemical engineering. Elsevier, Waltham. https://doi.org/10.1016/B978-0-12-409547-2.11133-3
24. Varadi M, Anyango S, Deshpande M et al. (2022) AlphaFold protein structure database: massively expanding the structural coverage of protein-sequence space with high-accuracy models. Nucleic Acids Res 50(D1): D439–D444. https://doi.org/10.1093/nar/gkab1061. PMID: 34791371; PMCID: PMC8728224
25. Bowie JU, Lüthy R, Eisenberg D (1991) A method to identify protein sequences that fold into a known three-dimensional structure. Science 253:164–170. https://doi.org/10.1126/science.1853201
26. Zhou H, Skolnick J (2013) FINDSITEcomb: a threading/structure-based, proteomic-scale virtual ligand screening approach. J Chem Inf Model 53:230–240. https://doi.org/10.1021/ci300510n
27. Teilum K, Olsen JG, Kragelund BB (2009) Functional aspects of protein flexibility. Cell Mol Life Sci 66:2231–2247. https://doi.org/10.1007/s00018-009-0014-6

28. Wlodawer A, Minor W, Dauter Z, Jaskolski M (2008) Protein crystallography for non-crystallographers, or how to get the best (but not more) from published macromolecular structures. FEBS J 275:1–21. https://doi.org/10.1111/j.1742-4658.2007.06178.x
29. Rupp B (2009) Biomolecular crystallography: principles, practice, and application to structural biology, 1st ed. Garland Science
30. Pozharski E, Weichenberger CX, Rupp B (2013) Techniques, tools and best practices for ligand electron-density analysis and results from their application to deposited crystal structures. Acta Crystallogr D Biol Crystallogr 69:150–167. https://doi.org/10.1107/S0907444912044423
31. Davis AM, St-Gallay SA, Kleywegt GJ (2008) Limitations and lessons in the use of X-ray structural information in drug design. Drug Discov Today 13:831–841. https://doi.org/10.1016/j.drudis.2008.06.006
32. Hu Y, Cheng K, He L et al (2021) NMR-based methods for protein analysis. Anal Chem 93:1866–1879. https://doi.org/10.1021/acs.analchem.0c03830
33. Sekhar A, Kay LE (2013) NMR paves the way for atomic level descriptions of sparsely populated, transiently formed biomolecular conformers. Proc Natl Acad Sci 110:12867–12874. https://doi.org/10.1073/pnas.1305688110
34. Van Drie JH, Tong L (2020) Cryo-EM as a powerful tool for drug discovery. Bioorg Med Chem Lett 30:127524. https://doi.org/10.1016/j.bmcl.2020.127524
35. Subramaniam S, Earl LA, Falconieri V et al (2016) Resolution advances in cryo-EM enable application to drug discovery. Curr Opin Struct Biol 41:194–202. https://doi.org/10.1016/j.sbi.2016.07.009
36. Cavasotto CN, Palomba D (2015) Expanding the horizons of G protein-coupled receptor structure-based ligand discovery and optimization using homology models. Chem Commun 51:13576–13594. https://doi.org/10.1039/C5CC05050B
37. Jumper J, Evans R, Pritzel A et al (2021) Highly accurate protein structure prediction with AlphaFold. Nature. https://doi.org/10.1038/s41586-021-03819-2
38. Baek M, DiMaio F, Anishchenko I et al (2021) Accurate prediction of protein structures and interactions using a three-track neural network. Science 373:871–876. https://doi.org/10.1126/science.abj8754
39. Lee C, Su B-H, Tseng YJ (2022) Comparative studies of AlphaFold, RoseTTAFold and modeller: a case study involving the use of G-protein-coupled receptors. Brief Bioinform bbac308. https://doi.org/10.1093/bib/bbac308
40. Heo L, Arbour CF, Feig M (2019) Driven to near-experimental accuracy by refinement via molecular dynamics simulations. Proteins Struct Funct Bioinforma 87:1263–1275. https://doi.org/10.1002/prot.25759
41. Zhang Y, Vass M, Shi D et al (2022) Benchmarking refined and unrefined AlphaFold2 structures for hit discovery. ChemRxiv. https://doi.org/10.26434/chemrxiv-2022-kcn0d-v2
42. Schiebel J, Krimmer SG, Röwer K et al. (2016) High-throughput crystallography: reliable and efficient identification of fragment hits. Structure 24(8): 1398–1409. ISSN 0969-2126, https://doi.org/10.1016/j.str.2016.06.010
43. Wu B, Barile E, De SK, Wei J, Purves A, Pellecchia M (2015) High-throughput screening by nuclear magnetic resonance (HTS by NMR) for the identification of PPIs antagonists. Curr Top Med Chem 15(20):2032–2042. https://doi.org/10.2174/1568026615666150519102459
44. Gorgulla C, Boeszoermenyi A, Wang ZF et al. (2020) An open-source drug discovery platform enables ultra-large virtual screens. Nature 580: 663–668. https://doi.org/10.1038/s41586-020-2117-z
45. Lipinski CA, Lombardo F, Dominy BW, Feeney PJ (1997) Experimental and computational approaches to estimate solubility and permeability in drug discovery and development settings. Adv Drug Deliv Rev 23:3–25. https://doi.org/10.1016/s0169-409x(00)00129-0
46. Baell JB, Holloway GA (2010) New substructure filters for removal of pan assay interference compounds (PAINS) from screening libraries and for their exclusion in bioassays. J Med Chem 53:2719–2740. https://doi.org/10.1021/jm901137j
47. Walters WP, Stahl MT, Murcko MA (1998) Virtual screening—an overview. Drug Discov Today 3:160–178. https://doi.org/10.1016/S1359-6446(97)01163-X

48. Erlanson DA, Fesik SW, Hubbard RE et al (2016) Twenty years on: the impact of fragments on drug discovery. Nat Rev Drug Discov 15:605–619. https://doi.org/10.1038/nrd.2016.109
49. John Harris C, Hill R, Sheppard D et al (2011) The design and application of target-focused compound libraries. Comb Chem High Throughput Screen 14:521–531
50. Moret N, Clark NA, Hafner M et al (2019) Cheminformatics tools for analyzing and designing optimized small-molecule collections and libraries. Cell Chem Biol 26:765-777.e3. https://doi.org/10.1016/j.chembiol.2019.02.018
51. Masimirembwa CM, Bredberg U, Andersson TB (2003) Metabolic stability for drug discovery and development. Clin Pharmacokinet 42:515–528. https://doi.org/10.2165/00003088-200342060-00002
52. Schnider P (2021) Overview of strategies for solving ADMET challenges. In: The medicinal chemist's guide to solving ADMET challenges. Royal Society of Chemistry, pp 1–15
53. Kumar S, Sharma R, Roychowdhury A (2012) Modulation of cytochrome-P450 inhibition (CYP) in drug discovery: a medicinal chemistry perspective. Curr Med Chem 19:3605–3621
54. Spyrakis F, Cozzini P, Kellogg GE (2010) Docking and scoring in drug discovery. Burger's Med Chem Drug Discov 601–684
55. Bitencourt-Ferreira G, de Azevedo WF (2019) Machine learning to predict binding affinity BT. In: de Azevedo Jr. WF (ed) Docking screens for drug discovery. Springer, New York, pp 251–273
56. Jones D, Kim H, Zhang X et al (2021) Improved protein-ligand binding affinity prediction with structure-based deep fusion inference. J Chem Inf Model 61:1583–1592. https://doi.org/10.1021/acs.jcim.0c01306
57. Thafar M, Bin RA, Albaradei S et al (2019) Comparison study of computational prediction tools for drug-target binding affinities. Front Chem. https://doi.org/10.3389/fchem.2019.00782
58. Dhakal A, McKay C, Tanner JJ, Cheng J (2022) Artificial intelligence in the prediction of protein–ligand interactions: recent advances and future directions. Brief Bioinform 23:bbab476. https://doi.org/10.1093/bib/bbab476
59. Dutta S, Bose K (2021) Remodelling structure-based drug design using machine learning. Emerg Top Life Sci 5:13–27. https://doi.org/10.1042/ETLS20200253
60. Chakraborti S, Hatti K, Srinivasan N (2021) 'All that glitters is not gold': high-resolution crystal structures of ligand-protein complexes need not always represent confident binding poses. Int J Mol Sci. https://doi.org/10.3390/ijms22136830
61. Blow DM (2002) Rearrangement of Cruickshank's formulae for the diffraction-component precision index. Acta Crystallogr Sect D 58:792–797. https://doi.org/10.1107/S0907444902003931
62. Cruickshank DWJ (1999) Remarks about protein structure precision. Acta Crystallogr Sect D 55:583–601. https://doi.org/10.1107/S0907444998012645
63. Deller MC, Rupp B (2015) Models of protein-ligand crystal structures: trust, but verify. J Comput Aided Mol Des 29:817–836. https://doi.org/10.1007/s10822-015-9833-8
64. Tickle IJ (2012) Statistical quality indicators for electron-density maps. Acta Crystallogr Sect D 68:454–467. https://doi.org/10.1107/S0907444911035918
65. Meyder A, Nittinger E, Lange G et al (2017) Estimating electron density support for individual atoms and molecular fragments in x-ray structures. J Chem Inf Model 57:2437–2447. https://doi.org/10.1021/acs.jcim.7b00391
66. Joosten RP, Joosten K, Murshudov GN, Perrakis A (2012) PDB_REDO: constructive validation, more than just looking for errors. Acta Crystallogr D Biol Crystallogr 68:484–496. https://doi.org/10.1107/S0907444911054515
67. Nishi H, Tyagi M, Teng S et al (2013) Cancer missense mutations alter binding properties of proteins and their interaction networks. PLoS ONE 8:e66273
68. Konc J, Česnik T, Konc JT et al (2012) ProBiS-database: precalculated binding site similarities and local pairwise alignments of PDB structures. J Chem Inf Model 52:604–612. https://doi.org/10.1021/ci2005687

69. Le Guilloux V, Schmidtke P, Tuffery P (2009) Fpocket: an open source platform for ligand pocket detection. BMC Bioinform 10:168. https://doi.org/10.1186/1471-2105-10-168
70. Halgren TA (2009) Identifying and characterizing binding sites and assessing druggability. J Chem Inf Model 49:377–389. https://doi.org/10.1021/ci800324m
71. McGovern SL, Shoichet BK (2003) Information decay in molecular docking screens against Holo, Apo, and modeled conformations of enzymes. J Med Chem 46:2895–2907. https://doi.org/10.1021/jm0300330
72. Koshland DE (1958) Application of a theory of enzyme specificity to protein synthesis*. Proc Natl Acad Sci 44:98–104. https://doi.org/10.1073/pnas.44.2.98
73. Kim K-H, Kim ND, Seong B-L (2010) Pharmacophore-based virtual screening: a review of recent applications. Expert Opin Drug Discov 5:205–222. https://doi.org/10.1517/17460441003592072
74. Sherman W, Day T, Jacobson MP et al (2006) Novel procedure for modeling ligand/receptor induced fit effects. J Med Chem 49:534–553. https://doi.org/10.1021/jm050540c
75. Hollingsworth SA, Dror RO (2018) Molecular dynamics simulation for all. Neuron 99:1129–1143. https://doi.org/10.1016/j.neuron.2018.08.011
76. Webb B, Sali A (2016) Comparative protein structure modeling using MODELLER. Curr Protoc Bioinforma 54:5.6.1–5.6.37. https://doi.org/10.1002/cpbi.3
77. Jacobson MP, Pincus DL, Rapp CS et al (2004) A hierarchical approach to all-atom protein loop prediction. Proteins Struct Funct Bioinforma 55:351–367. https://doi.org/10.1002/prot.10613
78. Schmid S, Hugel T (2020) Controlling protein function by fine-tuning conformational flexibility. Elife 9:e57180. https://doi.org/10.7554/eLife.57180
79. Möbitz H (2015) The ABC of protein kinase conformations. Biochim Biophys Acta - Proteins Proteomics 1854:1555–1566. https://doi.org/10.1016/j.bbapap.2015.03.009
80. Huse M, Kuriyan J (2002) The conformational plasticity of protein kinases. Cell 109:275–282. https://doi.org/10.1016/S0092-8674(02)00741-9
81. Wang X, Kim J (2012) Conformation-specific effects of Raf kinase inhibitors. J Med Chem 55:7332–7341. https://doi.org/10.1021/jm300613w
82. Tong M, Seeliger MA (2015) Targeting conformational plasticity of protein kinases. ACS Chem Biol 10:190–200. https://doi.org/10.1021/cb500870a
83. Kwarcinski FE, Brandvold KR, Phadke S et al (2016) Conformation-selective analogues of dasatinib reveal insight into kinase inhibitor binding and selectivity. ACS Chem Biol 11:1296–1304. https://doi.org/10.1021/acschembio.5b01018
84. Su M-G, Weng JT-Y, Hsu JB-K et al (2017) Investigation and identification of functional post-translational modification sites associated with drug binding and protein-protein interactions. BMC Syst Biol 11:132. https://doi.org/10.1186/s12918-017-0506-1
85. Liwo A, Czaplewski C, Ołdziej S, Scheraga HA (2008) Computational techniques for efficient conformational sampling of proteins. Curr Opin Struct Biol 18:134–139. https://doi.org/10.1016/j.sbi.2007.12.001
86. Haddad Y, Adam V, Heger Z (2020) Ten quick tips for homology modeling of high-resolution protein 3D structures. PLOS Comput Biol 16:e1007449
87. Hameduh T, Haddad Y, Adam V, Heger Z (2020) Homology modeling in the time of collective and artificial intelligence. Comput Struct Biotechnol J 18:3494–3506. https://doi.org/10.1016/j.csbj.2020.11.007
88. Sastry MG, Adzhigirey M, Day T et al (2013) Protein and ligand preparation: parameters, protocols, and influence on virtual screening enrichments. J Comput Aided Mol Des 27:221–234. https://doi.org/10.1007/s10822-013-9644-8
89. Vriend G (1990) WHAT IF: a molecular modeling and drug design program. J Mol Graph 8:52–56. https://doi.org/10.1016/0263-7855(90)80070-V
90. Kim S, Chen J, Cheng T et al (2018) PubChem 2019 update: improved access to chemical data. Nucleic Acids Res 47:D1102–D1109. https://doi.org/10.1093/nar/gky1033
91. Mendez D, Gaulton A, Bento AP et al (2018) ChEMBL: towards direct deposition of bioassay data. Nucleic Acids Res 47:D930–D940. https://doi.org/10.1093/nar/gky1075

92. Gilson MK, Liu T, Baitaluk M et al (2016) BindingDB in 2015: A public database for medicinal chemistry, computational chemistry and systems pharmacology. Nucleic Acids Res 44:D1045–D1053. https://doi.org/10.1093/nar/gkv1072
93. Cousins KR (2005) ChemDraw Ultra 9.0. CambridgeSoft, 100 CambridgePark Drive, Cambridge, MA 02140. www.cambridgesoft.com. See Web site for pricing options. J Am Chem Soc 127:4115–4116. https://doi.org/10.1021/ja0410237
94. Cole DJ, Horton JT, Nelson L, Kurdekar V (2019) The future of force fields in computer-aided drug design. Future Med Chem 11:2359–2363. https://doi.org/10.4155/fmc-2019-0196
95. O'Boyle NM, Banck M, James CA et al (2011) Open Babel: an open chemical toolbox. J Cheminform 3:33. https://doi.org/10.1186/1758-2946-3-33
96. Gohlke H, Klebe G (2002) Approaches to the description and prediction of the binding affinity of small-molecule ligands to macromolecular receptors. Angew Chemie Int Ed 41:2644–2676. https://doi.org/10.1002/1521-3773(20020802)41:15<2644::AID-ANIE2644>3.0.CO;2-O
97. Pantsar T, Poso A (2018) Binding affinity via docking: fact and fiction. Molecules 23:1899. https://doi.org/10.3390/molecules23081899
98. Plewczynski D, Łaźniewski M, Augustyniak R, Ginalski K (2011) Can we trust docking results? Evaluation of seven commonly used programs on PDBbind database. J Comput Chem 32:742–755. https://doi.org/10.1002/jcc.21643
99. van Gunsteren WF, Daura X, Fuchs PFJ et al (2021) On the effect of the various assumptions and approximations used in molecular simulations on the properties of bio-molecular systems: overview and perspective on issues. ChemPhysChem 22:264–282. https://doi.org/10.1002/cphc.202000968
100. Anighoro A (2020) Underappreciated chemical interactions in protein–ligand complexes BT—quantum mechanics in drug discovery. In: Heifetz A (ed). Springer US, New York, pp 75–86
101. Zhang X, Gong Z, Li J, Lu T (2015) Intermolecular sulfur···oxygen interactions: theoretical and statistical investigations. J Chem Inf Model 55:2138–2153. https://doi.org/10.1021/acs.jcim.5b00177
102. Ferreira LG, Dos Santos RN, Oliva G, Andricopulo AD (2015) Molecular docking and structure-based drug design strategies. Moleculee 20:13384–13421
103. Fan M, Wang J, Jiang H et al (2021) GPU-accelerated flexible molecular docking. J Phys Chem B 125:1049–1060. https://doi.org/10.1021/acs.jpcb.0c09051
104. Wang L, Chambers J, Abel R (2019) Protein-ligand binding free energy calculations with FEP+ BT. In: Bonomi M, Camilloni C (eds) Biomolecular simulations: methods and protocols. Springer, New York, pp 201–232
105. van der Kamp MW, Mulholland AJ (2013) Combined quantum mechanics/molecular mechanics (QM/MM) methods in computational enzymology. Biochemistry 52:2708–2728. https://doi.org/10.1021/bi400215w
106. Cao L, Ryde U (2018) On the difference between additive and subtractive QM/MM calculations. Front Chem. https://doi.org/10.3389/fchem.2018.00089
107. Ladbury JE (1996) Just add water! The effect of water on the specificity of protein-ligand binding sites and its potential application to drug design. Chem Biol 3:973–980. https://doi.org/10.1016/S1074-5521(96)90164-7
108. Zsidó BZ, Hetényi C (2021) The role of water in ligand binding. Curr Opin Struct Biol 67:1–8. https://doi.org/10.1016/j.sbi.2020.08.002
109. Klebe G (2011) On the validity of popular assumptions in computational drug design. J Cheminform 3:O18. https://doi.org/10.1186/1758-2946-3-S1-O18
110. Yang Y, Lightstone FC, Wong SE (2013) Approaches to efficiently estimate solvation and explicit water energetics in ligand binding: the use of WaterMap. Expert Opin Drug Discov 8:277–287. https://doi.org/10.1517/17460441.2013.749853
111. Cappel D, Sherman W, Beuming T (2017) Calculating water thermodynamics in the binding site of proteins—applications of WaterMap to drug discovery. Curr Top Med Chem 17:2586–2598

112. Lyne PD, Lamb ML, Saeh JC (2006) Accurate prediction of the relative potencies of members of a series of kinase inhibitors using molecular docking and MM-GBSA scoring. J Med Chem 49:4805–4808. https://doi.org/10.1021/jm060522a
113. Spyrakis F, BidonChanal A, Barril X, Javier Luque F (2011) Protein flexibility and ligand recognition: challenges for molecular modeling. Curr Top Med Chem 11:192–210
114. Gaudreault F, Chartier M, Najmanovich R (2012) Side-chain rotamer changes upon ligand binding: common, crucial, correlate with entropy and rearrange hydrogen bonding. Bioinformatics 28:i423–i430. https://doi.org/10.1093/bioinformatics/bts395
115. Alvarez-Garcia D, Barril X (2014) Relationship between protein flexibility and binding: lessons for structure-based drug design. J Chem Theory Comput 10:2608–2614. https://doi.org/10.1021/ct500182z
116. Lin X (2022) Applications of molecular dynamics simulations in drug discovery. In: Tripathi T, Dubey VK (eds) Advances in protein molecular and structural biology methods. Academic Press, pp 455–465
117. Amaro RE, Baudry J, Chodera J et al (2018) Ensemble docking in drug discovery. Biophys J 114:2271–2278
118. Tian S, Sun H, Pan P et al (2014) Assessing an ensemble docking-based virtual screening strategy for kinase targets by considering protein flexibility. J Chem Inf Model 54:2664–2679. https://doi.org/10.1021/ci500414b
119. Shaw DE, Dror RO, Salmon JK et al (2009) Millisecond-scale molecular dynamics simulations on anton. In: Proceedings of the conference on high performance computing networking, storage and analysis. Association for Computing Machinery, New York
120. Ngo VA, Garcia AE (2022) Millisecond molecular dynamics simulations of KRas-dimer formation and interfaces. Biophys J. https://doi.org/10.1016/j.bpj.2022.04.026
121. Pandey M, Fernandez M, Gentile F et al (2022) The transformational role of GPU computing and deep learning in drug discovery. Nat Mach Intell 4:211–221. https://doi.org/10.1038/s42256-022-00463-x
122. Puertas-Martín S, Banegas-Luna AJ, Paredes-Ramos M et al (2020) Is high performance computing a requirement for novel drug discovery and how will this impact academic efforts? Expert Opin Drug Discov 15:981–985. https://doi.org/10.1080/17460441.2020.1758664

Chapter 2
In Silico Discovery of Class IIb HDAC Inhibitors: The State of Art

Samima Khatun, Sk. Abdul Amin, Shovanlal Gayen, and Tarun Jha

Abstract HDAC6 and HDAC10 are class IIb HDAC isoenzymes. They have unique structural and physiological functions. They are key regulators of different physiological and pathological disease conditions. HDAC6 and HDAC10 are involved in different signaling pathways associated with several neurological disorders, various cancers at early as well as advanced stages, rare diseases, immunological conditions, etc. Thus, targeting these two enzymes has been found to be effective for various therapeutic purposes in recent years. More work is still needed to pinpoint the selectivity as well as potency of class IIb HDAC inhibitors (HDACi) for their clinical development. The present chapter deals with the structural biology of class IIb HDACs and discusses how in silico studies including the virtual screening approaches have been implemented to design HDAC6 and HDAC10 inhibitors. In addition, the interactions of class IIb HDACs with their inhibitors are also highlighted extensively to get a detail insight. This chapter offers understanding for designing newer class IIb HDAC inhibitors in future.

Keywords HDAC6 · HDAC10 · Drug design and discovery · QSAR · Molecular docking · MD simulation

2.1 Introduction

Epigenetic alterations caused by genetic flaws result in functional dysregulation of epigenetic regulators or proteins [1–4]. It eventually leads to changes in protein expression, which play a significant role in a variety of human diseases, including

S. Khatun · S. Gayen
Laboratory of Drug Design and Discovery, Department of Pharmaceutical Technology, Jadavpur University, Kolkata 700032, India

Sk. A. Amin · T. Jha (✉)
Natural Science Laboratory, Division of Medicinal and Pharmaceutical Chemistry, Department of Pharmaceutical Technology, Jadavpur University, Kolkata 700032, India
e-mail: tjupharm@yahoo.com

S. Kar and J. Leszczynski (eds.), *Current Trends in Computational Modeling for Drug Discovery*, Challenges and Advances in Computational Chemistry and Physics 35,
https://doi.org/10.1007/978-3-031-33871-7_2

various types of cancer, cardiovascular diseases, infections, inflammatory diseases, and neurological disorders [5–7]. A better understanding and application of epigenetics will aid in the identification of novel therapeutic treatments in the form of personalized medicine for various diseases [5]. Several histone post-translational modifications are critical epigenetic modulators and they are believed to influence gene expression [8]. Several post-translational modifications on histone proteins include: (1) the acetylation of specific lysine residues (by histone acetyltransferases), (2) the methylation of lysine and arginine residues (by histone methyltransferases), and (3) the phosphorylation of specific serine groups (by histone kinases) [9]. The amino termini of the histone proteins are utilized by the transcription regulators to carry out a number of post-translational modifications [10, 11].

Among the different post-translational modifications, the most researched process is the acetylation and deacetylation of histones. It takes place at the lysine amino termini and is controlled by both histone acetyltransferases (HATs) and histone deacetylases (HDACs) [12–14]. Due to the imbalance between HAT and HDAC, dysregulation of genetic expression results in chromatin instability and epigenetic diseases or disorders (Fig. 2.1). While HDAC inhibition results in continuous expression of the targeted gene, HAT inhibition results in the inexpression of the targeted gene [15–19]. It is well known that the overexpression of HDACs contributes to a variety of cancers as well as other neurological, autoimmune, inflammatory, cardiac, and pulmonary diseases [20–22]. HDACs are also found to deacetylate a number of non-histone proteins, including p53, E2F, α-tubulin, and Myo D. This leads to much more complex roles for HDACs in numerous other cellular processes. As a result, HDAC inhibition has drawn considerable interest and grown in importance as a drug target.

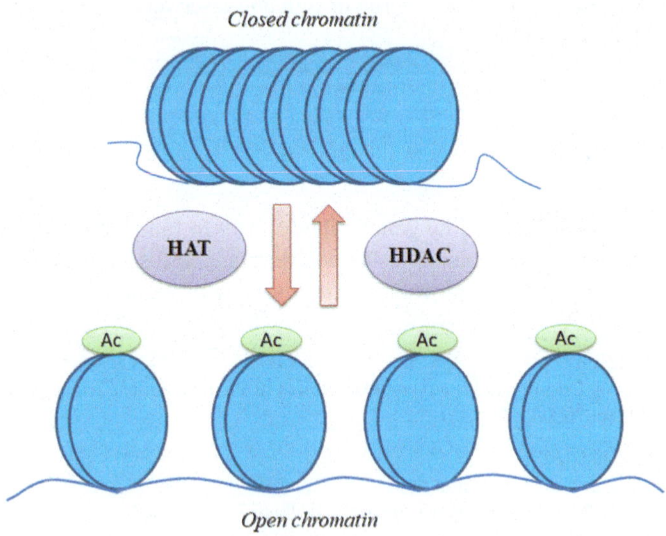

Fig. 2.1 Histone modification by HAT and HDAC

There are 18 isoforms of mammalian HDACs recognized until now. Based on their similarity with yeast protein and method of action, they are classified into four separate types [23–25]. The structure, enzymatic function, sub-cellular localization, and expression patterns of the four classes are unique [26]. The HDAC isozymes were numbered in the order in which they were discovered. HDAC1, HDAC2, HDAC3, and HDAC8 from class I have a sequence similarity to yeast- reduced potassium dependence (Rpd3)-like proteins [27–30]. They are mostly found in the nucleus and work primarily through histone proteins as their substrate. Class II HDACs have a similar amino acid sequence to yeast histone deacetylase 1 (Hda1). Based on sequence homology and domain structure, they were further categorized into two sub-classes, IIa (HDAC 4, 5, 7, and 9) and IIb (HDAC 6 and 10) [31–33]. Class IIa HDACs are located in the nucleus and they shuttle to the cytoplasm after being phosphorylated by kinases. Class IIb HDACs, on the other hand, are known to be cytoplasmic and act through diverse non-histone proteins as their substrates for deacetylase activity. Table 2.1 highlights the classification of class IIb HDAC isoforms as well as their cellular location and physiological roles. Class III HDACs or sirtuins, which include SIRT1, 2, 3, 4, 5, and 6, were so termed because they resemble the proteins that silence the yeast Sir2 gene [34]. Only HDAC11, which is known to have similarities to both class I and II catalytic domains of HDACs, belongs to class IV HDAC [35]. Class I, II, and IV HDACs are in the family of zinc-dependent HDAC isoforms, where the cofactor for the hydrolysis of acetylated substrates is the metal ion Zn^{2+}. Aside from that, class III HDACs depend on NAD+, which serves as a cofactor for their enzymatic activity [25].

A zinc-binding group (ZBG) that interacts with the zinc ion at the catalytic pocket, a cap group that interacts with the surface of the enzyme, and a linker that acts as a link between the cap and ZBG make up the canonical feature of HDAC inhibitors (HDACi) [36, 37]. Belinostat, panobinostat, romidepsin, and vorinostat (SAHA) have all received clinical approval for the treatment of lymphoma and multiple myeloma

Table 2.1 Classification of HDAC class IIb isoform, their cellular localization, and functions

Class	HDAC isoform	Chromosomal location	Amino acids no	Cellular localization	Location in body/ expression	Physiological function
IIb	HDAC6	Xp11.23	1215	Cytoplasm	Tissue specific	Regulation of protein degradation through aggresome pathway, Hsp90 chaperone activity, cytoskeletal dynamics, cell motility, angiogenesis
	HDAC10	2q13.33	669			Angiogenesis, autophagy, neurodegeneration

[38]. Due to their non-selectivity and broad-spectrum activity, these approved non-selective pan-HDACis are said to have a number of side effects, including exhaustion, nausea/vomiting, cardiotoxicity, etc. Therefore, there is a growing need for isoform-specific HDACi in order to reduce the side effects as well as apply them in more focused and selective treatments of a specific disease condition.

Until now, many HDAC isoforms have been investigated and their inhibitors are thoroughly defined. Since its first discovery in 1999, HDAC6, a class IIB HDAC isoform, has attracted also important attention among the different HDACs [39]. HDAC6 is a physically and functionally distinct cytoplasmic deacetylase. It is known for its deacetylase activity of certain cytosolic non-histone substrates like heat shock protein (Hsp90), cortactin, peroxiredoxin, α-tubulin, heat shock transcription facto-1 (HSF-1), etc. [40]. The first identified and extensively researched physiological substrate of HDAC6 is α-tubulin. HDAC6 controls the acetylation of lysine 40 in α-tubulin [41]. It is also known to participate in the tumorigenesis along with the development and metastasis through various pathways such as tubulin, Hsp90, and protein ubiquitination [42]. The effectiveness of selective HDAC6 inhibition in treating cancers such as bladder cancer, malignant melanoma, and lung cancer as well as neurodegenerative diseases such as Alzheimer's disease, Huntington's disease, and Parkinson's disease has also been extensively demonstrated in some studies and reports [43, 44]. The application of HDAC6i in rare disorders such as Rett syndrome, Charcot–Marie–Tooth disease, and amyotrophic lateral sclerosis has also been shown in recent investigations [45]. According to research using HDAC6 mutant mice, selective HDAC6 inhibitors are less cytotoxic to normal cells than pan-HDAC inhibitors, which mitigate their negative effects [46]. Tubacin, Tubastatin A, ACY-1215, ACY-241, and Nexturastat A are examples of specific HDAC6is. Recent studies on the X-ray crystal structures of HDAC6 CD2 and HDAC6i complexes have shed light on the structure and catalytic mechanism of the molecular characteristics that determines binding affinity for the target [46].

HDAC10 is also an important member in HDACs family and it shares structural similarities with HDAC6. The HDAC10 gene is located on chromosome 22 [47]. It is made up of 20 exons and two spliced transcripts. It comprises an N-terminal catalytic domain and a C-terminal leucine-rich domain. The expression of HDAC10, a member of the arginase/deacetylase superfamily, varies between the cytoplasm and the nucleus. Additionally, HDAC10 is expressed in the majority of human tissues, including the heart, liver, spleen, pancreas, placenta, kidney, and testicles. Recent research has shown that HDAC10 controls polyamine levels and functions as a polyamine deacetylase (PDAC) [47]. The pathogenesis of many malignancies is thought to include histone deacetylase 10 (HDAC10), and pharmacological blockade of this enzyme might aid in reversing the malignant phenotypes. A number of studies have highlighted HDAC10 inhibitors as possible anti-cancer agent [48–51]. However, the absence of the crystal structure of human HDAC10 hinders the structure-based rational drug design effort. HDAC10, the only other HDAC class IIb member, has received minimal attention from the medicinal chemistry community.

Prior to the 1960s, traditional drug discovery efforts in the pharmaceutical industry were largely committed to assess natural and synthetic chemicals against a specific biological endpoint [53–55]. After a potential drug or drug-like substance has been narrowed down from thousands of natural and synthetic compounds through an arduous process, medicinal chemists would then synthesize hundreds of related compounds (derivatives or analogs) to determine which molecule is both the safest and the most potent [56–60]. As a result, the costs and possible dangers with this approach dramatically increased.

The conventional drug design process underwent a significant change in the 1960s. The paradigms were shifted by a number of major publications released in the 1960s [53]. Since then, rational drug design (RDD) paradigms have attracted more attention for the design of new chemical entities as well as the optimization of chemical structures for improved biological activity. The development of computational chemistry, protein crystallography, and molecular biology in the early 1980s considerably benefited RDD paradigms in their efforts to improve the accuracy of binding affinity predictions. With fast advancing high throughput screening (HTS) and combinatorial chemical technology, computer-aided drug design (CADD) methodologies, and effective contributions, rational drug discovery has now become increasingly interdisciplinary [61].

In silico work such as quantitative structure–activity relationship (QSAR), pharmacophore mapping, virtual screening, homology modeling, molecular docking and molecular dynamic (MD) simulation studies applied to design class IIb HDAC inhibitors is the main emphasis of this chapter. The present chapter is broken into two sections. The first section presents an overview of class IIb HDACs (HDAC6 and HDAC10) as well as their structural biology, functions, and mechanism of action. The next section discusses about the different approaches related to in silico discovery of class IIb HDAC inhibitors. This chapter may be useful in the future for designing highly active as well as selective HDAC class IIb inhibitors.

2.2 Structural Biology of HDAC6

Among the different HDACs, HDAC6 has distinctive structural and functional characteristic. Its localization is mainly in the cytoplasm and primarily responsible for deacetylation of non-histone proteins like α-tubulin of microtubule. Unlike other HDACs, HDAC6 has two catalytic deacetylase domains in its structure as shown in Fig. 2.2. It has also an ubiquitin binding domain that is responsible to form aggresomes for degradation of polyubiquitinated misfolded proteins [62]. The N-terminal domain of HDAC6 is comprised two domains, namely nuclear localized signal (NLS) and nuclear export signal (NES). NLS is rich in arginine and lysine amino acids, while NES is rich in leucine amino acid. The two catalytic domains 1 and 2 are comprised amino acids 88–447 and 482–800, respectively. The dynein motor binding region (DMB) is connecting the two catalytic domains. The SE14 domain which is a Ser-Glu tetrapeptide domain sequence (SE14) is important for intracellular retention and tau interaction of HDAC6.

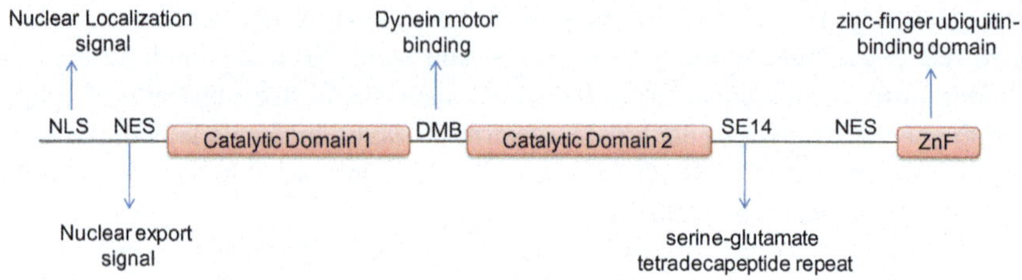

Fig. 2.2 Different domains in HDAC6

Hubbert et al. [41] first reported the in vivo and in vitro tubulin deacetylation for HDAC6. From the study, they have proved that HDAC6 is responsible for microtubule-dependent cell motility. Among the two catalytic domains, only one catalytic domain binds to the tubulin and is important for deacetylation of tubulin [63]. It is documented that CD2 of HDAC6 has tubulin deacetylation function and the function of CD1 is not revealed [64]. Several post-translational modifications including acetylation [65], sumoylation, ubiquitination [66], phosphorylation, etc. are responsible for the regulation of deacetylase activity of HDAC6.

2.2.1 Insight into HDAC6 Crystal Structures

Several X-ray crystal structures of HDAC6 were reported from *Homo sapiens* (human) and *Danio rerio* (zebrafish), and the structures give important insights in ligand receptor interactions for different HDAC6 inhibitors. The crystal structures of both the catalytic domains have been studied extensively and reported [67]. The crystal structures of both catalytic domains from zebrafish HDAC6 in complex with inhibitors were also reported [68].

Figure 2.3 shows the important amino acids responsible for ligand–receptor interactions for HDAC6 with enantiomers of trichostatin A (TSA). The isolated catalytic domain 1 contains (*R*)-TSA in its catalytic center (Fig. 2.3a). The crystal structure of catalytic domain 2 with (*S*)-TSA bound is shown in Fig. 2.3b. The analysis of the two structures revealed that the backbone structures of both catalytic domains are very similar. The ligand binding sites in the two structures are highly conserved and clearly point out the importance of the narrow hydrophobic channel in the binding site. P83, F202, W261, etc. are important residues in case of catalytic domain 1, and F643, L712, etc. are important in catalytic domain 2. Several important residues are found in both the structures for interaction with the Zn^{2+} ions. An important difference found between the two structures is the presence of bulkier amino acid W261 in catalytic domain 1, whereas in catalytic domain 2, the amino acid is F643. Thus, (*S*)-TSA can selectively bind to HDAC6 over other HDACs in that its cap group interacts with F463. It has been also proved that catalytic domain of HDAC6 can accommodate different substrates, whereas catalytic domain 1 is highly specific

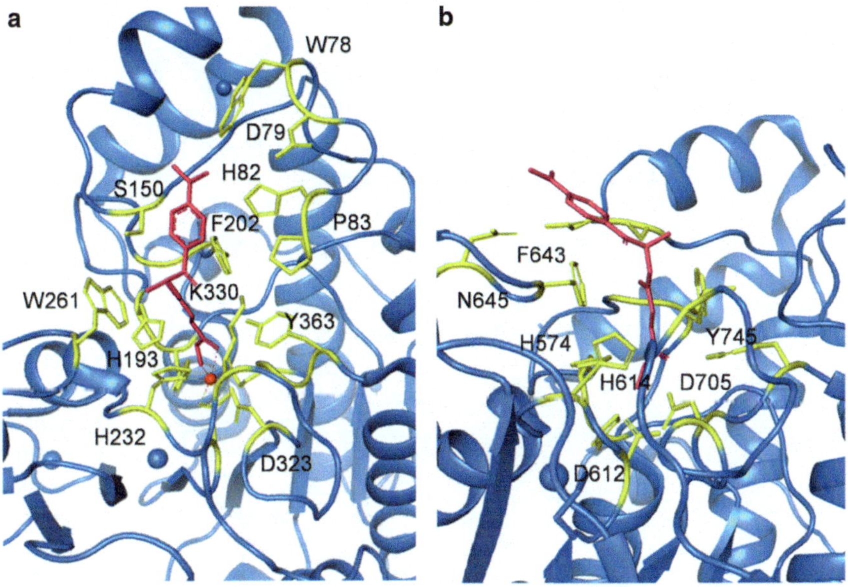

Fig. 2.3 Ligand–receptor interactions for HDAC6 with (*R*) and (*S*)-enantiomers of trichostatin A (TSA)

for the hydrolysis of C-terminal of acetyl-lysine residues. In different HDACs, the inhibitor is involved in bidentate coordination with the Zn^{2+} ion in its ligand–receptor interactions, whereas a specific inhibitor of HDAC6 is involved in monodentate coordination with the Zn^{2+} revealing unique enzyme specificity of HDAC6 enzyme. The detailed interaction showing important amino acids in ligand–receptor interactions for RTS-V5 is shown in Fig. 2.4a. The hydroxamate moiety of RTS-V5 interacts with the active site Zn^{2+} in monodentate fashion. Beyond the Zn^{2+} coordination, there are other amino acids found important for its selectivity and specificity for HDAC6. The aromatic ring of phenyl hydroxamate of the inhibitor is very close to the amino acid F643. In the complex, it has been found that S531 forms hydrogen bond with the inhibitor which may be important also for its selectivity. Lastly, typical bidentate and monodentate interactions with different inhibitors are highlighted in Fig. 2.4b, c, respectively.

From the analysis of different ligand–receptor complexes, it has been found that the structure of the linker is important to maintain the orientation of the cap group of the inhibitor toward the loop. The specific inhibitor of HDAC6 generally shows important interaction with the amino acids like F583 and F643. In case of RTS-V5, it has been found that the linker is very close to the amino acid F643. This interaction is unique for HDAC6-specific inhibitor. In general, the binding site of HDAC6 is wider than other class I HDACs. This feature allows the binding of inhibitors with bulky cap groups as well as aromatic or heteroaromatic linker features to the active site of HDAC6. This feature can be taken into account for the design of selective HDAC6 inhibitor.

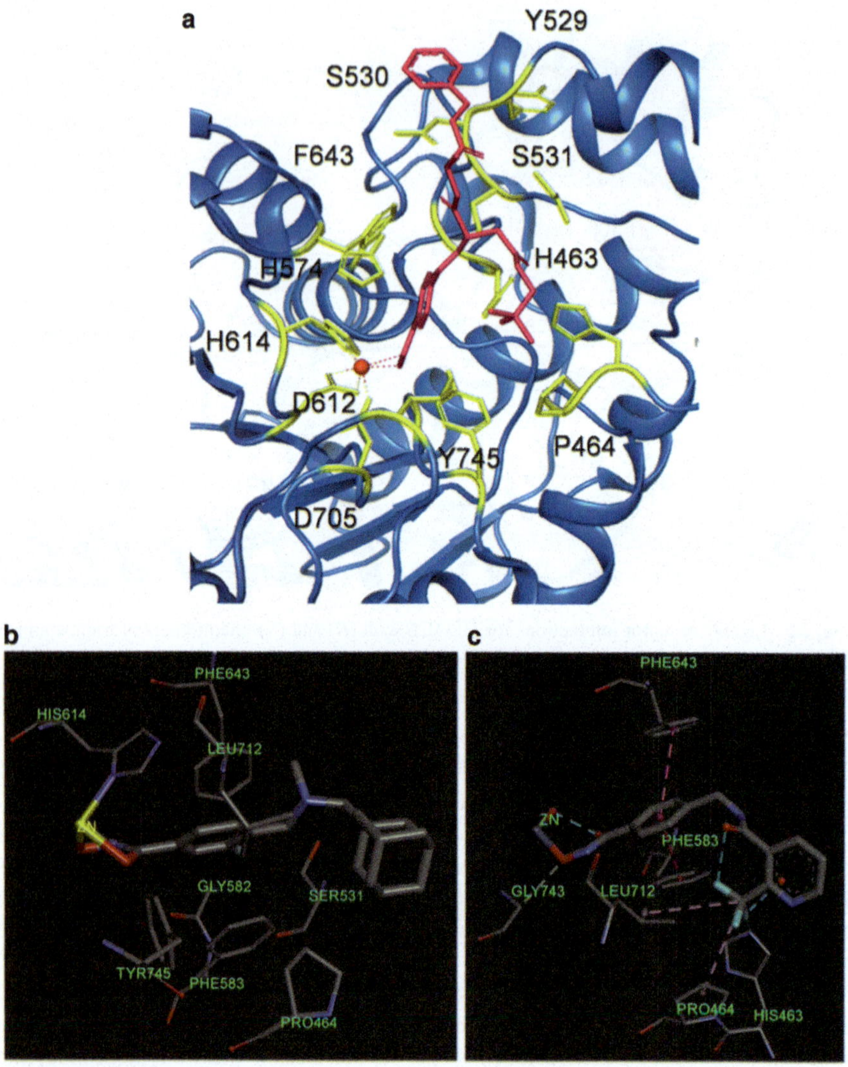

Fig. 2.4 **a** Ligand–receptor interaction for the inhibitor RTS-V5 with *dr*HDAC6 CD2 binding site (PDB: 6CW8). **b** Bidentate (PDB: 6DVO), **c** monodentate (PDB: 6PZO) interactions of the inhibitors with the HDAC6 receptor

2.2.2 Insight into HDAC10 Crystal Structures

The HDAC10 crystal structure was determined at 2.85 Å resolutions for Y307F zHDAC10 complexed with the trifluoromethyl ketone inhibitor. The structure has a butterfly-like architecture where each domain adopted the α/β fold observed in other HDAC proteins (Fig. 2.5). The structure is comprised amino terminal polyamine

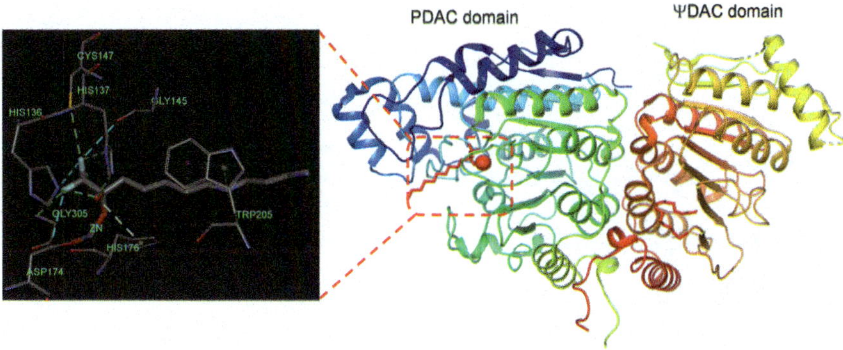

Fig. 2.5 Structure of Y307F zHDAC10 complexed with the trifluoromethyl ketone inhibitor

deacetylase (PDAC) domain as well as the C-terminal pseudodeacetylase (ΨDAC) domain. PDAC domain of HDAC10 is catalytically active, whereas in case of HDAC6, both domains are catalytically active. It was revealed that the tertiary structure of the PDAC domain is almost similar to the catalytic domain 1 and catalytic domain 2 of HDAC6 [69–71]. The ligand binding site was situated at the base of the active site tunnel of HDAC10 where the inhibitor bound in an extended conformation. The trifluoromethyl ketone moiety of the inhibitor is making asymmetric interactions with the Zn^{2+} ion in HDAC10 ligand binding site. Two histidine residues H136 and H137 are making important hydrogen bond interactions with the receptor. A close-up view of the active site of the enzyme HDAC10 revealed that the active site is much constricted than other HDACs like HDAC6. In the active site, an amino acid E274 is acting as gatekeeper residue and its electrostatic interactions with the ligand may be responsible for specificity of the enzyme activity. Another structural features present in the PDAC domain is that presence of 3_{10} helix having a consensus sequence P_{23}(E,A)CE_{26} sterically constricts the binding site of HDAC10. This allows the binding of long slender polyamines in the HDAC10 binding site [2]. Thus, unique structural features present in HDAC10 can be exploited for the design of the selective HDAC10 inhibitors.

In order to get more structural insights with the catalytic mechanism involving HDAC10, X-ray crystallographic study with the intact substrates into the active site of "humanized" D. rerio (zebrafish) HDAC10 having A24E and D94A substitutions was performed [63]. The structure gives important insights into substrate recognition process in HDAC10 as well as stabilization mechanism of transition states in the catalysis process. The studies highlight the importance of Y307 to assist the Zn^{2+} ion in polarizing the substrate carbonyl and stabilize the negative charge in transition state complexes [70]. Recently, the crystal structure of HDAC10-Tubastatin A was solved at 2.00 Å resolutions (Fig. 2.6). The structure shows that the hydroxamate moiety of Tubastatin A forms complex with the Zn^{2+} ion in the active site of HDAC10 [71]. There is also a specific hydrogen bond formation with the ligand carbonyl group to the Y307. The important histidine's H136 and H137 form important hydrogen

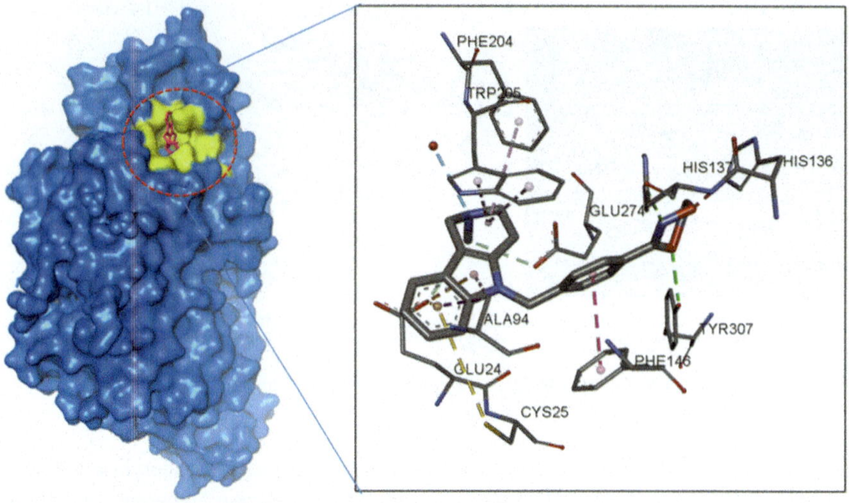

Fig. 2.6 Crystal structure of HDAC10-tubastatin A complex

bond also with the ligand. The role of the histidine dyad in the HDAC10 reaction mechanism bears some resemblance to that of HDAC6.

The phenyl group of the ligand forms aromatic interaction with the W205 and F146. However, this aromatic interaction is not contributing to the selectivity of HDAC10 as similar aromatic cavity is also seen in case of HDAC6. The tricyclic tetrahydro-γ-carboline group present in the ligand acts as a capping group of HDAC10 inhibitors. This is mainly interacting with the indole moiety of W205. E24 and E274 form important electrostatic interaction with the ligand as shown in Fig. 2.6. These interactions will be very helpful to guide the selective HDAC10 inhibitors. However, there is no structure available for human HDAC10 with the inhibitors, which may guide more effectively the design of selective HDAC10 inhibitors.

2.3 Different Tools of in Silico Drug Discovery and Its Applications

Numerous CADD applications are utilized at almost early phases of the drug discovery cascades. Thus, CADD can be described as a method to accelerate and economize the method of the drug development process [61, 72–75]. It allows better engaging on experiments and subsequently reduces the cost as well as time to find new drugs. CADD comprises (i) in silico design and prediction of novel compounds by making the drug discovery and development process faster, (ii) identifying and optimizing new compounds by the aid of computational approach, and (iii) eliminate

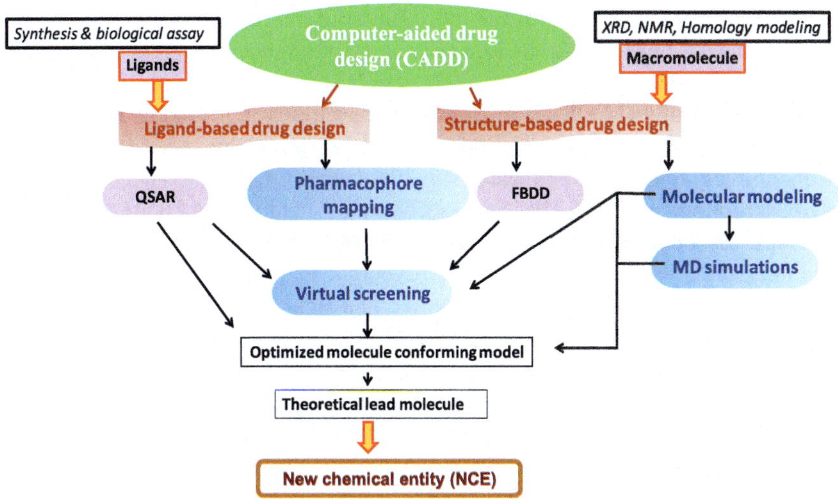

Fig. 2.7 Different tools of in silico drug discovery describing about structure- or ligand-based approaches of lead optimization

compounds with undesirable properties and selecting candidates with more chances for success.

Pharmacophore-based techniques currently are an integral part of many CADD workflows (Fig. 2.7) [76–78] and have been extensively employed for many assignments such as de novo design, virtual screening, and lead optimization. Pharmacophore model can be generated from both receptor-based and ligand-based techniques (Fig. 2.7).

Similarly, molecular docking [77, 81–85] and molecular dynamic (MD) simulations allow to understand the three-dimensional binding mode of a given molecule in the binding site of a macromolecule (protein/DNA). The binding affinity can also be quantitatively predicted by a docking score, and the stability of the protein–ligand complex is judged by proper MD simulation studies [86, 87]. More interestingly, pharmacophore-based virtual screening when combined with docking analyses provides great chance of acceptability [87].

2.3.1 Design Strategies for HDAC6 Inhibitors

To examine the potential of achieving several dimensions of isoform selectivity in the inhibition of HDACs, Kozikowski et al. in 2008 synthesized a series of structurally distinct HDAC inhibitors and applied QSAR modeling strategies to explain the potency of HDAC6 inhibitory activities as well as its selectivity over HDAC1, HDAC2, HDAC8, and HDAC10. The inhibitors are having the 2,4′-diaminobiphenyl group, which is suitably decorated with an amino acid residue at the *o*-amino group

[88]. The amino acid acts as a potential isoform differentiating, surface recognition element, and it is linked to a hydroxamate or mercaptoacetamide moieties that chelate to the catalytic zinc ion.

Different significant QSAR models were developed individually for HDAC6, HDAC1, HDAC2, HDAC8, and HDAC10. These models highlighted the importance of lipophilicity (clogP) and indicator variables I-NHCOCH2SH, I-Thiazole for the HDAC inhibitory activities. The result nicely explained the higher HDAC6 inhibitory activities for phenylthiazoles (compound **2**, compound **3**) and lower HDAC6 inhibitory activities for biphenyl mercaptoacetamides (compound **1**) (Fig. 2.8). The cap group of the inhibitor is not contributing in a significant way for the inhibitory activities as explained in Sect. 2.2.1. The QSAR models were also developed for the selectivity of HDAC6 over other HDACs. These models explained the effects of different structural and physicochemical properties of the inhibitors for its selective HDAC6 inhibition. In summary, these modeling strategies nicely correlated the experimental and predicted HDAC6 activities. In addition, cell-based experiments were carried out to determine the possible isoform and tissue selectivity of these novel inhibitors. Finally, this study drew attention to the fact that certain mercaptoacetamides do show useful levels of HDAC6 selectivity. Most importantly, the current research has identified two hydroxamates bearing meta-substituted phenylthiazole CAPs (compound **2**, compound **3**) that have IC_{50} values < 0.2 nM in in vitro HDAC6 inhibition studies. Moreover, several phenylthiazoles were found to exhibit submicromolar to low nanomolar IC_{50} values in the pancreatic cancer cell proliferation studies.

Tang et al. used a combinatorial QSAR approach to build models for 59 chemically diverse HDAC inhibitors [89]. The studies identified a novel HDAC6 inhibitor that signifies the power of QSAR-based virtual screening strategies in HDAC-targeted drug discovery. The variable selection methods of *k* nearest neighbor (*k*NN) and support vector machines (SVM) are used in QSAR model building independently by the use of Molconn Z and MOE chemical descriptors [90]. Highly predictive QSAR models were developed with leave-one-out cross-validated (LOO-CV) q^2 and external R^2 values as high as 0.80 and 0.87, respectively, utilizing the kNN/Molconn Z approach. Extensive external validations on both *k*NN and SVM models were conducted using two external datasets as described in Fig. 2.9. The Y-randomization test was run in addition to external validation to determine the model's robustness. A rigorously validated QSAR models were used for virtual screening (VS) of an in-house database collection of over 9.5 million compounds compiled from the ZINC7.0 database, the ASINEX Synergy libraries, the World Drug Index (WDI) database, and other commercial databases. The study yielded 45 novel putative HDAC inhibitors. These computational hits contained several unique structural features that were absent in the original dataset. Four computational hits with interesting chemical features were evaluated, out of which one compound was identified as selective HDAC6 inhibitor (compound **4**).

Zhao et al. developed the models of two classes of HDAC inhibitors (HDAC1 and HDAC6) in 2013 [90]. The selectivity and activity of HDAC inhibitors were studied

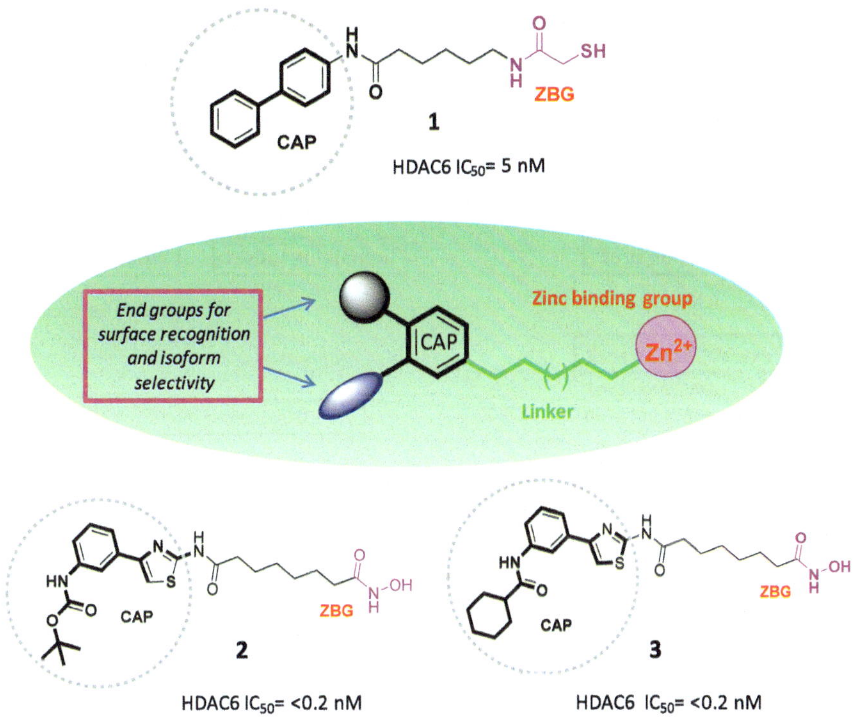

Fig. 2.8 Model structure containing a ZBG, linker, and a CAP region with end groups for surface recognition and isoform selectivity for HDAC6 inhibition. Compound 1 contains a biphenyl cap and it is linked to a mercaptoacetamide moiety that chelates to the catalytic zinc ion. Compounds 2 and 3 are two hydroxamates bearing *meta*-substituted phenylthiazole CAPs that exhibit < 0.2 nM IC_{50} values in the in vitro HDAC6 inhibition

using a two-step modeling approach. A schematic representation of the novel QSAR approach is depicted in Fig. 2.10.

First, a binary classification model was built to classify two types of inhibitors based on their activity against HDAC1 and HDAC6. Then, for each subclass, two continuous models were created to predict the activity value of HDAC1 and HDAC6 inhibitors. All three models were developed using the GA-kNN method and dragon descriptors. External validation was performed using an external prediction set and Y-randomization test. For each of the three datasets, highly predictive models were constructed. The classification accuracies of the models for the external test set were as high as 100% for the classification model. External R^2 values for HDAC1 and HDAC6 inhibitor consecutive models were 0.947 and 0.911, respectively. The outcomes validated the models' accuracy. All of the models were used to screen 1,000 compounds from the PubMed dataset. Virtual screening yielded 13 structurally diverse consensus hits as HDAC6 inhibitors.

Pham-The et al. in 2017 explored diverse machine learning (ML) techniques for the development of reliable QSAR models capable of distinguishing HDAC6 to

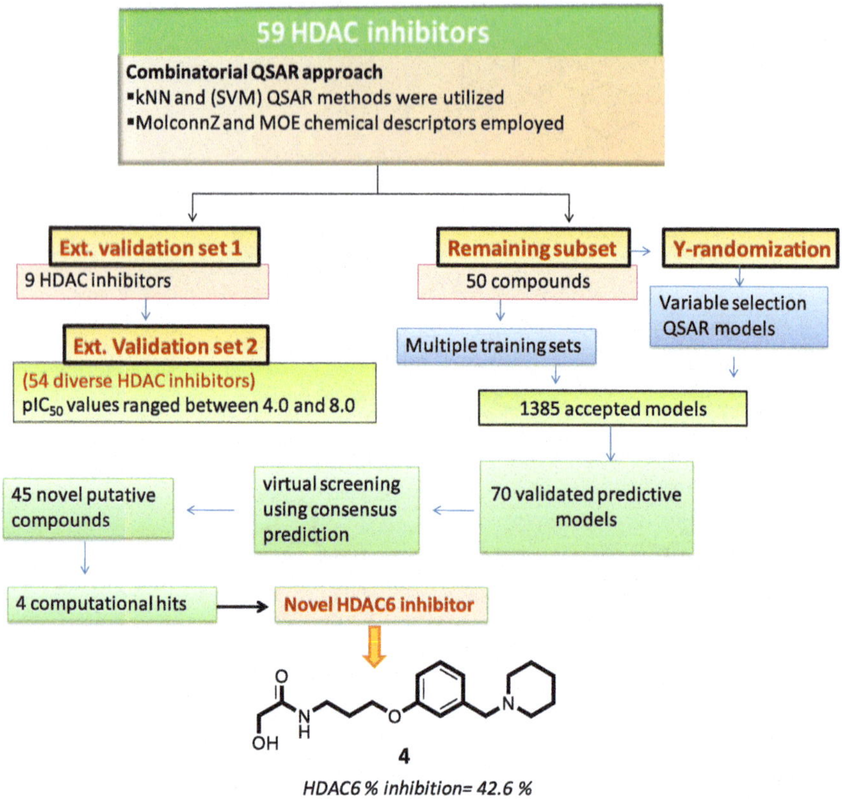

Fig. 2.9 Identification of novel inhibitor (compound **4**) for HDAC6 by QSAR modeling of known inhibitors, virtual screening, and experimental validation

HDAC2 inhibitors [91]. The ChEMBL (https://www.ebi.ac.uk/chembl/) and Drug-Bank databases were used to curate a large, structurally diverse collection of chemicals. The database contains 191 compounds as HDAC6 inhibitor/HDAC2 non-inhibitor and 95 compounds as HDAC6 non-inhibitor/HDAC2 inhibitor. The study pointed out several important compounds such as quinazoline-4-one derivatives (**5**), tetrahydro-1H-benzazepines (**6**), biphenyl hydroxylpyridin-2-thiones (**7**), 3-hydroxypyridine-2-thiones (**8**), and phenyl hydroxamic acids (**9**) as potential HDAC6 inhibitors (Fig. 2.11).

Zeb et al. in 2018 designed a study to investigate non-hydroxamate HDAC6 inhibitors [92]. Ligand-based pharmacophore was established from a training set of 26 compounds of HDAC6 inhibitors. A lowest total cost of 115.63, highest cost difference of 135.00, lowest RMSD of 0.70, and highest correlation of 0.98 were the statistical parameters of pharmacophore (Hypo1). Fischer's randomization and test set validation methods were used to validate the pharmacophore, which was then used as a screening tool for chemical databases. The pharmacophore model (Fig. 2.12)

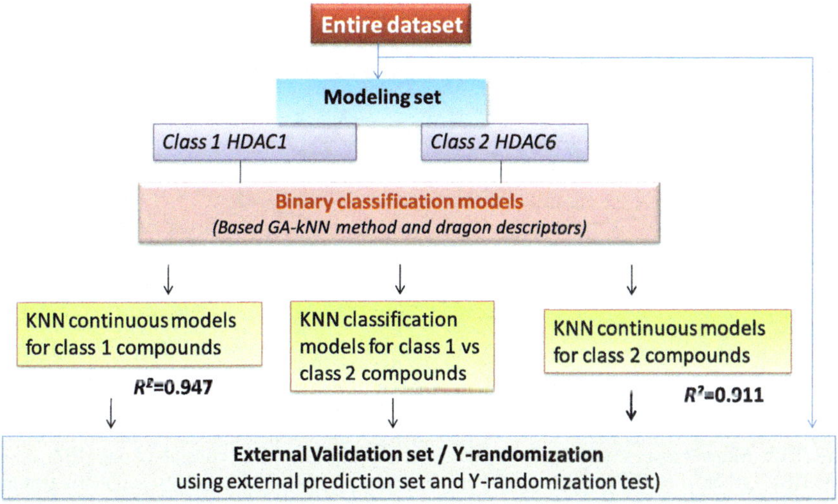

Fig. 2.10 Workflow assigned for two-step QSAR approach

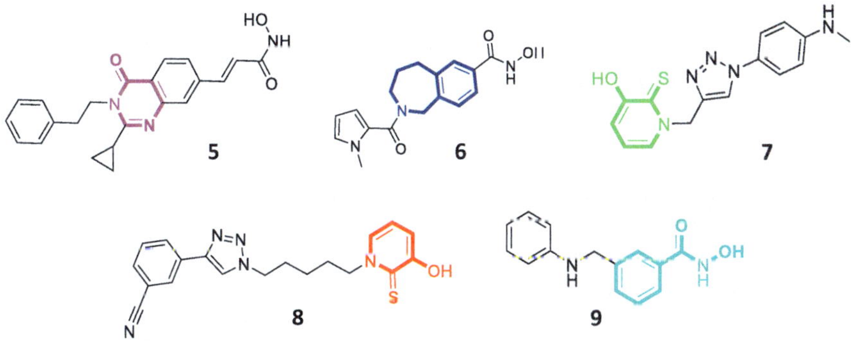

Fig. 2.11 Different HDAC6-selective inhibitors pointed out in the study. The colored region indicates important scaffolds like quinazoline-4-one moiety (pink), tetrahydro-1H-benzazepine (blue), biphenyl hydroxylpyridin-2-thiones (green), 3-hydroxypyridine-2-thione (red), and phenyl hydroxamic acid (sky blue) present in compounds 5, 6, 7, 8, and 9, respectively

indicates four features like HBA, HBD, RA, and HYP which are important for the design of HDAC6 inhibitor.

The pharmacophore-based screening methods were applied to identify novel HDAC6 inhibitors. To identify drug-like compounds, the screened compounds were analyzed using fit value (> 10.00), estimated Inhibitory Concentration (IC_{50}) (< 0.459), Lipinski's Rule of Five, and ADMET Descriptors. In addition, the drug-like hit compounds were docked into the active site of HDAC6 (PDB ID: 5EDU) using GOLD software. The best docked compounds were selected on the basis of goldfitness score > 66.46 and chemscore < 28.31 and hydrogen bonds with catalytic

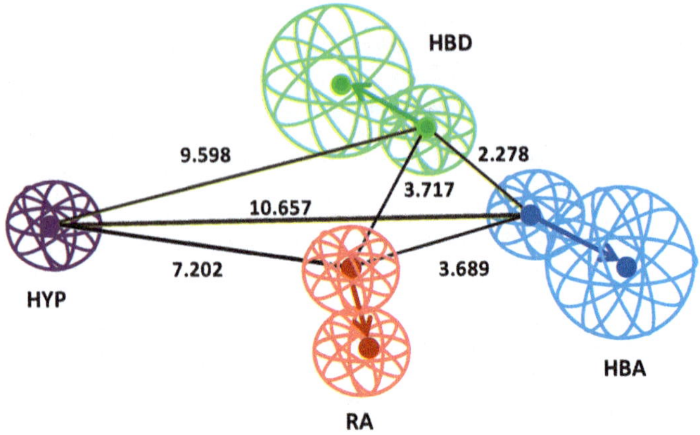

Fig. 2.12 Manual representation of pharmacophore model. The model consists of one HBA (blue), one HBD (green), one RA (magenta), and one HYP (purple) as important pharmacophoric features

active residues. The binding modes of the final three hit compounds were investigated by also using a 20-ns MD simulation. The MD simulation results showed that the hit compounds formed several interactions including π-π—stacking, hydrogen bonds, π-cation, π-sulfur, and hydrophobic interactions with the active site residues of HDAC6. Furthermore, docking analysis was used to assess the proposed specificity of the newly discovered hits against HDAC8. The final hit molecules (**10**, **11**, and **12**) have been proposed as promising platforms for the development of novel HDAC6 inhibitors (Fig. 2.13).

Debnath et al. generated a number of five featured pharmacophore hypotheses to identify selective HDAC6 inhibitors [93]. The study involved a combination of pharmacophore-based virtual screening, molecular docking, 3D-QSAR, absorption, distribution, metabolism, excretion, and toxicity (ADMET) study, and in vitro HDAC6 inhibitory activity assay of identified hits. Thirty-two known HDAC inhibitors were selected from the literature and a common pharmacophore hypothesis was generated. The best hypotheses ADDRR4 was composed of five features: one hydrogen bond acceptors (A2), two hydrogen bond donors (D3, D4), and two aromatic rings (R7 and R8). The 3D-QSAR model developed from the pharmacophore ADDRR4 was utilized to search the Phase database. The ligand pharmacophore mapping process was used to identify compounds in the database that shared at least four pharmacophoric features. This step yielded 500 top-scoring hits with fitness scores ≥ 1.0. These hit molecules were subjected to ADME filtration followed by molecular docking against HDAC6 (PDB ID: 5WPB) to predict the binding affinity toward the HDAC6. Generated pharmacophores were employed to match the final five hits. In vitro HDAC inhibitory activity clearly demonstrated that compound **13** showed marginal selectivity for HDAC6 ($IC_{50} = 0.62$ nM). It was investigated that the hit compound **13** preferentially binds to the catalytic domain 2 rather than

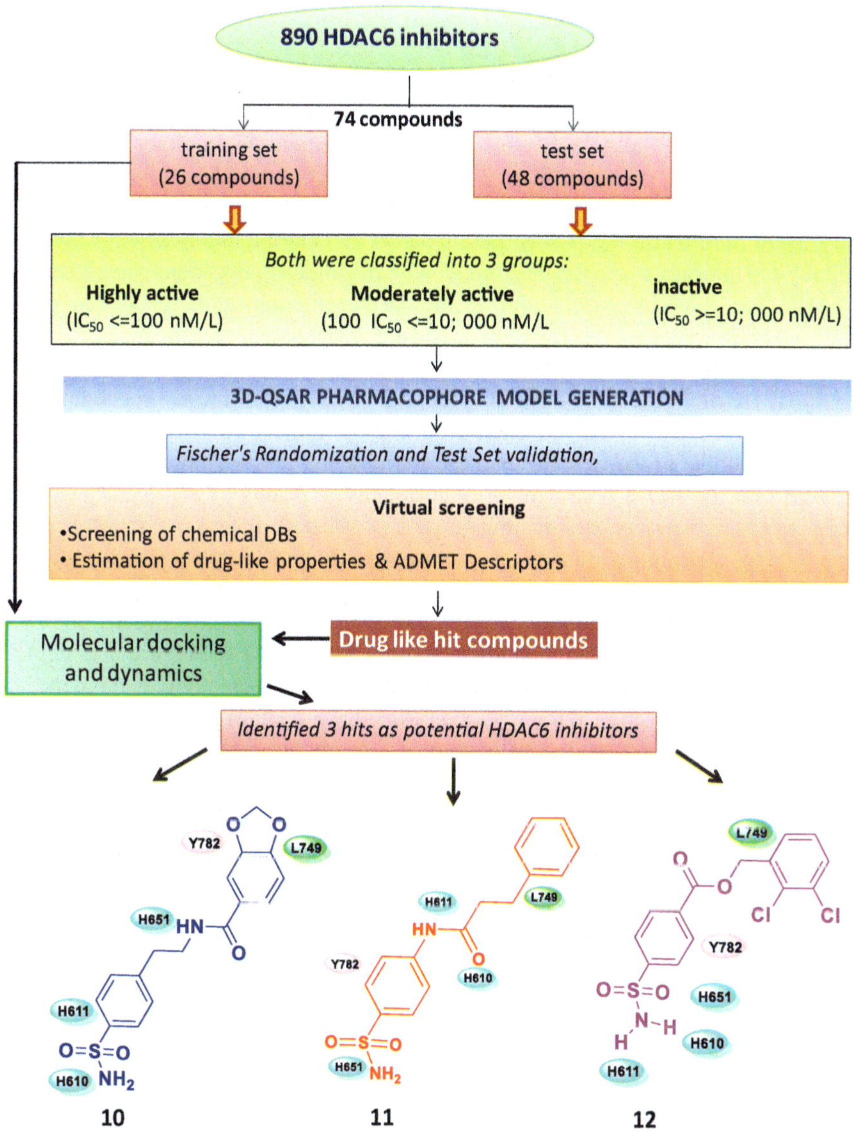

Fig. 2.13 Investigation of non-hydroxamate-based HDAC6 inhibitors using pharmacophore modeling, molecular docking, and molecular dynamics simulation approach

the zinc finger domain. Figure 2.14 depicts the workflow for the identification of novel HDAC6 inhibitor compound **13**.

Sharma et al. carried out a 3D-QSAR study on structurally diverse HDAC6 inhibitors [94]. The inhibitory activities of the diverse set of ligands were correlated with different field descriptors (steric, electrostatic, hydrophobic, donor, and acceptor) by using common feature-based pharmacophore alignment for the 3D-QSAR CoMFA, CoMSIA, docking, and MD simulations. The HipHop module in Discovery Studio (DS v2.5) was used to select and align a highly selective dataset of HDAC6 inhibitors, and the best model with the highest rank score was chosen for CoMFA and CoMSIA model building. Internal and external validation methods were used to validate the models. The models had a good statistical significance and a high correlation of 0.978 and 0.991 for the best CoMFA and CoMSIA models, respectively. These models could be used to predict the activities of the test set compounds as well as to derive useful information about the molecules' steric, electrostatic, and hydrophobic properties. Further, the training and test set molecules were then docked into the HDAC6 active site to correlate the results of 3D-QSAR with the molecular interactions within the binding site of HDAC6. In addition to this, molecular dynamics was used to indicate structural requirements for new inhibitor design. The docked poses were compared with the contour maps obtained from CoMFA/CoMSIA. For HDAC6 inhibitory action, the contour map studies revealed the relative relevance of steric, electrostatic, donor, and acceptor fields. Figure 2.15 depicts the pharmacophoric characteristics that are critical for the selectivity and potency of HDAC6 inhibitory action. This work confirms the soundness and predictive power of the proposed models with excellent consistency between molecular docking and

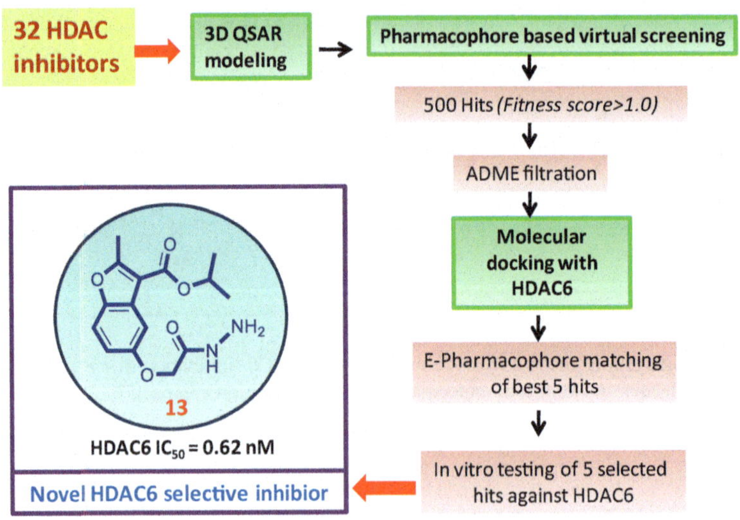

Fig. 2.14 Screening workflow for the novel HDAC6-selective inhibitor **13**. Inhibitory activity of compound 13 against HDAC6 was 0.62 nM

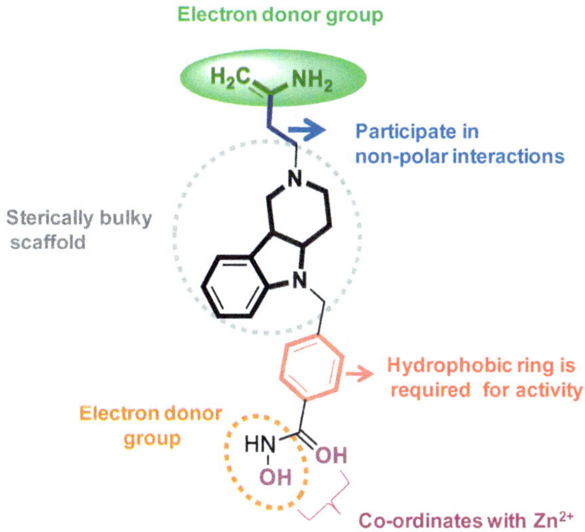

Fig. 2.15 Pharmacophoric features essential for HDAC6 inhibitory activity [94]

CoMFA and CoMSIA contour maps and the also provides valuable insights for a deeper comprehension of the basic physicochemical and structural features for designing novel HDAC6 inhibitors for drug development.

To define structural motifs which aid HDAC6 selectivity, Ruzic et al. developed valid and reliable 3D-QSAR models in order to define structural features that influence the inhibitory potency of hydroxamic acid-based HDAC1 and HDAC6 inhibitors [95]. Virtual docking studies using crystal structures of human HDAC6 isoforms were performed for conformer generation of dataset compounds. An attempt was made to eliminate bias in conformer synthesis by employing geometrical optimization, and as a result, using this innovative approach, descriptors for 3D-QSAR study related to virtually bioactive conformations were developed. The obtained conformers were utilized for generation of specific molecular descriptors (Grid-Independent Descriptors–GRIND) and 3D-QSAR modeling. A comparative analysis of the descriptors was used, to select specific structural determinants of the compounds for selective HDAC6 inhibition. GRIND descriptors were utilized using a combination of ligand-based and fragment-based approaches to define the differences in structural determinants critical for HDAC6 inhibition, as well as to provide directions for designing novel HDAC6-selective inhibitors. The potential new ligands were chosen for further research based on their projected HDAC6 selectivity, pharmacokinetic profile, synthetic tractability, and in silico cytotoxicity against a variety of human cancer cell lines. TIP-TIP GRIND variables were used to define the overall molecular shape of the HDAC inhibitors. The previously stated variables agreed with the DRY-O GRIND variable (var303) in the HDAC6 3D-QSAR model, indicating important hydrophobic interactions of the CAP region with the outside rim of HDAC6 second catalytic pocket. These variables showed that the distance between the CAP group and the hydroxamic acid (linker length) should be increased (number of carbon atoms equal or higher than 6). Fragment-based approach indicated 14 compounds

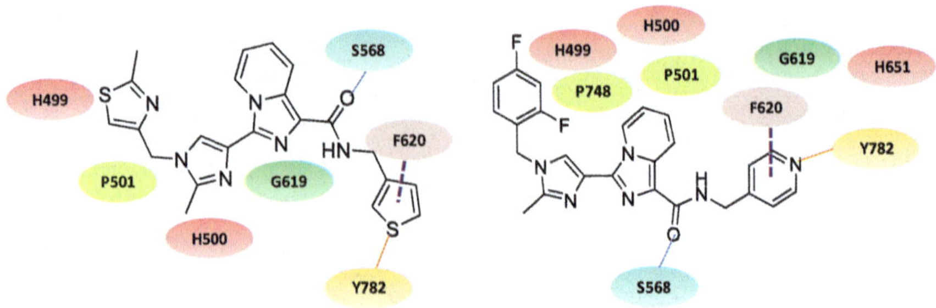

14 HDAC6 %inhibition (20μM)= 81%

15 HDAC6 %inhibition (20μM)= 61%

Fig. 2.16 Structures of highly selective HDAC6 inhibitors **14** and **15**

with increased selectivity toward HDAC6 isoform. The leverage values were used to populate the novel developed compounds into the expected applicability domains, and hence, the *pKi* values of the designed compounds were regarded accurately predicted. Finally, the proposed compounds were chosen for synthesis and in vitro testing because they had a superior anticipated HDAC6 selectivity profile, enhanced in silico ADME properties, and lower in silico toxicological concerns than clinically approved HDACi. The rational drug design technique used in this study would aid in the search for new chemically appealing CAP groups that could behave as unique interaction groups with the outer rim of HDAC6 CD2 binding pocket.

In order to discover novel inhibitor for targeting HDAC6 selectively for clinical applications, Yan et al. [96] employed virtual screening methods through the pharmacophore generation and pharmacophore-based virtual screening, molecular docking, PAINS remover, and MD simulations. Following virtual screening, 15 molecules were chosen for biological activity research. Two of the hits (compounds **14** and **15**) demonstrated inhibitory efficacy against HDAC6 after in vitro bioassays, with compound **14** inhibiting HDAC6 by 81% at a dose of 20 μM. Furthermore, inhibitory efficacy showed that these compounds were highly selective to HDAC6 (Fig. 2.16). The investigation of binding modalities of these hit molecules can provide a point of reference for the future development of highly active HDAC6 inhibitors.

2.4 Design Strategies for HDAC 10 Inhibitors

Histone deacetylase 10 (HDAC10) has been linked to the pathophysiology of several malignancies and neurological illnesses, making the finding of new inhibitors of the isoform a crucial undertaking. However, the lack of a crystallographic structure of human HDAC10 (hHDAC10) hinders efforts to develop structure-based drugs. Furthermore, from the standpoints of medicinal chemistry and structure-based drug design, HDAC10 has received little attention as a cancer therapeutic target so far.

Uba et al. in 2019 employed the X-ray crystal structure of *Danio rerio* (zebrafish) HDAC10 (PDB ID: 5TD7) as a template structure to model the 3D structure of human HDAC10 due to their high sequence similarity, especially around their catalytic domain [97]. The best model (M0017) was evaluated by computing its z-score (a measure of the deviation of the total energy of the structure with respect to an energy distribution derived from random conformations) and by docking of a series of known HDAC10 inhibitors to its catalytic cavity. Moreover, ligand-based virtual screening against the ZINC database was used to identify potential HDAC10-selective inhibitors. The screening identified the top 100 hits with scaffolds similar to quisinostat. M0017, the top-ranked built homology model, and its complexes with quisinostat, as well as the best hit, compound **16**, were subjected to molecular dynamics simulations. CHARMM-GUI server employing CHARMM36 force field (http://www.charmm-gui.org/) was used to generate the input files for nanoscale molecular dynamics (NAMD). M0017's catalytic domain was thus extracted and aligned to the catalytic domain of *Danio rerio* HDAC10's experimental crystal structure. Comparative analysis of root mean-squared deviation (RMSD), root-mean-squared fluctuation (RMSF), potential energy, and radius of gyration (Rg) of these systems revealed that HDAC10 compound **16** complex remained the most stable over time. Therefore, M0017 could be used as a structure-based inhibitor against HDAC10, and compound **16** (Fig. 2.17) could be used as a lead compound and a scaffold for further optimization in the direction of selective HDAC10 inhibitor design.

In 2020, Uba et al. performed the first molecular dynamics study of *h*HDAC10 developed using the crystallographic structure of Danio rerio (zebrafish) HDAC10 (zHDAC10) as a template [98]. The objective behind exploring the two homologs, despite the fact that they came from different organisms, was to allow for robust comparisons that led to the inference of reliable *h*HDAC10 conformation from *z*HDAC10. Both *h*HDAC10 and zHDAC10, as well as their complexes

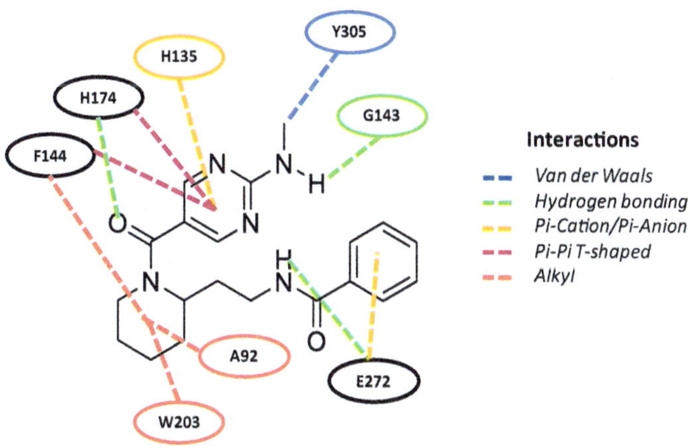

Fig. 2.17 Different interactions shown by hit compound **16** in the active site of HDAC10

with trichostatin A (TSA), quisinostat, and the natural ligand (5TD7), 7-[(3-aminopropyl)amino]-1,1,1-trifluoroheptane-2,2-diol (PDB ID; FKS), were subjected to 100 ns unrestrained MD simulations. Comparative analyses of the MD trajectories revealed that *z*HDAC10 and its complexes were more stable over time than *h*HDAC10 and its associated complexes. Nonetheless, docking of active and inactive set molecules demonstrated that more stable conformations of *h*HDAC10 might be obtained over time [98]. The current findings suggest that the initial predicted structure of *h*HDAC10 may not accurately reflect the active conformation in the real biological environment. The experimentally determined crystallographic structure, *z*HDAC10, demonstrated greater structural stability than the homology-modeled structure, *h*HDAC10. Further docking of inactive and active set molecules indicated that the modeled structure could be utilized to develop structural based inhibitors with confidence. Overall, this research could offer more information on the reliability of the HDAC10-predicted structure for application in selective inhibitor design.

Geraldy et al. discovered tubastatin A, an HDAC6 inhibitor, to potently bind with HDAC10 pocket [99]. Docking into human HDAC10 homology models revealed that a hydrogen bond between cap group nitrogen atoms is responsible for strong HDAC10 binding. The gatekeeper residue Glu272 is the single element that confers HDAC10 PDAC over HDAC activity (Fig. 2.18). This selectivity is due to the polar interaction between glutamic acid and a positively charged polyamine. Different homology models of *h*HDAC10 were built and docking studies were performed to investigate the binding of inhibitors into HDAC10 pocket. Initially, two *h*HDAC10 homology models Model I and Model II were prepared, based on a crystal structure of *z*HDAC10 (PDB ID 5td7) and *h*HDAC6 (PDB ID 5edu), respectively. Docking of tubastatin A was performed using both Model I and Model II. Model I failed to demonstrate reliable docking poses, whereas Model II demonstrated high-scoring docking poses of tubastatin A, with a hydrogen bond to Glu272, and has an L1 loop that is more similar to its parent *h*HDAC6 structure (Fig. 2.18). Further, to investigate the loop flexibility, two additional models (Model III and Model IV) were built. Surprisingly, docking of tubastatin A into these models was successful, leading to hydrogen-bonded docking poses similar to what was found in Model II. To summarize, the four models only revealed reliable docking poses when the L1 loop adopted a conformation different from that of the *z*HDAC10 structure, indicating that the loop flexibility of L1 is required for tubastatin A binding.

To accelerate the drug design efforts, Herbst-Gervasoni et al. described the preparation of "humanised" zebrafish HDAC10, which has two amino acid substitutions, A24E and D94A that provide an active site outline more comparable to that of human HDAC10 [71]. The X-ray crystal structures of this HDAC10 variant complexed with tubastatin A and indole analogs with pendant tertiary amines show that inhibitors capable of hydrogen bonding with gatekeeper E274 have a high affinity and selectivity for HDAC10 over HDAC6 [71] (Fig. 2.19).

Ukey et al. used computational methods such as homology modeling, docking, pharmacophore analysis, and molecular dynamic (MD) simulations to decipher the specificity rendering interaction features of HDAC 10 and other ZnHDACs [100]. Eight approved pan-HDAC inhibitors (PHIs) and experimentally validated 91

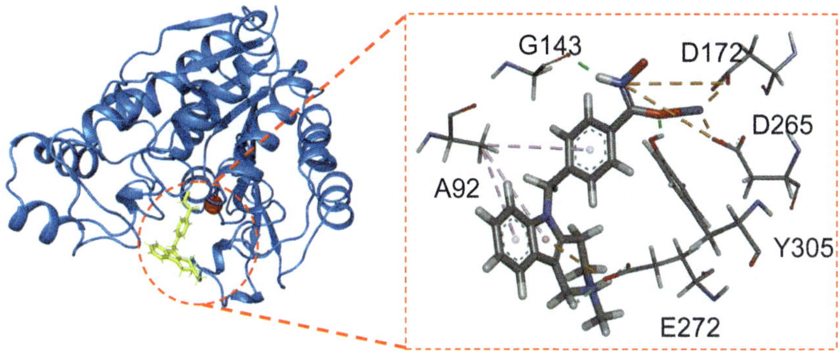

Fig. 2.18 Docking pose of tubastatin A in *h*HDAC10

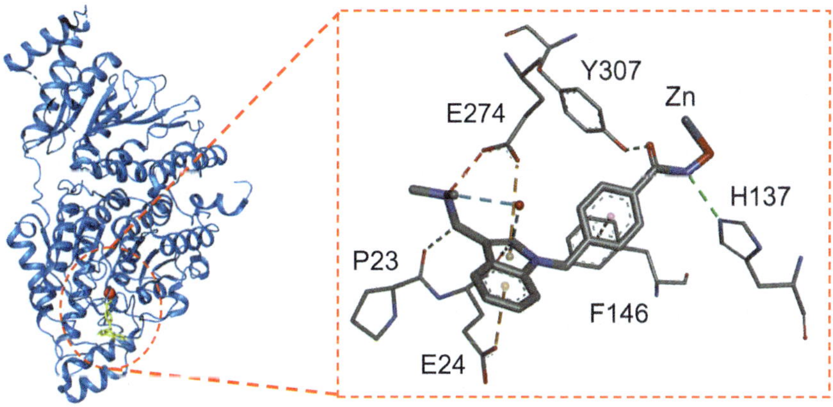

Fig. 2.19 Interaction of HDAC10 (Pdb: 6wdv) with indole analogue

reported subtype-specific inhibitors (SSIs) (activity/1000 nM) of various ZnHDACs were docked to 40 MD generated conformations of each ZnHDAC after which MM-GBSA binding energy estimates were made. To pinpoint distinct subtype-specific interaction features for each Class II ZnHDAC, the sequences, structures, physiochemical properties, and patterns of interaction of the binding sites procured from docking were thoroughly compared. A 20-ns MD simulation was run on 12 complexes (each Class II ZnHDAC bound to one SSI and one PHI) in explicit water models to further confirm the stabilities of these features. Despite high sequence similarity, different pharmacophoric patterns were seen in the binding pockets of each subtype. Due to subtype-specific protein–ligand interactions, the presence of amides, ketone, hydroxyl, carboxyl groups, moieties occupying additional sub-pockets and interacting with Zn^{2+}, etc., in the SSIs affects the orientations of the binding site

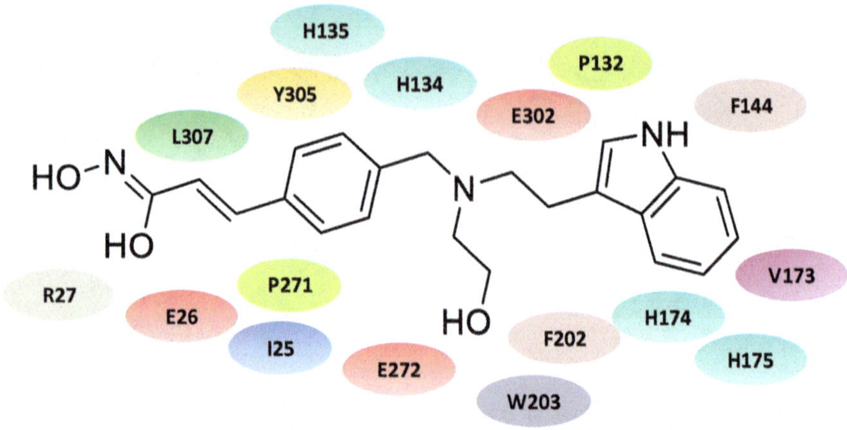

Fig. 2.20 Interactions (green: π-stacking; red: cation-π; magenta: H-bond) of SSIs and PHIs with the relevant class II ZnHDACs BS residues

residues (BSRs). Such distinct interaction features and pharmacophoric patterns (Fig. 2.20) can be used to develop subtype-specific ZnHDAC inhibitors.

2.5 Conclusion

HDAC6 is the only HDAC known to have two catalytic domains and to be predominately cytoplasmic in its structure. This distinguishing feature allowed HDAC6 to target specific substrates involved in proteasomal degradation, cell shape and migration, microtubule dynamics, apoptosis, axonal growth abnormalities, and different signaling pathways that contribute to the pathogenic response of many diseases. This broad variety of roles and activity of HDAC6 is well employed in many malignancies, neurological illnesses, epigenetic rare diseases, and inflammatory diseases. Utilizing a variety of selective HDAC6*i*, various researches have been published to date examining and comprehending the cellular and physiological roles of HDAC6. Although numerous studies described the preclinical testing of HDAC6 isoform inhibitors, only **ACY-1215** (ricolinostat) and **ACY-241** (citarinostat) (Fig. 2.21) have so far entered phase II clinical trials. Designing efficient and selective HDAC6*i* must undoubtedly take druggability and poor bioavailability into consideration. Since the bulky cap group is made up of extremely hydrophobic moieties and the linker function is mostly made up of linear elongation alkyl functions, the hydrophobicity combined with the flexibility provided by linear functionalities may result in varying interaction with HDAC enzymes.

Like HDAC6, HDAC10 demonstrates a variety of intricate functions and clinical implications. The theoretical groundwork for clinical application is laid by its functions, which include DNA damage repair, autophagy, gene transcription and

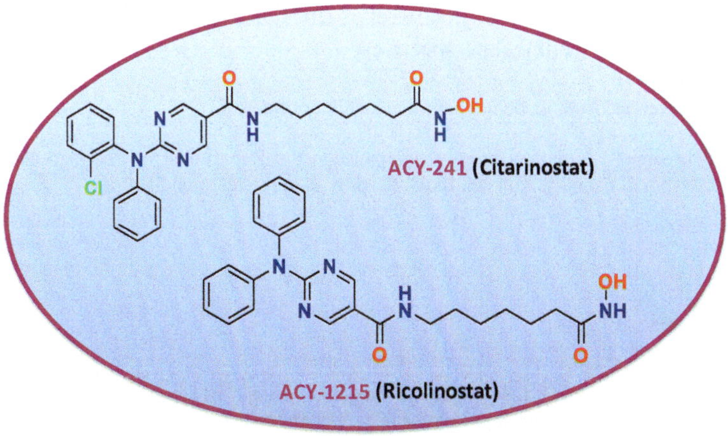

Fig. 2.21 Structures of HDAC6-selective inhibitors that have entered phase II clinical trials

HDAC10-mediated tumor cell proliferation, migration, invasion, apoptosis, metastasis, angiogenesis, and immunological modulation. There are still difficulties to be overcome, though. In order to provide a more comprehensive understanding of HDAC10 and other class IIb enzyme, further information on the mechanisms of aberrant HDAC regulation is first and foremost necessary. HDAC10 may become a crucial anti-tumor target and contribute to therapeutic applications with more in-depth research.

So far, the majority of HDAC6- and HDAC10-selective inhibitors have hydroxamic acids as ZBG with different structural changes on the cap and linker regions. Despite the fact that numerous other ZBGs with higher potencies have been identified, none of them displayed substantial cellular activity. Nonetheless, the greater zinc-binding ability of various ZBGs, such as hydroxamate, may be a serious issue in terms of selective binding to specific HDAC, resulting in undesired toxicities. As a result, the main difficulty hampered by these functions is the selectivity of HDAC Class IIb inhibitors. Moreover, these moieties are not only responsible for enzyme binding, but also for druggability and bioavailability. Aside from that, crystallographic data can be used to examine the binding mode of contacts, which can assist to discover essential functional aspects as well as vital amino acid residues. These are useful in the development of selective HDAC*i*. Furthermore, the time and effort required in designing selective inhibitors prior to synthesis and biological screening may be lessened by the initial ligand-docking interactions with the inhibitors of HDAC6 and 10 followed by ADMET screening using various modeling techniques. Taking these critical features into account, it is anticipated that selective and effective HDAC6 and HDAC10 specific inhibitors may be designed in future.

Overall, this chapter outlines the distinctive structural features of class IIb HDACs, as well as their varied interactions with inhibitors. This would assist in designing and developing more effective and selective class IIb HDACi, overcoming current hurdles and broadening their clinical perspectives.

Conflict of Interest Authors do not have any conflict of interest.

Acknowledgements Authors sincerely acknowledge the Department of Pharmaceutical Technology, Jadavpur University, Kolkata, India, for providing the research facilities.

References

1. Wang Y, Xie Q, Tan H, Liao M, Zhu S, Zheng LL, Huang H, Liu B (2021) Targeting cancer epigenetic pathways with small-molecule compounds: therapeutic efficacy and combination therapies. Pharmacol Res. https://doi.org/10.1016/j.phrs.2021.105702
2. Prachayasittikul V, Prathipati P, Pratiwi R, Phanus-Umporn C, Malik AA, Schaduangrat N, Seenprachawong K, Wongchitrat P, Supokawej A, Prachayasittikul V, Wikberg JE, Nantasenamat C (2017) Exploring the epigenetic drug discovery landscape. Expert Opin Drug Discov 12:345–362. https://doi.org/10.1080/17460441.2017.1295954
3. Miranda Furtado CL, Dos Santos Luciano MC, Silva Santos RD, Furtado GP, Moraes MO, Pessoa C (2019) Epidrugs: targeting epigenetic marks in cancer treatment. Epigenetics 14:1164–1176. https://doi.org/10.1080/15592294.2019.1640546
4. https://www.nature.com/scitable/topicpage/epigenetic-influences-and-disease-895/. Accessed on 19 June 2021
5. Yu B, Ouyang L (2019) Epigenetic regulation and drug discovery for cancer therapy. Curr Top Med Chem 19:971. https://doi.org/10.2174/156802661912190730153906
6. Shetty MG, Pai P, Deaver RE, Satyamoorthy K, Babitha KS (2021) Histone deacetylase 2 selective inhibitors: a versatile therapeutic strategy as next generation drug target in cancer therapy. Pharmacol Res 170:105695. https://doi.org/10.1016/j.phrs.2021.105695
7. Nebbioso A, Tambaro FP, Dell'Aversana C, Altucci L (2018) Cancer epigenetics: moving forward. PLoS Genet 14:e1007362. https://doi.org/10.1371/journal.pgen.1007362
8. Rasool M, Malik A, Naseer MI, Manan A, Ansari SA, Begum I, Qazi MH, Pushparaj PN, Abuzenadah AM, Al-Qahtani MH, Kamal MA, Gan SH (2015) The role of epigenetics in personalized medicine: challenges and opportunities. BMC Med Genomics 8:S5. https://doi.org/10.1186/1755-8794-8-S1-S5
9. Biel M, Wascholowski GA (2005) Epigenetics—an epicenter of gene regulation: histones and histone-modifying enzymes. Angew Chem Int Ed 44:3186–3216. https://doi.org/10.1002/anie.200461346
10. Glozak MA, Sengupta N, Zhang X, Seto E (2005) Acetylation and deacetylation of non-histone proteins. Gene 363:15–23. https://doi.org/10.1016/j.gene.2005.09.010
11. Lawlor L, Yang XB (2019) Harnessing the HDAC-histone deacetylase enzymes, inhibitors and how these can be utilised in tissue engineering. Int J Oral Sci 11:20. https://doi.org/10.1038/s41368-019-0053-2
12. Hull EE, Montgomery MR, Leyva KJ (2016) HDAC inhibitors as epigenetic regulators of the immune system: impacts on cancer therapy and inflammatory diseases. Biomed Res Int 2016:8797206. https://doi.org/10.1155/2016/8797206
13. Amin SA, Trivedi P, Adhikari N, Routholla G, Vijayasarathi D, Das S, Ghosh B, Jha T (2021) Quantitative activity–activity relationship (QAAR) driven design to develop hydroxamate derivatives of pentanoic acids as selective HDAC8 inhibitors: synthesis, biological evaluation and binding mode of interaction studies. New J Chem 45:17149. https://doi.org/10.1039/d1nj02636d

14. Bourguet E, Ozdarska K, Moroy G, Jeanblanc J, Naassila M (2018) Class I HDAC inhibitors: potential new epigenetic therapeutics for alcohol use disorder (AUD). J Med Chem 61:1745–1766. https://doi.org/10.1021/acs.jmedchem.7b00115
15. Wang N, Peng YJ, Su X, Prabhakar NR, Nanduri J (2021) Histone deacetylase 5 is an early epigenetic regulator of intermittent hypoxia induced sympathetic nerve activation and blood pressure. Front Physiol 12:688322. https://doi.org/10.3389/fphys.2021.688322
16. Cao H, Li L, Yang D, Zeng L, Yewei X, Yu B, Liao G, Chen J (2019) Recent progress in histone methyltransferase (G9a) inhibitors as anticancer agents. Eur J Med Chem 179:537–546. https://doi.org/10.1016/j.ejmech.2019.06.072
17. Vaidya GN, Rana P, Venkatesh A, Chatterjee DR, Contractor D, Satpute DP, Nagpure M, Jain A, Kumar D (2021) Paradigm shift of "classical" HDAC inhibitors to "hybrid" HDAC inhibitors in therapeutic interventions. Eur J Med Chem 209:112844. https://doi.org/10.1016/j.ejmech.2020.112844
18. Amin SA, Adhikari N, Jha T (2017) Is dual inhibition of metalloenzymes HDAC-8 and MMP-2 a potential pharmacological target to combat hematological malignancics? Pharmacol Res 122:8–19. https://doi.org/10.1016/j.phrs.2017.05.002
19. Amin SA, Adhikari N, Jha T (2017) Structure-activity relationships of hydroxamate-based histone deacetylase-8 inhibitors: reality behind anticancer drug discovery. Future Med Chem 9:2211–2237. https://doi.org/10.4155/fmc-2017-0130
20. Atadja P (2009) Development of the pan-DAC inhibitor panobinostat (LBH589): successes and challenges. Cancer Lett 280:233–241. https://doi.org/10.1016/j.canlet.2009.02.019
21. Osko JD, Christianson DW (2020) Structural determinants of affinity and selectivity in the binding of inhibitors to histone deacetylase 6. Bioorg Med Chem Lett 30:127023. https://doi.org/10.1016/j.bmcl.2020.127023
22. Rajan PK, Udoh UA, Sanabria JD, Banerjee M, Smith G, Schade MS, Sanabria J, Sodhi K, Pierre S, Xie Z, Shapiro JI, Sanabria J (2020) The role of histone acetylation-/methylation-mediated apoptotic gene regulation in hepatocellular carcinoma. Int J Mol Sci 21:8894. https://doi.org/10.3390/ijms21238894
23. de Ruijter AJM, van Gennip AH, Caron HN, Kemp S, van Kuilenburg ABP (2003) Histone deacetylases (HDACs): characterization of the classical HDAC family. Biochem J 370:737–749. https://doi.org/10.1042/bj20021321
24. Wang X-X, Wan R-Z, Liu Z-P (2018) Recent advances in the discovery of potent and selective HDAC6 inhibitors. Eur J Med Chem 143:1406–1418. https://doi.org/10.1016/j.ejmech.2017.10.040
25. Marmorstein R, Roth SY (2001) Histone acetyltransferases: function, structure, and catalysis. Curr Opin Genet Dev 11:155–161. https://doi.org/10.1016/s0959-437x(00)00173-8
26. Haberland M, Montgomery RL, Olson EN (2009) The many roles of histone deacetylases in development and physiology: implications for disease and therapy. Nat Rev Genet 10:32–42. https://doi.org/10.1038/nrg2485
27. Taunton J, Hassig CA, Schreiber SL (1996) A mammalian histone deacetylase related to the yeast transcriptional regulator Rpd3p. Science 272:408–411. https://doi.org/10.1126/science.272.5260.408
28. Yang WM, Inouye C, Zeng Y, Bearss D, Seto E (1996) Transcriptional repression by YY1 is mediated by interaction with a mammalian homolog of the yeast global regulator RPD3. PNAS 93:12845–12850. https://doi.org/10.1073/pnas.93.23.12845
29. Yang WM, Yao YL, Sun JM, Davie JR, Seto E (1997) Isolation and characterization of cDNAs corresponding to an additional member of the human histone deacetylase gene family. J Biol Chem 272:28001–28007. https://doi.org/10.1074/jbc.272.44.28001
30. Hu E, Chen Z, Fredrickson T, Zhu Y, Kirkpatrick R, Zhang GF, Johanson K, Sung CM, Liu R, Winkler J (2000) Cloning and characterization of a novel human class I histone deacetylase that functions as a transcription repressor. J Biol Chem 275:15254–15264. https://doi.org/10.1074/jbc.M908988199
31. Kao HY, Downes M, Ordentlich P, Evans RM (2000) Isolation of a novel histone deacetylase reveals that class I and class II deacetylases promote SMRT-mediated repression. Genes Dev 14:55–66

32. Kao HY, Lee CH, Komarov A, Han CC, Evans RM (2002) Isolation and characterization of mammalian HDAC10, a novel histone deacetylase. J Biol Chem 277:187–193. https://doi.org/10.1074/jbc.M108931200
33. Zhou X, Richon VM, Rifkind RA, Marks PA (2000) Identification of a transcriptional repressor related to the noncatalytic domain of histone deacetylases 4 and 5. Proc Natl Acad Sci 97:1056–1061. https://www.ncbi.nlm.nih.gov/pmc/articles/PMC15519/. Accessed 2 May 2020
34. Seto E, Yoshida M (2014) Erasers of histone acetylation: the histone deacetylase enzymes. Cold Spring Harb Perspect Biol 6:a018713. https://doi.org/10.1101/cshperspect.a018713
35. Gao L, Cueto AF, Atadja P (2002) Cloning and functional characterization of HDAC11, a novel member of the human histone deacetylase family. J Biol Chem 277:25748–25755. https://doi.org/10.1074/jbc.M111871200
36. Finnin MS, Donigian JR, Cohen A, Richon VM, Rifkind RA, Marks PA, Breslow R, Pavletich NP (1999) Structures of a histone deacetylase homologue bound to the TSA and SAHA inhibitors. Nature 401:188–193. https://doi.org/10.1038/43710
37. Wang DF, Helquist P, Wiech NL, Wiest O (2005) Toward selective histone deacetylase inhibitor design: homology modeling, docking studies, and molecular dynamics simulations of human class I histone deacetylases. J Med Chem 48:6936–6947. https://doi.org/10.1021/jm0505011
38. Adhikari N, Amin SA, Trivedi P, Jha T, Ghosh B (2018) HDAC3 is a potential validated target for cancer: an overview on the benzamide-based selective HDAC3 inhibitors through comparative SAR/QSAR/QAAR approaches. Eur J Med Chem 157:1127–1142. https://doi.org/10.1016/j.ejmech.2018.08.081
39. Grozinger CM, Hassig CA, Schreiber SL (1999) Three proteins define a class of human histone deacetylases related to yeast Hda1p. Proc Natl Acad Sci 96:4868–4873. https://doi.org/10.1073/pnas.96.9.4868
40. Li Y, Shin D, Kwon SH (2013) Histone deacetylase 6 plays a role as a distinct regulator of diverse cellular processes. FEBS J 280:775–793. https://doi.org/10.1111/febs.12079
41. Hubbert C, Guardiola A, Shao R, Kawaguchi Y, Ito A, Nixon A, Yoshida M, Wang XF, Yao TP (2002) HDAC6 is a microtubule-associated deacetylase. Nature 417:455–458. https://doi.org/10.1038/417455a
42. Shen S, Svoboda M, Zhang G, Cavasin MA, Motlova L, McKinsey TA, Eubanks JH, Bařinka C, Kozikowski AP (2020) Structural and in vivo characterization of tubastatin A, a widely used histone deacetylase 6 inhibitor. ACS Med Chem Lett 11:706–712. https://doi.org/10.1021/acsmedchemlett.9b00560
43. Simões-Pires C, Zwick V, Nurisso A, Schenker E, Carrupt PA, Cuendet M (2013) HDAC6 as a target for neurodegenerative diseases: what makes it different from the other HDACs? Mol Neurodegener 8:1–16. https://doi.org/10.1186/1750-1326-8-7
44. Li T, Zhang C, Hassan S, Liu X, Song F, Chen K, Zhang W, Yang J (2018) Histone deacetylase 6 in cancer. J Hematol Oncol 11:1–10. https://doi.org/10.1186/s13045-018-0654-9
45. Brindisi M, Saraswati AP, Brogi S, Gemma S, Butini S, Campiani G (2020) Old but gold: tracking the new guise of histone deacetylase 6 (HDAC6) enzyme as a biomarker and therapeutic target in rare diseases. J Med Chem 63:23–39. https://doi.org/10.1021/acs.jmedchem.9b00924
46. Osko JD, Porter NJ, Narayana Reddy PA, Xiao YC, Rokka J, Jung M, Hooker JM, Salvino JM, Christianson DW (2020) Exploring structural determinants of inhibitor affinity and selectivity in complexes with histone deacetylase 6. J Med Chem 63:295–308. https://doi.org/10.1021/acs.jmedchem.9b01540
47. Cheng F, Zheng B, Wang J, Zhao G, Yao Z, Niu Z, He W (2021) Histone deacetylase 10, a potential epigenetic target for therapy. Biosci Rep. 41:BSR20210462. https://doi.org/10.1042/BSR20210462
48. Pojani E, Barlocco D (2022) Selective inhibitors of histone deacetylase 10 (HDAC-10). Curr Med Chem 29:2306–2321. https://doi.org/10.2174/0929867328666210901144658

49. Morgen M, Steimbach RR, Géraldy M, Hellweg L, Sehr P, Ridinger J, Witt O, Oehme I, Herbst-Gervasoni CJ, Osko JD, Porter NJ, Christianson DW, Gunkel N, Miller AK (2020) Design and synthesis of dihydroxamic acids as HDAC6/8/10 inhibitors. ChemMedChem 15:1163–1174. https://doi.org/10.1002/cmdc.202000149
50. Oehme I, Lodrini M, Brady NR, Witt O (2013) Histone deacetylase 10-promoted autophagy as a druggable point of interference to improve the treatment response of advanced neuroblastomas. Autophagy 9:2163–2165. https://doi.org/10.4161/auto.26450
51. Lakshmaiah KC, Jacob LA, Aparna S, Lokanatha D, Saldanha SC (2014) Epigenetic therapy of cancer with histone deacetylase inhibitors. J Cancer Res Ther 10:469–478. https://doi.org/10.4103/0973-1482.137937
52. Bugide S, Gupta R, Green MR, Wajapeyee N (2021) EZH2 inhibits NK cell-mediated antitumor immunity by suppressing CXCL10 expression in an HDAC10-dependent manner. Proc Natl Acad Sci USA 118:e2102718118. https://doi.org/10.1073/pnas.2102718118
53. Muratov EN, Bajorath J, Sheridan RP, Tetko IV, Filimonov D, Poroiko V, Oprea TI, Baskin II, Varnek A, Roitberg A, Isayev O (2020) QSAR without borders. Chem Soc Rev 49:3525–3564. https://doi.org/10.1039/d0cs00098a
54. Amin SA, Gayen S (2016) Modelling the cytotoxic activity of pyrazolo-triazole hybrids using descriptors calculated from the open source tool "PaDEL-descriptor." J Taibah Univ Sci 10:896–905. https://doi.org/10.1016/j.jtusci.2016.04.009
55. Lombardino JG, Lowe JA (2004) The role of the medicinal chemist in drug discovery–then and now. Nat Rev Drug Discov 3:853–862. https://doi.org/10.1038/nrd1523
56. Hughes JP, Rees S, Kalindjian SB, Hughes PKL (2011) Principles of early drug discovery. Br J Pharmacol 162:1239–1249. https://doi.org/10.1111/j.1476-5381.2010.01127.x
57. Bajorath J (2002) Integration of virtual and high-throughput screening. Nat Rev Drug Discov 1:882–894. https://doi.org/10.1038/nrd941
58. Amin SA, Adhikari N, Jha T, Gayen S (2016) First molecular modeling report on novel arylpyrimidine kynurenine monooxygenase inhibitors through multi-QSAR analysis against Huntington's disease: a proposal to chemists! Bioorg Med Chem Lett 26:5712–5718. https://doi.org/10.1016/j.bmcl.2016.10.058
59. Kiriiri GK, Njogu PM, Mwangi AN (2020) Exploring different approaches to improve the success of drug discovery and development projects: a review. Futur J Pharm Sci 6:27. https://doi.org/10.1186/s43094-020-00047-9
60. Shaker B, Ahmad S, Lee J, Jung C, Na D (2021) In silico methods and tools for drug discovery. Comput Biol Med 137:104851. https://doi.org/10.1016/j.compbiomed.2021.104851
61. Radaeva M, Dong X, Cherkasov A (2020) The use of methods of computer-aided drug discovery in the development of topoisomerase II inhibitors: applications and future directions. J Chem Inf Model 60:3703–3721. https://doi.org/10.1021/acs.jcim.0c00325
62. Boyault C, Sadoul K, Pabion M, Khochbin S (2007) HDAC6, at the crossroads between cytoskeleton and cell signaling by acetylation and ubiquitination. Oncogene 26:5468–5476. https://doi.org/10.1038/sj.onc.1210614
63. Haggarty SJ, Koeller KM, Wong JC, Grozinger CM, Schreiber SL (2003) Domain-selective small-molecule inhibitor of histone deacetylase 6 (HDAC6)-mediated tubulin deacetylation. PNAS 100:4389–4394. https://doi.org/10.1073/pnas.0430973100
64. Zhang X, Yuan Z, Zhang Y, Yong S, Salas-Burgos A, Koomen J, Olashaw N, Parsons JT, Yang XJ, Dent SR, Yao TP, Lane WS, Seto E (2007) HDAC6 modulates cell motility by altering the acetylation level of cortactin. Mol Cell 27:197–213. https://doi.org/10.1016/j.molcel.2007.05.033
65. Han Y, Jeong HM, Jin YH, Kim YJ, Jeong HG, Yeo CY, Lee KY (2009) Acetylation of histone deacetylase 6 by p300 attenuates its deacetylase activity. Biochem Biophys Res Commun 383:88–92. https://doi.org/10.1016/j.bbrc.2009.03.147
66. Hook SS, Orian A, Cowley SM, Eisenman RN (2002) Histone deacetylase 6 binds polyubiquitin through its zinc finger (PAZ domain) and copurifies with deubiquitinating enzymes. PNAS 99:13425–13430. https://doi.org/10.1073/pnas.172511699.

67. Liu Y, Li L, Min J (2016) HDAC6 finally crystal clear. Nat Chem Biol 12:660–661. https://doi.org/10.1038/nchembio.2158
68. Miyake Y, Keusch JJ, Wang L, Saito M, Hess D, Wang X, Melancon BJ, Helquist P, Gut H, Matthias P (2016) Structural insights into HDAC6 tubulin deacetylation and its selective inhibition. Nat Chem Biol 12:748–754. https://doi.org/10.1038/nchembio.2140
69. Hai Y, Shinsky SA, Porter NJ, Christianson DW (2017) Histone deacetylase 10 structure and molecular function as a polyamine deacetylase. Nat commun 8:1–9. https://doi.org/10.1038/ncomms15368
70. Herbst-Gervasoni CJ, Christianson DW (2021) X-ray crystallographic snapshots of substrate binding in the active site of histone deacetylase 10. Biochem 60:303–313. https://doi.org/10.1021/acs.biochem.0c00936
71. Herbst-Gervasoni CJ, Steimbach RR, Morgen M, Miller AK, Christianson DW (2020) Structural basis for the selective inhibition of HDAC10, the cytosolic polyamine deacetylase. ACS Chem Biol 15:2154–2163. https://doi.org/10.1021/acschembio.0c00362
72. Amin SA, Jha T (2020) Fight against novel coronavirus: a perspective of medicinal chemists. Eur J Med Chem 201:112559. https://doi.org/10.1016/j.ejmech.2020.112559
73. Amin SA, Adhikari N, Gayen S, Jha T (2019) Reliable structural information for rational design of benzoxazole type potential cholesteryl ester transfer protein (CETP) inhibitors through multiple validated modeling techniques. J Biomol Struct Dyn 37:4528–4541. https://doi.org/10.1080/07391102.2018.1552895
74. Seidel T, Schuetz DA, Garon A, Langer T (2019) The pharmacophore concept and its applications in computer-aided drug design. Prog Chem Org Nat Prod 110:99–141. https://doi.org/10.1007/978-3-030-14632-0_4
75. Macalino SJ, Gosu V, Hong S, Choi S (2015) Role of computer-aided drug design in modern drug discovery. Arch Pharm Res 38:1686–1701. https://doi.org/10.1007/s12272-015-0640-5
76. Choudhury C, Sastry GN (2019) Pharmacophore modelling and screening: concepts, recent developments and applications in rational drug design. In: Structural bioinformatics: applications in preclinical drug discovery process, Springer, Cham, pp 25–53
77. Kontoyianni M (2017) Docking and virtual screening in drug discovery. Methods Mol Biol 1647:255–266. https://doi.org/10.1007/978-1-4939-7201-2_18
78. Voet A, Zhang KY (2012) Pharmacophore modelling as a virtual screening tool for the discovery of small molecule protein-protein interaction inhibitors. Curr Pharm Des 18:4586–4598. https://doi.org/10.2174/138161212802651616
79. Yang SY (2010) Pharmacophore modeling and applications in drug discovery: challenges and recent advances. Drug Discov Today 15:444–450. https://doi.org/10.1016/j.drudis.2010.03.013
80. Kutlushina A, Khakimova A, Madzhidov T, Polishchuk P (2018) Ligand-based pharmacophore modeling using novel 3d pharmacophore signatures. Molecules 23:3094. https://doi.org/10.3390/molecules23123094
81. Guedes IA, de Magalhães CS, Dardenne LE (2014) Receptor-ligand molecular docking. Biophys Rev 6:75–87. https://doi.org/10.1007/s12551-013-0130-2
82. Cournia Z, Allen B, Sherman W (2017) Relative binding free energy calculations in drug discovery: recent advances and practical considerations. J Chem Inf Model 57:2911–2937. https://doi.org/10.1021/acs.jcim.7b00564
83. Huang SY, Zou X (2010) Advances and challenges in protein-ligand docking. Int J Mol Sci 11:3016–3034. https://doi.org/10.3390/ijms11083016
84. Maia EHB, Assis LC, de Oliveira TA, Da Silva AM, Taranto AG (2020) Structure-based virtual screening: from classical to artificial intelligence. Front Chem 8:343. https://doi.org/10.3389/fchem.2020.00343
85. Lin X, Li X, Lin X (2020) A review on applications of computational methods in drug screening and design. Molecules 25:1375. https://doi.org/10.3390/molecules25061375
86. Amin SA, Banerjee S, Singh S, Qureshi IA, Gayen S, Jha T (2021) First structure-activity relationship analysis of SARS-CoV-2 virus main protease (Mpro) inhibitors: an endeavor on COVID-19 drug discovery. Mol Divers 25:1827–1838. https://doi.org/10.1007/s11030-020-10166-3

87. Scior T, Bender A, Tresadern G, Medina-Franco JL, Martínez-Mayorga K, Langer T, Cuanalo-Contreras K, Agrafiotis DK (2012) Recognizing pitfalls in virtual screening: a critical review. J Chem Inf Model 52:867–881. https://doi.org/10.1021/ci200528d
88. Kozikowski AP, Chen Y, Gaysin AM, Savoy DN, Billadeau DD, Kim KH (2008) Chemistry, biology, and QSAR studies of substituted biaryl hydroxamates and mercaptoacetamides as HDAC inhibitors nanomolar-potency inhibitors of pancreatic cancer cell growth. ChemMedChem 3:487–501. https://doi.org/10.1002/cmdc.200700314
89. Tang H, Wang XS, Huang XP, Roth BL, Butler KV, Kozikowski AP, Jung M, Tropsha A (2009) Novel inhibitors of human histone deacetylase (HDAC) identified by QSAR modeling of known inhibitors, virtual screening, and experimental validation. J Chem Inf Model 49:461–476. https://doi.org/10.1021/ci800366f
90. Zhao L, Xiang Y, Song J, Zhang ZA (2013) A novel two-step QSAR modeling work flow to predict selectivity and activity of HDAC inhibitors. Bioorg Med Chem Lett 23:929–933. https://doi.org/10.1016/j.bmcl.2012.12.067
91. Pham-The H, Casañola-Martin G, Diéguez-Santana K, Nguyen-Hai N, Ngoc NT, Vu-Duc L, Le-Thi-Thu H (2017) Quantitative structure–activity relationship analysis and virtual screening studies for identifying HDAC2 inhibitors from known HDAC bioactive chemical libraries. SAR QSAR in Environ Res 28:199–220. https://doi.org/10.1080/1062936X.2017.1294198
92. Zeb A, Park C, Son M, Rampogu S, Alam SI, Park SJ, Lee KW (2018) Investigation of non-hydroxamate scaffolds against HDAC6 inhibition: a pharmacophore modeling, molecular docking, and molecular dynamics simulation approach. J Bioinform Comput Biol 16:1840015. https://doi.org/10.1142/S0219720018400152
93. Debnath S, Debnath T, Bhaumik S, Majumdar S, Kalle AM, Aparna V (2019) Discovery of novel potential selective HDAC8 inhibitors by combine ligand-based, structure-based virtual screening and in-vitro biological evaluation. Sci Rep 9:1–14. https://doi.org/10.1038/s41598-019-53376-y
94. Sharma M, Jha P, Verma P, Chopra M (2019) Combined comparative molecular field analysis, comparative molecular similarity indices analysis, molecular docking and molecular dynamics studies of histone deacetylase 6 inhibitors. Chem Biol Drug Des 93:910–925. https://doi.org/10.1111/cbdd.13488
95. Ruzic D, Petkovic M, Agbaba D, Ganesan A, Nikolic K (2019) Combined ligand and fragment-based drug design of selective histone deacetylase-6 inhibitors. Mol Inform 38:1800083. https://doi.org/10.1002/minf.201800083
96. Yan G, Li D, Zhong X, Liu G, Wang X, Lu Y, Qin F, Guo Y, Duan S, Li D (2021) Identification of HDAC6 selective inhibitors: pharmacophore based virtual screening, molecular docking and molecular dynamics simulation. J Biomol Struct Dyn 39:1928–1939. https://doi.org/10.1080/07391102.2020.1743760
97. Uba AI, Yelekçi K (2019) Homology modeling of human histone deacetylase 10 and design of potential selective inhibitors. J Biomol Struct Dyn 37:3627–3636. https://doi.org/10.1080/07391102.2018.1521747
98. Uba AI, Yelekçi K (2020) Crystallographic structure versus homology model: a case study of molecular dynamics simulation of human and zebrafish histone deacetylase 10. J Biomol Struct Dyn 38:4397–4406. https://doi.org/10.1080/07391102.2019.1691658
99. Géraldy M, Morgen M, Sehr P, Steimbach RR, Moi D, Ridinger J, Oehme I, Witt O, Malz M, Nogueira MS, Koch O (2019) Selective inhibition of histone deacetylase 10: hydrogen bonding to the gatekeeper residue is implicated. J Med Chem 62:4426–4443. https://doi.org/10.1021/acs.jmedchem.8b01936
100. Ukey S, Choudhury C, Sharma P (2021) Identification of unique subtype-specific interaction features in Class II zinc-dependent HDAC subtype binding pockets: A computational study. J Biosci 46:1–22. https://doi.org/10.1007/s12038-021-00197-9

Chapter 3
Role of Computational Modeling in Drug Discovery for Alzheimer's Disease

Mange Ram Yadav, Prashant R. Murumkar, Rahul Barot, Rasana Yadav, Karan Joshi, and Monica Chauhan

Abstract Researchers are striving hard for the last two decades to discover new therapeutically effective molecules for the treatment of Alzheimer's disease (AD). Unfortunately, the exact etiology of Alzheimer's disease is not yet fully known, which is proving to be the main hurdle for the discovery of drugs to treat the disease. Several factors are involved in the etiology of AD, like oxidative stress, low levels of acetylcholine (ACh), β-amyloid aggregation, and tau protein phosphorylation in the brain. But unfortunately, no single drug has proved clinically effective till today to prevent or stop progression of the disease. The existing drugs simply improve the worsening clinical symptoms of AD and help in delaying the process of progression of the disease to a fully blown state. Currently, available drugs in the market for the treatment of AD include donepezil, galantamine and rivastigmine, the three acetylcholinesterase inhibitors (AChEIs), and memantine an *N*-methyl-*D*-aspartate receptor (NMDAR) antagonist. These drugs are used mainly to alleviate mild cognitive impairment (MCI) providing temporary relief from the symptoms. This chapter discusses about the application of various computational tools for compounds containing different heterocyclic moieties like quinoline, pyridine, pyrimidine, coumarine, chromane, indole, etc., which could serve as potential leads to design potent novel anti-Alzheimer's agents.

Keywords Alzheimer's disease · QSAR · Molecular modeling · Docking · Molecular dynamics · Drug research

---

M. R. Yadav (✉)
Centre of Research for Development, Parul University, Waghodia Road, Vadodara, Gujarat 391760, India
e-mail: mryadav11@yahoo.co.in

P. R. Murumkar · R. Barot · R. Yadav · K. Joshi · M. Chauhan
Faculty of Pharmacy, The Maharaja Sayajirao University of Baroda, Vadodara, Gujarat 390001, India

S. Kar and J. Leszczynski (eds.), *Current Trends in Computational Modeling for Drug Discovery*, Challenges and Advances in Computational Chemistry and Physics 35,
https://doi.org/10.1007/978-3-031-33871-7_3

3.1 Introduction

Alzheimer's disease (AD) is a progressive, incurable, and debilitating neurodegenerative disease that eventually leads to a state of dementia. Currently, around 50 million people are affected by dementia worldwide, and by 2050 that number may rise to over 150 million making AD the seventh major cause of deaths [1, 2]. The early signs of Alzheimer's include the presence of β-amyloid (Aβ) plaques and neurofibrillary tangles (NFTs) in the central nervous system (CNS). Two main types of Alzheimer's disease exist, the early-onset familial Alzheimer's (FAD), and the sporadic Alzheimer's (SAD) which has late onset. FAD is hereditary and typically begins early in life. It tends to affect the first-degree relatives of the patients. It is believed to be caused by mutations in three genes involved in the expression of amyloid precursor protein (APP), presenilin 1 (PSEN1), and presenilin 2 (PSEN2). SAD, on the other hand, is considered to be caused by genetic and environment factors and generally begins in later part of life. Mainly, Eε4 (ε4 allele of ApoE gene), a gene apolipoprotein, is responsible for SAD, while FAD has a strong genetic link to chromosomes 1, 14, 19, and 21. The progression of Alzheimer's is divided into three stages: preclinical AD, AD with mild cognitive impairment (MCI), and AD causing dementia. The dementia phase of Alzheimer's is further divided into three levels of mild, moderate, and severe, which indicate the manifestation of the degree of symptoms affecting a person's ability to perform daily activities [3, 4].

Low levels of the neurotransmitter acetylcholine (ACh) due to high levels of cholinesterase (ChE) enzymes and hyperactivity of *N*-methyl-*D*-aspartate (NMDA) glutamatae receptors in the brain are the two main biochemical reasons for cognitive impairment in AD patients. Increased free radicals and an increased influx of ions caused by hyperactivity of glutamate NMDA receptors in the brain prove detrimental to the neurons [3].

Hypothesis Behind the Development of AD

Several hypotheses have been proposed by the researchers for explaining the genesis of AD in human beings over the years. These hypotheses are mainly classified as follows:

3.1.1 The Cholinergic Hypothesis

Acetyl choline, a neurotransmitter plays important role in cognitive processes. As per the cholinergic hypothesis, synaptic acetylcholine deficit and impairment of cholinergic neurons lead to the indisputable causes of AD. These lead to decrease in memory and learning abilities and neurobehavioral actions such as agitation and aggression. AChE and BuChE are responsible for decreased levels of ACh in the brain as they rapidly hydrolyze the neurotransmitter into acetate and choline [5]. In AD, the levels of both AChE and BuChE are increased, which cause decrease in the levels of

ACh. Additionally, studies have shown that both of the cholinesterases interact with amyloid-β (Aβ) to form aggregates that cause greater neurotoxicity than the free Aβ. The AChE-Aβ complex causes an uncontrolled Ca^{2+} influx into the neurons' interior chamber that leads to dysfunction in mitochondria and loss of membrane potential. Thus, inhibition of the cholinesterase enzymes remains the most common approach for AD treatment [6].

3.1.2 The Amyloid Hypothesis

β-amyloid has emerged as the most focused target for the treatment of AD because synaptic plasticity is significantly disrupted by Aβ that induces the formation of reactive oxygen species (ROS), which are responsible for the disability of mitochondrial function and neuro-inflammation [6]. Studies suggest that Aβ is involved in many processes like formation of extracellular plaques and disruption of neuronal cell structure, leading to the aggregation and accumulation of Aβ fibrils in the neurons [7, 8].

3.1.3 Tau Protein Hypothesis

Tau protein is another crucial protein which is responsible for the genesis of AD. Recent studies have shown that in sporadic AD, the tau hypothesis may be the initiator of the events. Post-translational hyper-phosphorylation of tau protein leads to formation of aggregates similar to Aβ, known as neurofibrillary tangles (NFTs). However, it needs to be clarified that Aβ and NFTs are not exclusively correlated to AD progression, because tau hyper-phosphorylation and tau production abnormalities have also been observed in other types of neurodegenerative disorders.

For the treatment of Alzheimer's disease, three drugs are acetylcholinesterase inhibitors, i.e., donepezil, galantamine, and rivastigmine, while memantine is an *N*-methyl-*D*-aspartate (NMDA) receptor antagonist. Cholinesterase inhibitors are advocated in mild to moderate cases of Alzheimer's disease while in moderate to severe cases, memantine is recommended. Combination of donepezil and memantine has given better results than the use of either of the drugs all alone. A new drug, Aducanumab, developed by Biogen, is the first human monoclonal antibody to be approved that targets Aβ directly. It binds to both Aβ fibrils as well as the soluble oligomers [9, 10].

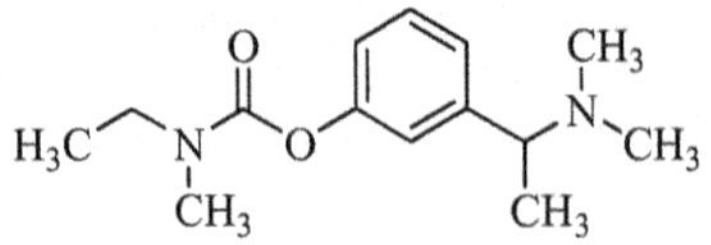

(Rivastigmine)

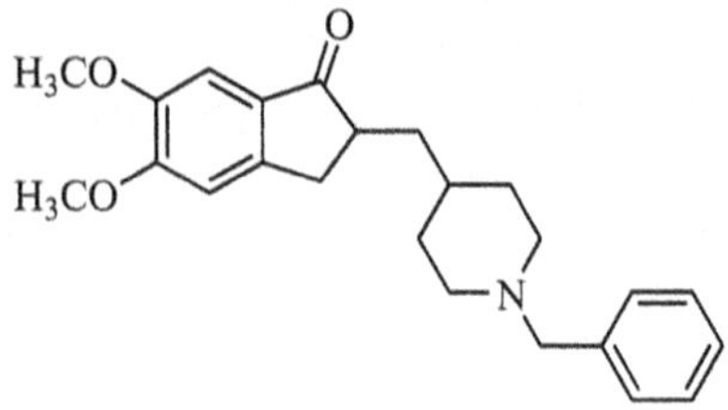

(Donepezil)

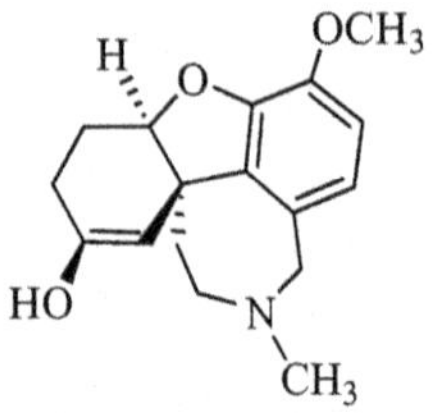

(Galantamine)

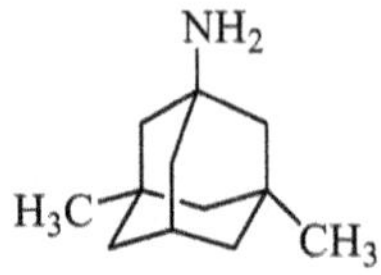

(Memantine)

Treatment Strategies for AD

Multi-target drugs (MTDs) approach is the buzzword used widely these days to treat multifactorial diseases like AD. MTDs are defined as molecules in which two or more pharmacophoric features are incorporated in a single molecular entity. This single molecule can target different biological targets. Development of MTDs is considered to be a promising strategy, but any MTD has yet to reach the clinician's desk and see the light of the day for the treatment of AD. More than 140 MTDs eliciting different mechanism of action are currently in clinical trials in different phases for AD treatment. Currently in the AD interventional studies, there are 111 new chemical entities in clinical trials either in phase 2 or phase 3 and 29 in phase 3. In order to reverse the progression of AD, a majority of the MTDs are in different phases of clinical trials target neuro-inflammation, oxidative stress, mitochondrial dysfunction, metal dysregulation, tau proteins, amyloids (both AChE and BACE), or some miscellaneous targets [11–13] (Fig. 3.1).

3.2 Role of Computational Studies in the Designing of Anti-Alzheimer's Agents

In this chapter, we have endeavored to discuss about the role of various computational techniques for the designing of new anti-AD agents. Rational drug design and discovery is highly dependent on computer-aided drug design techniques. There are two important drug designing techniques that are mainly used in computer-aided drug design, i.e., ligand and structure-based drug design techniques. Ligand-based

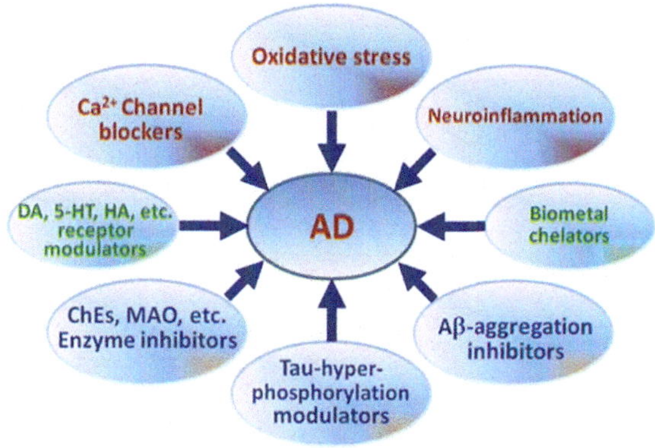

Fig. 3.1 Various targets exploited for developing anti-AD agents (reused with permission *source* Fig. 1, Oset-Gasque and Marco-Contelles [13])

designing technique is also called as indirect drug designing technique wherein only the structures of the ligand molecules are considered [14], whereas the structure-based drug design or direct drug designing technique additionally requires the three dimensional structure of the target enzyme/protein [15, 16]. In our own lab, we have used CADD techniques for designing of TACE inhibitors [17–27] and anti-AD agents [28–37].

This book chapter describes some recent applications of various computational techniques to discover or design new potential anti-Alzheimer's agents, by applying different approaches like molecular docking, quantitative structure–activity relationship (QSAR), and molecular dynamics simulations. Since a huge quantum of literature related to development of anti-AD agents is available in the public domain, we have covered the literature reported from the year 2018 onward in this chapter, related to some potent small molecules having activity in nanomolar range against the main AD targets like AChE, BuChE, PDE, BACE1, amyloid β, GSK-3β, and also having antioxidant and metal chelating properties. Chemical moieties falling under different chemical classes are discussed in the following sections.

3.2.1 *Tacrine-Based Scaffolds as Anti-AD Agents*

Tacrine was the first US-FDA approved drug made available for the treatment of mild to moderate cases of AD, but later on, it was withdrawn due to its severe side effects like gastrointestinal discomfort and hepatotoxicity. But, some features of tacrine like decrease in Aβ peptide-induced apoptosis in cortical neurons were observed to be positive for the treatment of AD. Based on these observations, Przybyłowska et al. designed novel *N*- and *O*-phosphorylated tacrine derivatives [38]. Among the compounds of the reported series, compound **(1)** exhibited the highest inhibitory activity with an IC_{50} value of 6.11 nM against AChE, while compound **(2)** offered an IC_{50} of 1.97 nM against BuChE, proving compound **(1)** and compound **(2)** to be 6 and 12-fold more potent than tacrine, respectively.

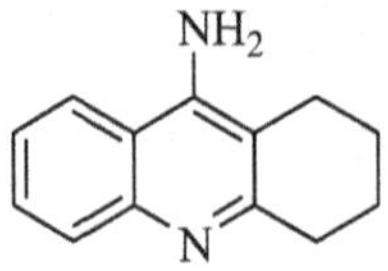

Tacrine

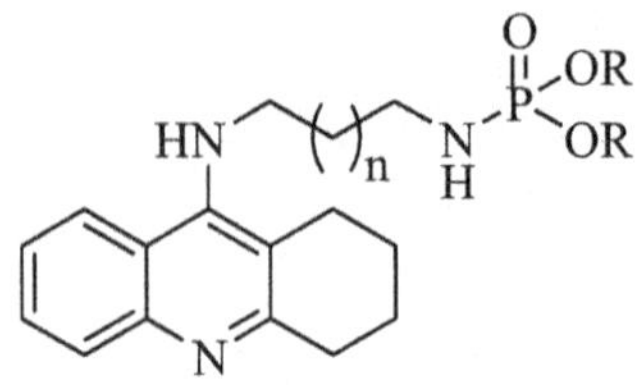

1. n = 8, R = Et
2. n = 12, R = Ph
3, n = 6, R = Ph

Molecular modeling studies revealed that all the compounds offered satisfactory free binding energies, for AChE in the range of −9.2 kcal/M and for BuChE −7.8 to −10.92 kcal/M compared to the reference compound tacrine which yielded binding energies of −9.0 kcal/M for AChE and −8.2 kcal/M for BuChE. Compound (**3**) proved to be the best fitting molecule for inhibiting the catalytic activity in the cavity of AChE and BuChE with docking energies of −12.5 and −10.7 kcal/M, respectively. In the CAS region of both of the enzymes, aromatic ring of tacrine showed π–π stacking interaction with Phe330 and Trp84 (AChE), and Trp82 (BuChE), and H-bonding between the nitrogen atom of tacrine and His440 and His438 of AChE and BuChE, respectively. Additionally, the phosphoramide group formed H-bond with the Tyr121 and Thr110 in the mid-gorge regions of AChE and BuChE, respectively, while non-covalent interactions were observed with Trp279 of AChE.

In the pathology of AD, hyper-phosphorylation of tau protein plays a significant role resulting into microtubule polymerization and NFT generation that lead to necrosis and degeneration of the neuronal cells, leading to progressive deterioration of the patient's condition. Glycogen synthase kinase 3 beta (GSK-3β) catalyzes the disordered phosphorylation of tau protein. Therefore, tau protein signaling pathway involving GSK-3β has become a vital target for designing of new anti-AD compounds. Based on this philosophy, Yao et al. reported novel tacrine-pyrimidone hybrids as potent dual GSK-3β and AChE inhibitors as anti-AD agents [39]. In the reported series of compounds, **4** exhibited excellent dual GSK-3 and AChE inhibition activities with IC_{50} values of 89.3 and 51.1 nM, respectively.

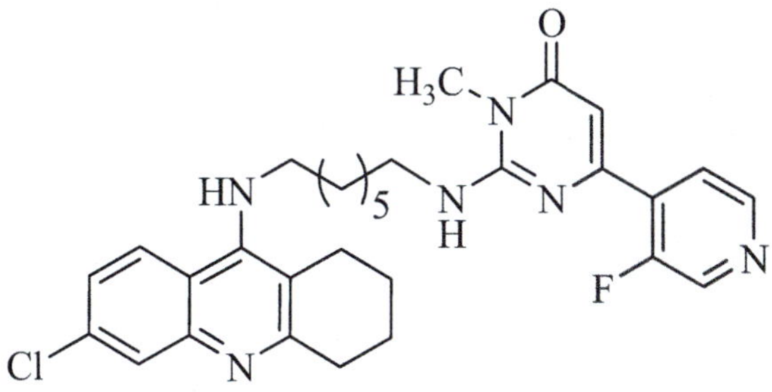

(4)

As per the docking studies, compound **(4)** was perfectly fitting in the catalytic anionic site (CAS). The tacrine fragment was positioned in the interior gorge, while the pyrimidone fragment was located in the exterior gorge of the peripheral anionic site (PAS) of AChE. Five non-bonding interactions were observed between compound **(4)** and the binding gorge of the enzyme. H-bond between chloro group of compound **(4)** and oxygen atom of Tyr72 (3.5 Å) drives the tacrine unit to bind in the reverse way, costing one π–π interaction with Trp86, but additionally making two π–π stacking interactions with Trp286, and H-bonding with Ser293 with a distance of 2.2 Å (Fig. 3.2).

Pyrimidone fragment of compound **(4)** intruded into GSK-3β protein's interior part of the binding pocket, while the tacrine unit got stuck to the exterior portion of the active pocket (Fig. 3.3a, b). The hybrid molecule **(4)** made a π-cation interaction (4.5 Å) with Arg141, and two H-bonds with Val-135 and Lys-85 with distances of 2.6 and 2.7 Å, respectively, on the opposite side of the peripheral groove (Fig. 3.3c). Compound **(4)** was fitting into the binding pockets of GSK-3β and AChE and showed good affinity for both the targets by interacting with several amino acid residues making secondary bonds through the cooperation of the tacrine unit, pyrimidone

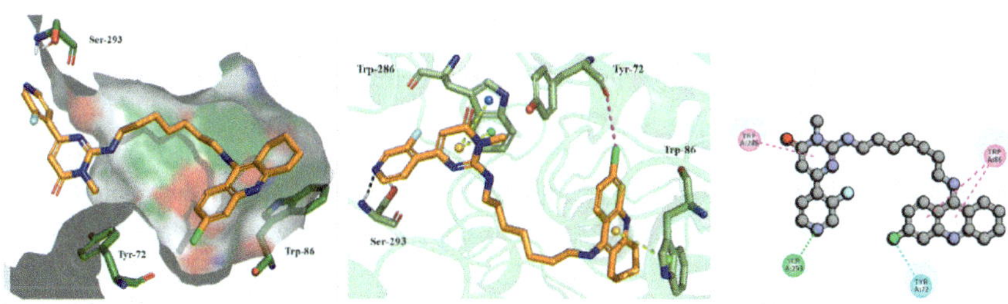

(a) (b) (c)

Fig. 3.2 Binding of compound (**4**) with AChE. **a**, **b** are shown using the PDB code 4EY7. The position of compound (**4**) in brown is depicted within the pocket of AChE, and the interactions are shown between compound (**4**) and the amino acid residues. In **c**, the 2D binding mode of compound (**4**) with AChE is depicted, hydrogen bonds shown in green, the π–π stacking interaction, shown in purple, and halogen bond, shown in cyan (reused with permission *source* Fig. 3, Yao et al. [39])

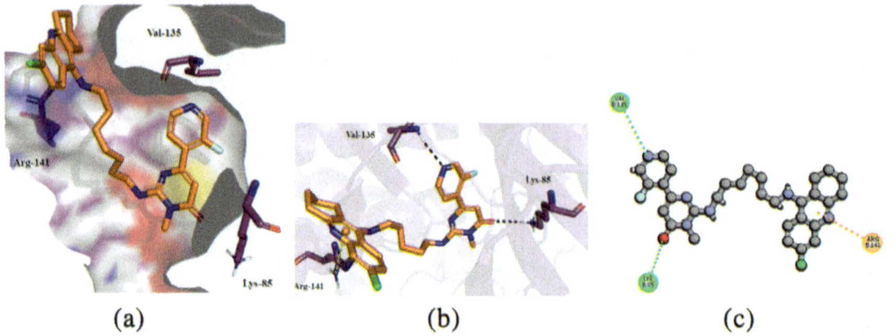

Fig. 3.3 Interaction of compound (4) with GSK-3β, **a**, **b** display the location of compound (4) in the active site of the protein, and **c** shows a 2D representation via hydrogen bonds and π-cation bond (reused with permission *source* Fig. 3, Yao et al. [39])

moiety, and the alkylamine linker catapulting it an excellent dual GSK-3β and AChE inhibitor.

Furthermore, Ozten et al. reported carbamate derivatives of tacrine as potent cholinesterase inhibitors. Compound **(5)** showed the best inhibitory activity against both the enzymes BuChE and AChE with IC_{50} values of 16.96 and 22.15 nM, respectively [40].

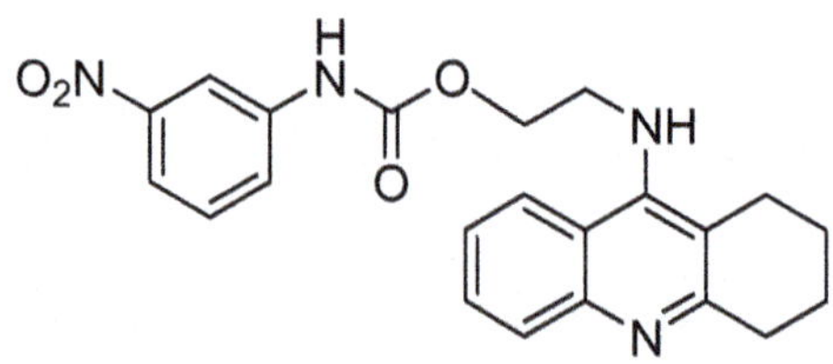

(5)

In the modeling research, it has been noticed that compound **(5)** interacts with AChE and BuChE enzymes through π-cation and π–π stacking interactions with the phenyl ring, in a similar manner to tacrine and its derivatives. Compound **(5)** forms strong hydrogen bonds to AChE, with residues His478 in the catalytic site, Glu233 in the choline binding region, and Gly153 in the acyl-binding region and also has a π-cation interaction in the acyl-binding site with Phe369 through its -NO_2 group. Additionally, compound **(5)** interacts with water molecules and forms stable bridges with Tyr360 in the BuChE binding cavity and Glu225 in the choline binding site. The residues from the acyl-binding region also play a role in stabilizing compound **(5)** in the BuChE binding site through hydrogen bonding with Gly143 and Gly145. Although some non-covalent interactions were seen to increase the affinity of compound **(5)** temporarily for AChE and BuChE, but these were not sustained during the entire MD simulation. The MD study suggests that compound (**5**) forms covalent bonds with residues Ser226 and Ser234 for AChE and BuChE, respectively.

Chufarova et al. have reported hybrid molecules of tacrine and acridine as potential anti-acetylcholinesterase and anti-butyrylcholinesterase enzymes [41]. Compound (**6**) was found to be the most potent inhibitor having IC_{50} values of 1.7 and 7.6 pM against BuChE and AChE enzymes, respectively. Compound (**6**) was found to be having better activity than the standard tacrine without any significant cytotoxic effect.

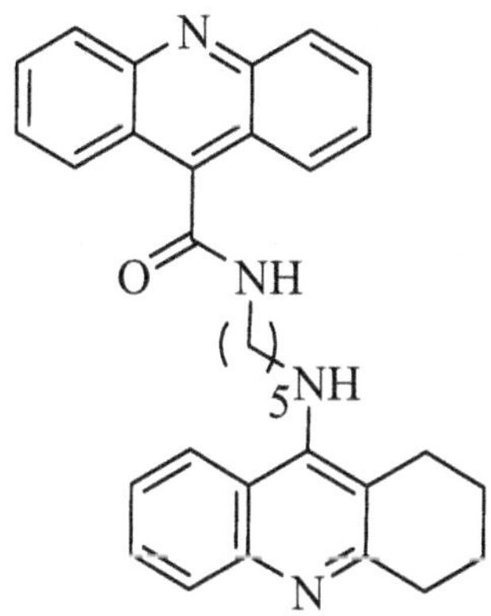

(6)

Modeling studies revealed significant interactions by the potent compounds in the active site of the enzyme when compared to tacrine. Compound (**6**) was found to be having dual-binding mode (bound to PAS and CAS both). It was oriented in the opposite direction of tacrine making π–π stacking with residues Tyr341 and Trp286 along with the two T-shaped aromatic interactions with residues Tyr124 and Tyr72 in the PAS site of AChE. Similarly in CAS site, it showed T-shaped interaction with residue Tyr124 and π–π stacking with Trp86 residue. Whereas, in BuChE, the compound showed very limited interactions. Tacrine moiety interacted with residue Tyr332 and acridine moiety interacted with Phe329 residue by two π-stacking interactions. There were no T-shaped interactions observed in the BuChE site.

Li et al. reported a novel series of tacrine-phenolic acid and tacrine-phenolic acid-ligustrazine, dihybrids, and trihybrids, respectively, as acetylcholinesterase inhibitors [42]. Among them, compound (**7**) displayed potent activity as acetylcholinesterase inhibitor with IC_{50} value of 3.9 nM; *hAChE* IC_{50} value of 65.2 nM, along with the blocking of β-amyloid self-aggregation effect at 20 μM.

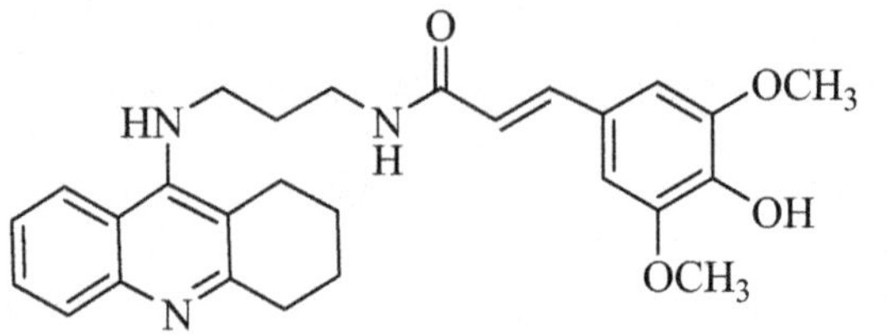

(7)

In order to see the molecular interactions, the most potent compound (**7**) was subjected to docking studies in the active site of the enzyme cholinesterase. Docking studies using *Tc*AChE (PDB ID 3CKM) revealed that tacrine component of compound (**7**) interacted within the CAS site with phenyl ring of Phe330 residue and indole moiety of Trp84 via aromatic π–π stacking, whereas the sinapic acid component interacted via hydrophobic interactions with residues Ser286 and Asp285 within the PAS site. Similarly, using AChE (PDB ID: 4EY7) active site, these studies revealed that sinapic acid component made aromatic π–π stacking interaction with Trp286 residue in the PAS site, and the tacrine moiety was involved through the hydrophobic interactions with residue Tyr124 that offered stacking interaction with indole ring of Trp86. These docking interactions in the CAS and PAS sites elicited the active binding of compound (**7**) with *Tc*AChE and *h*AChE. Compound (**7**) has shown π–π stacking interaction between Trp82 residue and tacrine moiety in the active site of *h*BuChE (PDB ID 1P0I) and hydrogen bonding interactions between Ser198 residue and hydroxyl group.

Derabli et al. disclosed poly-functionalized tacrine-derived compounds via Friedlander reaction and evaluated them for cholinesterase (BuChE and AChE) inhibitory activity [43]. Among the reported derivatives, compound (**8**) containing 4-chlorophenyl and *S*-ethyl moieties yielded the most potent compounds against AChE (IC_{50} value of 4.32 μM), while compound (**9**) elicited higher potency against BuChE with IC_{50} value of 2.74 μM.

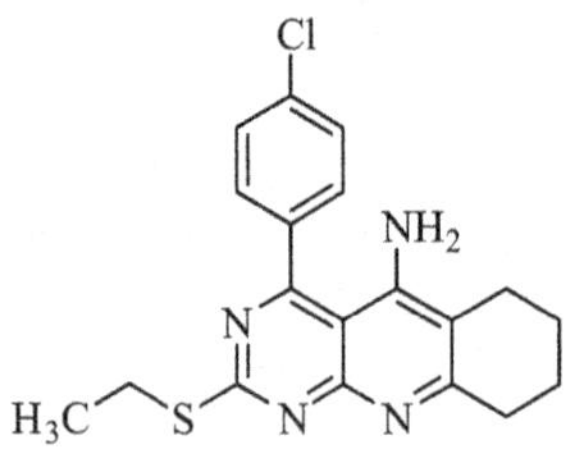

(8)

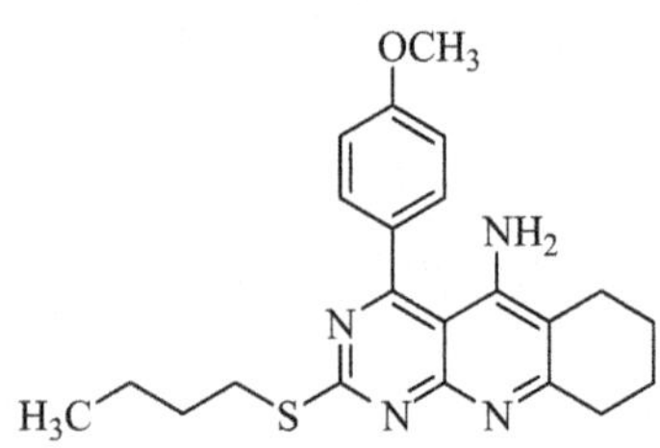

(9)

Molecular modeling studies were performed to find out the binding interactions of both of the compounds (**8** and **9**), in the active sites of BuChE (PDB ID: 5dyw) and AChE (PDB ID: 4EY7). Biological evaluation confirmed that compound **(8)** was more potent against AChE than compound **(9)**, which was more potent than compound (**8**) against BuChE. In the AChE active site, compound **(8)** exhibited hydrophobic interactions with Trp8 residue and hydrogen interactions with Tyr124, as the heterotacrine moiety of the compound (**8**) was much better oriented to offer a good binding affinity. Whereas with BuChE, compound **(9)** showed pi-interactions with phenyl ring of Tyr332 residue and hydrogen bond interaction with Asp70 residue.

Zhu et al. synthesized a series of tacrine-ferulic acid hybrids as multi-target-directed ligands [44]. Among the reported series of compounds, compound **(10)** displayed promising activity as BuChE inhibitor with IC_{50} value of 101.40 nM, and as AChE inhibitor with IC_{50} value of 37.02 nM. The compound **(10)** also exhibited potent amyloid β-protein self-aggregation inhibitory activity.

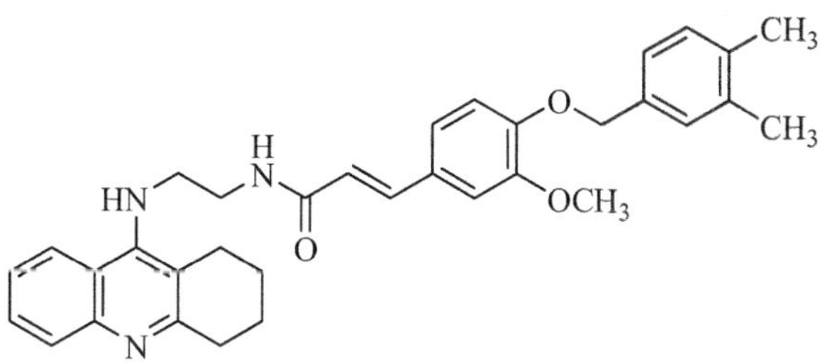

(10)

To investigate the intermolecular interactions of compound (**10**) in the *hu*AChE (PDB ID: 4EY7) and *hu*BuChE (PDB ID: 4TPK) active sites, molecular docking studies were performed. From these studies, it was revealed that in *hu*AChE 1,2,3,4-tetrahydroacridine moiety of compound **(10)** made π–π stacking interactions with residue Trp86 and hydrogen bonding with residue Tyr337 in the CAS binding site of the enzyme. Whereas the ferulic acid phenyl ring interacted with Trp286 and Tyr341 residues via π–π stacking interactions within the PAS site. This phenyl ring also interacted with residues Leu289, Ser293, and Tyr124 via van der Waals interactions (Fig. 3.4).

Więckowska et al. have carried out the synthesis of multi-target-directed ligands and reported a compound **(11)** displaying potent activity as an antagonist of 5-HT_6 receptor and non-competitive inhibitor of cholinesterase (IC_{50}, *h*AChE = 14 nM, IC_{50}, *eq*BuChE = 22 nM) and amyloid-β anti-aggregation activity (IC_{50} = 1.27 uM) [45]. From the docking studies, it has also been revealed that in the active site of the enzyme, the short chains (2–3 carbon atom linkers) and long chains (6–8 carbon atom linkers) exhibit distinct binding modes. Short chain derivatives of the tacrine fragment interact with Phe330 and Trp84 residues of the catalytic binding site of AChE, whereas π–π stacking interaction is formed with the 1-(3-(benzyloxy)-2-methylphenyl)piperazine fragment and methylphenyl or indole fragment of Trp279 residue of the peripheral site of AChE, while π-cation interaction takes place between

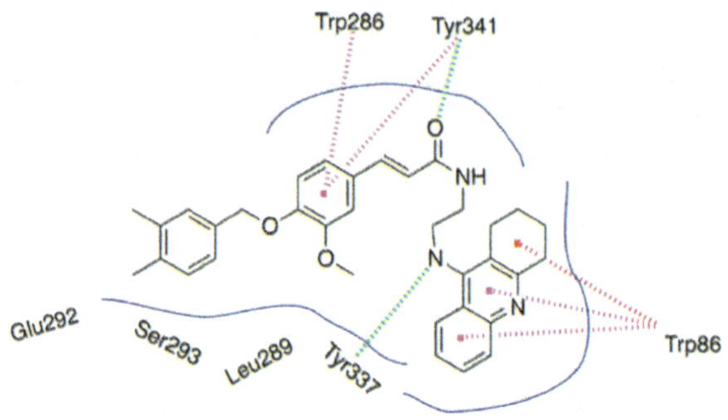

Fig. 3.4 Binding mode of compound (**10**) with *huAChE*. Binding interactions are indicated in different colors as dashed lines. Red: π–π T-shaped, purple: π–π stacking and green color: hydrogen bond interaction (reused with permission *source* Fig. 1, Zhu et al. [44])

Tyr334 residue and piperazine. Similarly, long chain compounds show same interactions but piperazine interacts with Phe331 instead of Tyr334 and forms cation-π interaction, whereas in short chain compounds, tacrine component forms π–π stacking interaction with Tyr334, and rest of the molecule exists in perpendicular positioning the catalytic site. Compounds with 5 carbon linkers were showing poor binding affinity as they were unable to make interactions with residues Trp279, Tyr334 or Phe331.

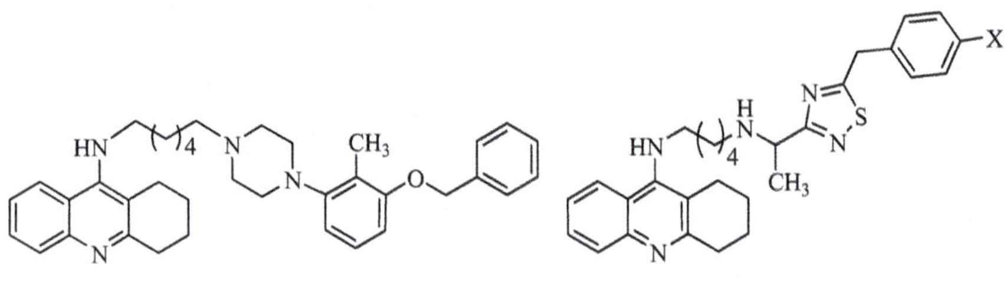

(**11**) (**12**) X = Cl; (**13**) X = F

Makhaeva et al. reported a series of conjugates of tacrine-1,2,4-thiadiazole derivatives using quantum mechanical characterization followed by molecular docking which revealed exceptional binding affinity of these compounds for the PAS and CAS active sites of BuChE and AChE both. The in vitro assay results justified the in silico predictions for these compounds. The compounds displaced propidium dye from PAS, indicating that the compounds were blocking AChE-induced β-amyloid aggregation [46]. Molecular docking studies were performed using PDB ID: 4EY7 for AChE and 1POI for BuChE using AutoDock 4.2.6. Molecular docking studies predicted dual binding in both of the active sites of AChE in which the tacrine part formed π-cation interactions and π–π stacking with Trp86, and ionic interactions with Glu2020 along with hydrogen bonding stabilized these interactions. The pentamine in the structures (**12, 13**) additionally produced ionic and π-cation interactions between the amino acid residues in the gorge, mainly with Tyr341 and the charged amino group of these compounds. Molecular docking studies predicted that the substituents on the aromatic group on the thiadiazole moiety did not produce substantial anti-AChE effect because of small gorge cavity area in the target enzyme, which was not the case with BuChE, as the compounds could bind tightly in the active site in folded conformation. Compound **(12)** containing pentamine fragment achieved better results compared to pentamine conjugates of compound **(13)**. The compounds (**12** & **13**) offered in vitro IC_{50} values of 0.00380 ± 0.00029 μg/mol and 0.0725 ± 0.0016 μg/mol, respectively, as anti-BuChE agents, when compared with the standard drug tacrine (anti-BuChE IC_{50} value of 0.0295 ± 0.0020 μg/mol).

3.2.2 *Indole-Based Anti-AD Agents*

In the year 2021, a series of sulfonamide group containing dihydropyranoindoles were reported by Sarfaraz S et al. along with their in vitro AChE activity. All the synthesized compounds showed AChE (IC_{50} = 0.41–8.79 μM) and BuChE (IC_{50} = 1.17–30.17 μM) inhibitory activity [47]. Among the compounds of the reported series, compound **(14)** showed the highest inhibitory activity for BuChE (IC_{50} = 1.17 μM) as well as AChE (IC_{50} = 0.41 μM) when compared with galantamine and rivastigmine as the standard drugs. Moreover, the enzyme kinetics revealed that the compound **(14)** possessed mixed type of inhibition, and it was bound to catalytic sites (CAS) as well as to the peripheral anionic sites (PAS) of both the enzymes. Compound **(14)** was docked in the active site of AChE (Fig. 3.5) to understand its binding mode. The modeling studies revealed that it interacted with active sites of both the enzymes BuChE and AChE. Compound **(14)** showed multiple H-bonding interactions with Asp72 residue of PAS and through Ser200 residue with CAS and acyl-binding pocket with Phe288 residue. Two π–π interactions were observed with Trp334 and Tyr279 residues in the PAS while one π–π interaction was formed at anionic sub-site with Phe330 residue. For comparison purposes, galantamine and rivastigmine were also docked into the active site of AChE wherein interactions similar to those observed with compound **(14)** were seen. The orientations of compound **(14)** galantamine and rivastigmine with AChE are shown in Fig. 3.5. In case of BuChE, compound **(14)** formed two H-bonding interactions with Ser287 residue in the acyl pocket and Glu197 residue in the PAS. It also displayed π–π stacking interactions with His438 of CAS, Trp82 of PAS and Phe329 of the anionic sub-site. These results indicate mixed type of inhibition by compound **(14)** of BuChE. Compound **(14)** exactly fits into BuChE binding pocket with similar interactions as seen in the case of standard drugs.

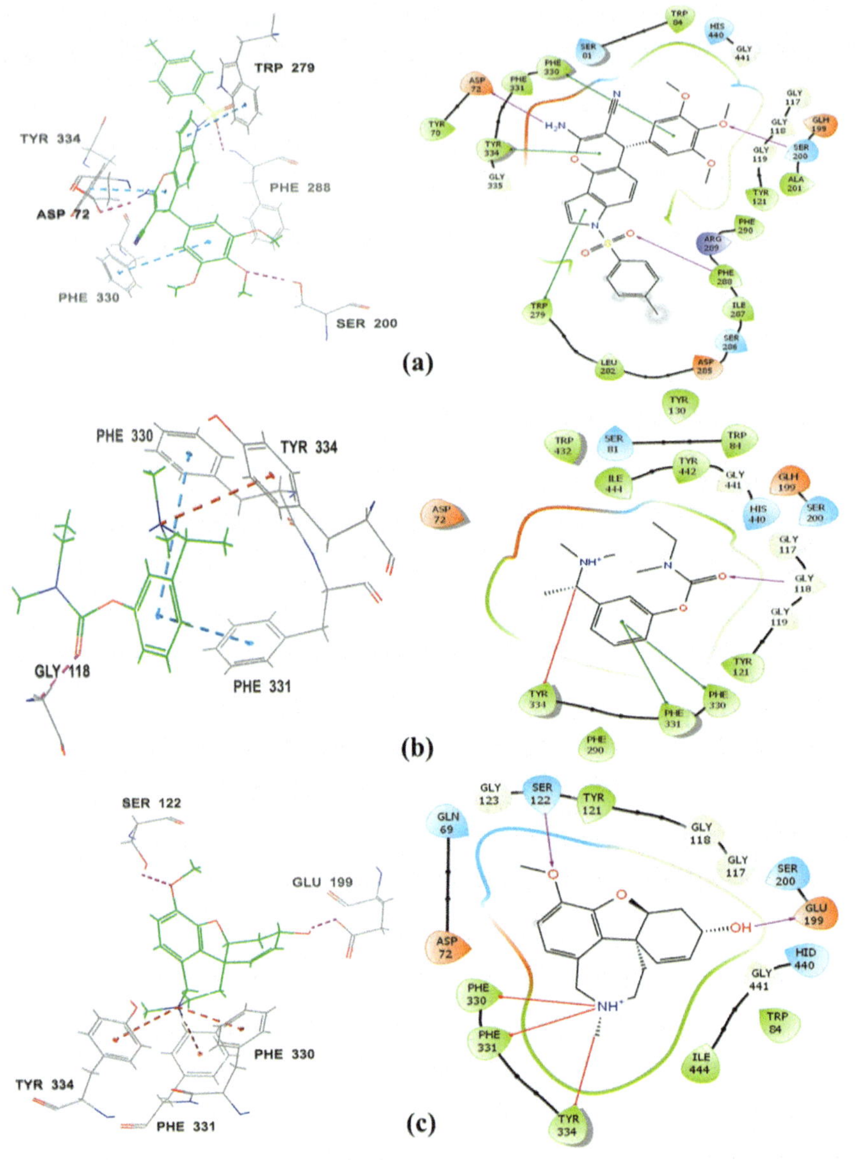

Fig. 3.5 Docking interaction of **a** compound (**14**), **b** rivastigmine, and **c** galantamine within the active site of AChE (reused with permission *source* Fig. 6, Shaikh et al. [47])

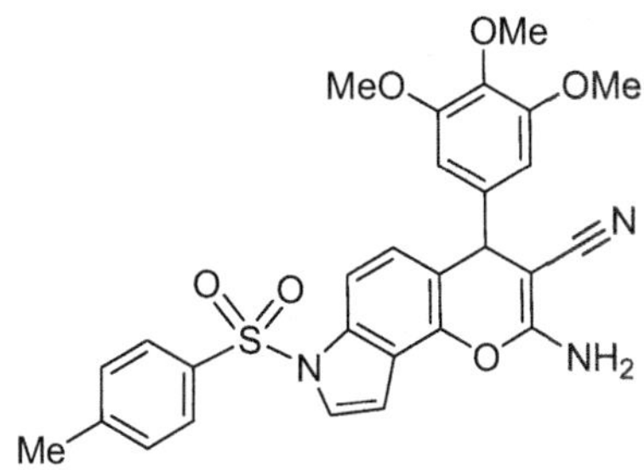

(14)

He et al. found that melatonin-ferulic acid derivatives can act as selective inhibitors of HDAC6, an enzyme involved in memory, neurodevelopment, and cognitive processes. Inhibitors of HDACs have been recognized as a potential treatment for AD type neurodegenerative disorders, wherein HDAC6 plays an important role in deacetylation of tau protein that leads to aggregation of phosphorylated tau protein into neurofibrillary tangles. The study showed that the melatonin-ferulic acid derivatives have immunomodulatory and neuroprotective effects [48].

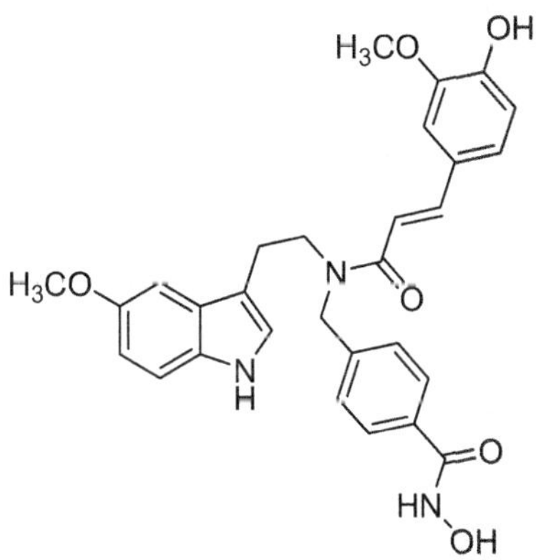

(15)

Compound **(15)** showed selective inhibition of HDAC6 among the reported series of compounds, with an IC_{50} value of 30.7 nM, and it exhibited more than 25-fold higher selectivity than the other reported compounds from the series. Moreover, compound **(15)** was comparable to ferulic acid in demonstrating DPPH radical scavenging activity. To identify putative binding modes of compounds **(15)** with HDAC6, computational docking studies were performed using PDB: 6DVM. For the capping groups, two binding modes could be envisaged, i.e., the binding groups could be flanked by loops L1 and L2 above the pocket. In one of the binding modes, the indole moiety showed aromatic interactions with His463, whereas van der Waals interactions were seen between hydroxymethoxyphenyl ring and Leu712 residue.

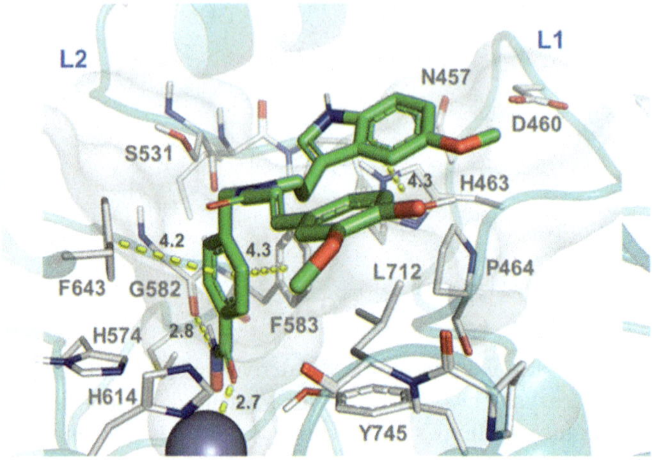

Fig. 3.6 Interaction of compound **(15)** with HDAC6 (PDB: 6DVM) (reused with permission *source* Fig. 2, He et al. [48])

As far as the switched orientation is concerned, it was found less favorable as the indole ring was placed above Leu712, and the -OH group of the ferulic acid-based moiety approached Asp460 for H-bond interaction in the docking studies. On the basis of these observations, it was assumed that the likely mode of binding of the compound (**15**) was binding of both of the capping groups into the L1 loop pocket of HDAC6, offering selectivity to the compound for the enzyme HDAC6 (Fig. 3.6).

Ghamari et al. disclosed a series of H3R antagonist and anti-cholinesterase dual acting compounds. They have identified compound (**16**), as the lead molecule using computational studies [49]. The molecular docking studies showed its effective binding interactions within the active sites of the enzymes. For cholinesterase inhibitory activity, PDB IDs selected for AChE were 4EY7 and for BuChEit were 4TPK. GOLD software was used to perform the docking studies. In modeling studies, it was seen that compound (**16**) exhibited one π–π stacking with the residue Trp286, two hydrophobic interactions with Tyr337, Tyr341, Trp86, and His447, and one hydrogen bonding with the amino acid Tyr124 within the active site of AChE; whereas in BuChE active site, this compound has shown hydrophobic interactions with residues Trp82, Tyr128, and Glu197 forming a hydrophobic pocket and a hydrogen bonding interaction with residue Tyr332. It was revealed from the in vitro studies that the compounds of the series possessed potent H3R antagonist and cholinesterase inhibitory activities. Compound (**16**) showed H3R antagonistic activity with EC_{50} value of 0.73 nM, AChE inhibitory activity with IC_{50} value of 9.09 μM, and BuChE inhibitory activity with IC_{50} value of 21.1 μM.

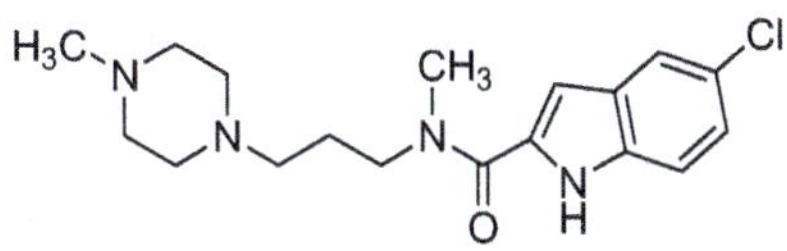

(16)

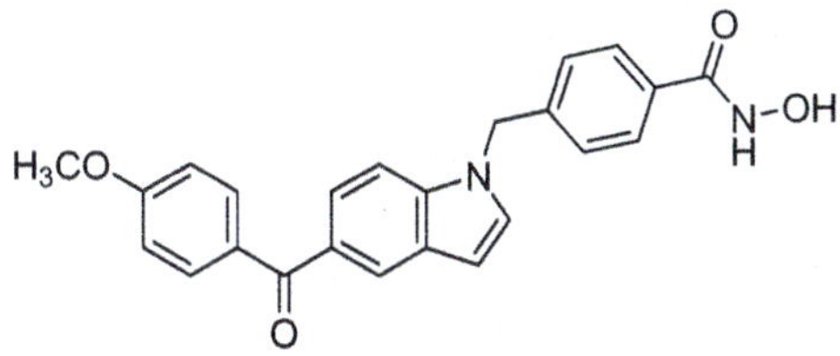

(17)

Lee et al. disclosed a series of 5-aroylindolyl-substituted hydroxamic acids. Among them, compounds **(17)** displayed the most potent inhibitory activity against histone deacetylase 6 (HDAC6) along with the reduction of tau protein phosphorylation and aggregation [50]. The phosphorylated tau protein was transferred to the Hsp70/CHIP complex resulting into the ubiquitination of the phosphorylated tau protein. Its inhibitory activity also results into increased levels of acetylated Hsp90 that affects the binding of Hsp90 to HDAC6.

Docking studies were also undertaken along with the in vitro experimentation, to elucidate the selectivity of the synthesized compounds. Different crystal structures of histone deacetylase like HDAC1 (PDB ID: 5ICN), HDAC2 (PDB ID: 5IX0), HDAC3 (PDB ID: 4A69), HDAC6 (PDB ID: 5EDU), and HDAC8 (PDB ID: 1W22) were used for docking of compound **(17)** in order to determine its selectivity. It was revealed that compound **(17)** was bound in a specific pocket of protein (5EDU) which is not present in any other histone deacetylase and its anisole moiety played a very important role in its selectivity as it interacted with the residues N494, D496, and W497 to form van der Waals interactions and with residue N494 to forms hydrogen bonding (Fig. 3.7).

Glycogen synthase kinase 3β (GSK-3β) has been identified as a novel target to treat Alzheimer's disease. A series of GSK 3β inhibitors has been explored by Lozinskaya et al., A compound **(18)** bearing 3-arylidene-2-oxindole scaffold has been reported as a novel GSK-3β inhibitor with an IC_{50} value of 4.19 nM for the enzyme

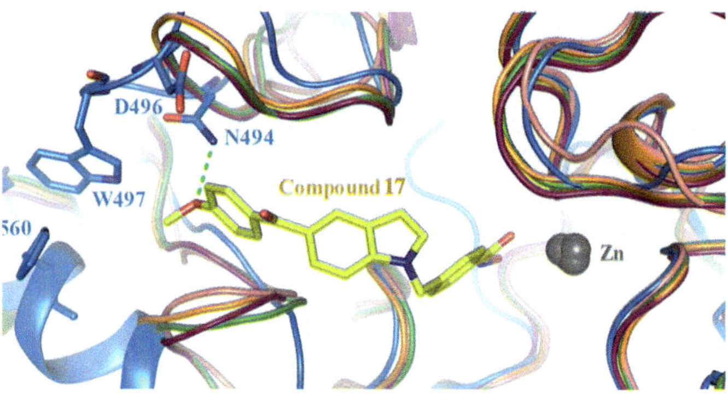

Fig. 3.7 Interaction of compound **(17)** with 5EDU (reused with permission *source* Fig. 4a, Lee et al. [50])

[51]. Docking studies were performed using Autodock Vina 1.1.2 using the protein structure of GSK-3β (PDB ID: 3SD0 and 4J1R). The oxindole moiety of compound (**18**) occupied the hydrophobic cleft of the allosteric site of the ATP-binding site of GSK-3β in the docking studies. It exhibited tight alkyl-π interaction with the lipophilic side chains of Ala83, Val110, Leu132, Leu188, and Cys199 for both of the geometric isomers. A characteristic feature of the allosteric GSK-3β inhibitors, i.e., hydrogen bond interaction between NH and Asp133 was also observed.

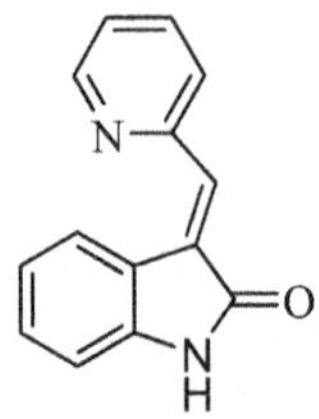

(18)

3.2.3 Pyridine and Pyrimidine-Based Scaffolds as Anti-AD Agents

Shoaib et al. designed and synthesized a novel series of phenylsulfonyl-pyrimidine carboxylate MTDLs for the treatment of AD. Compound (**19**) displayed potent inhibitory activity against AChE having IC_{50} value of 47.33 ± 0.02 nM among the reported series of compounds. Compound (**19**) displayed non-competitive inhibition of AChE in the enzyme kinetics study with Ki value of 8 nM. AChE-induced $A\beta_{1\text{-}42}$ aggregation was also inhibited by the compound at a concentration of 10 μM in the thioflavin T-assay [52]. All the compounds of the series displayed high stability in CAS site of AChE enzyme with binding free energy ranging from −12.38 to −9.69 kcal/M through π–π interactions and H-bondings in the

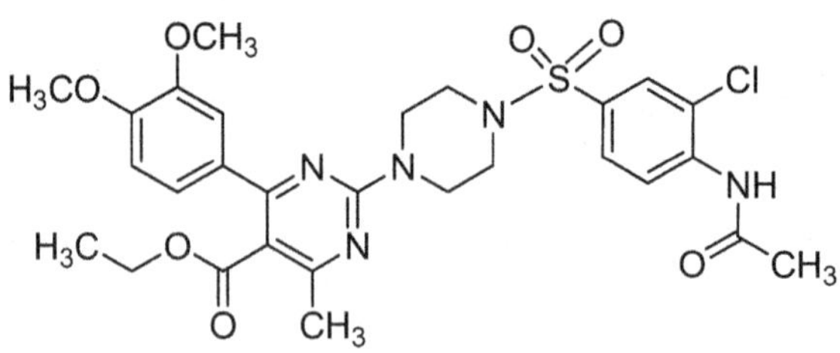

(19)

docking studies. The phenyl ring of the 3,4-dimethoxyphenyl component of compound (**19**) displayed π–π stacking interactions (distance 3.8 Å) with Trp279 residue in the PAS sub-site (Fig. 3.8). Indole ring system of Trp84 residue was involved in sandwich-type π–π stacking with the aromatic ring of *o*-chlorophenylcetamido group (distance 4.1 Å) of compound (**19**) in the CAS sub-site of the enzyme. Methoxy group of compound (**19**) offered strong H-bonding with Arg289 residue at a distance of 1.8 Å. A hydrogen bond (3.1 Å) was formed between pyrimidine ring nitrogen and Tyr121 residue. Additionally, methoxyl group of compound (**19**) formed H-bonding with Ser286 (3.2 Å). Carbonyl group was also involved in forming H-bond with Ser124 and Gly123 at distances of 3.3 and 2.7 Å, respectively. Based on the docking results, it could be concluded that the selected compound possessed dual-binding inhibitory activity for AChE by targeting PAS and CAS sub-sites. Similarly, using AutoDock software, docking studies for compound (**19**) were further extended on *h*BuChE employing the crystal structure of *hBuChE* (PDB ID: 4TPK) as shown in Fig. 3.9. It displayed better interactions (binding free energy −8.7 kcal/M) with the enzyme. In the CAS sub-site, Trp82 residue of the enzyme interacted with the aromatic ring through π–π stacking interaction. Ser70 fragment of the enzyme interacted with the methoxyl group through H-bonding, while the NH-group of acetamido function made the strongest bond with Glu115 fragment of the enzyme. The docking results clearly indicate that these pyrimidine-phenylsulfonyl derivatives target both the sub-sites, i.e., PAS and CAS of AChE as well as of BuChE.

Zhang et al. reported a novel series of pyrazole-pyrimidinone derivatives as PDE9A inhibitors having additional antioxidant activity, as anti-AD agents in the year 2018. Out of 14 compounds, 12 compounds showed IC_{50} values below 200 nM along with good antioxidant activity. Compound (**20**) among the reported series of compounds offered an IC_{50} value of 56 nM, the lowest one making it the most potent compound of the series [53]. Compound (**20**) proved to be safe toward human neuroblastoma (SH-SY5Y) cells in the toxicity studies. From the structure–activity

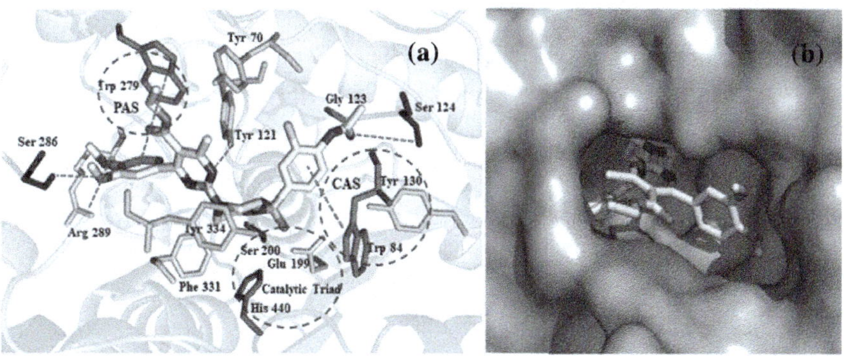

Fig. 3.8 Docking interactions of compound (**19**) within the active site of AChE (PDB code: 1EVE) (reused with permission *source* Fig. 3, Manzoor et al. [52])

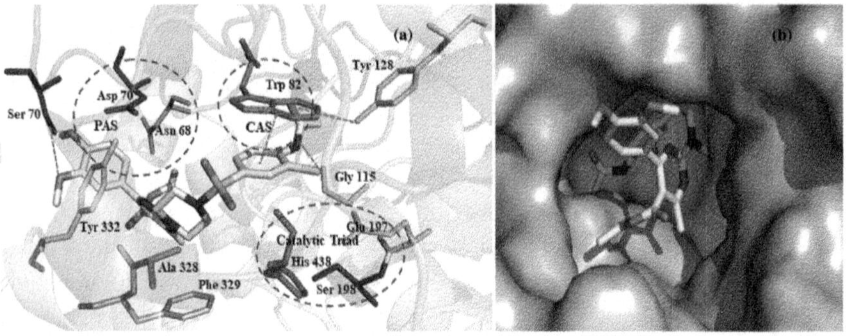

Fig. 3.9 Docking interactions of compound (**19**) within the active site of BuChE (PDB code: 4TPK) (*Source* Fig. 4, Manzoor et al. [52])

relationship and molecular docking studies, it was revealed that for good selectivity for PDE9A, the compounds needed to show interactions with Tyr424 as it is unique in PDE9. Along with this, the compounds also needed to interact with Gln453 and Phe456.

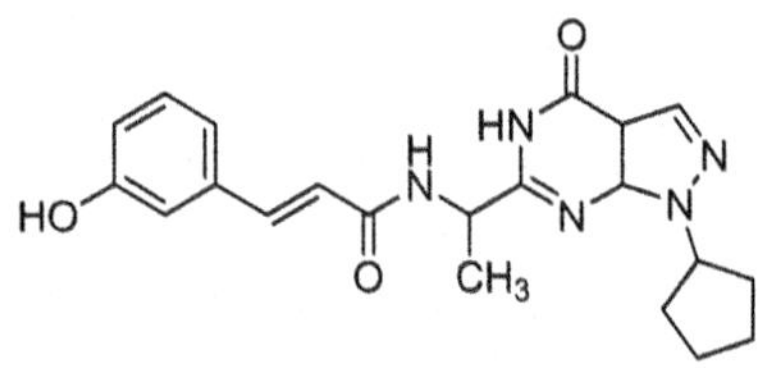

(20)

In 2018, Ghobadian et al. reported a series of some carbazole-benzyl-pyridine-based BChE inhibitors [54]. From the in vitro activity, it was found that all these derivatives displayed potent and selective anti-BuChE activity having strong BuChE inhibition in the range of 0.073 to 1.6 μM. Compound (**21**) exhibited mixed type of inhibition with the highest selectivity and potency with an IC_{50} value of 0.073 μM. Docking studies performed on the compound showed that it was bound to the catalytic site, peripheral site, the choline site, and the acyl pocket of BuChE. It was bound to the enzyme with a free binding energy of −8.68 kcal/M. BuChE choline binding site was fully occupied by the planar part of the carbazole moiety. No violation of Lipinski's rule of five was observed during the screening of these derivatives. Additionally, the compound (**21**) was found to be neuroprotective at 10 μM. It was also observed that the compound could inhibit self-induced and AChE-induced aggregations at concentrations of 10 and 100 μM, respectively.

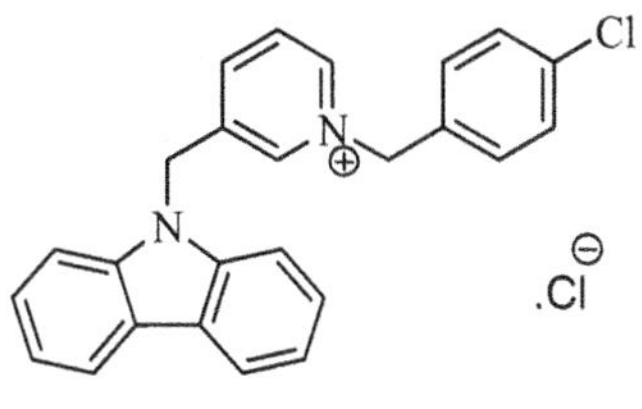

(21)

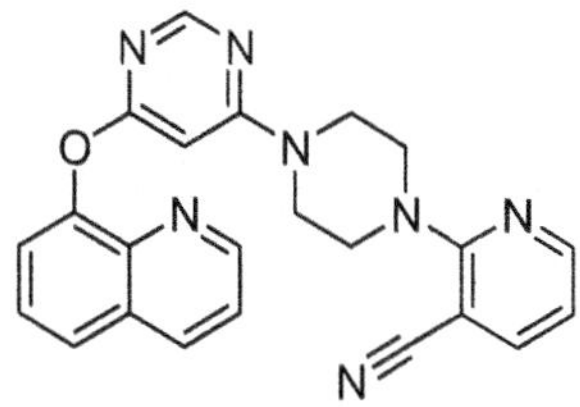

(22)

A hybrid series of triazolopyrimidine and pyrimidine derivatives was reported by Kumar et al. for the treatment of Alzheimer's disease as potent acetylcholinesterase inhibitors [55]. 2-(4-(6-(Quinolin-8-yloxy)pyrimidin-4-yl)piperazin-1-yl)nicotinonitrile **(22)** with an IC_{50} value of 36 nM was found to be the most potent compound of the series. Using the crystal structures of *Tc*AChE and *rh*AChE, docking studies were undertaken, in order to see the interactions between the hybrid molecules and the binding residues in the active site of AChE. Docking studies revealed that compound (**22**) interacted at three different sites of *rh*AChE, namely PAS, CAS, and catalytic site in its active binding domain. The cyanopyridine and 8-oxyquinoline moieties of compound (**22**) interacted with the active site residues Trp86 and Trp286 of the enzyme via π–π stacking.

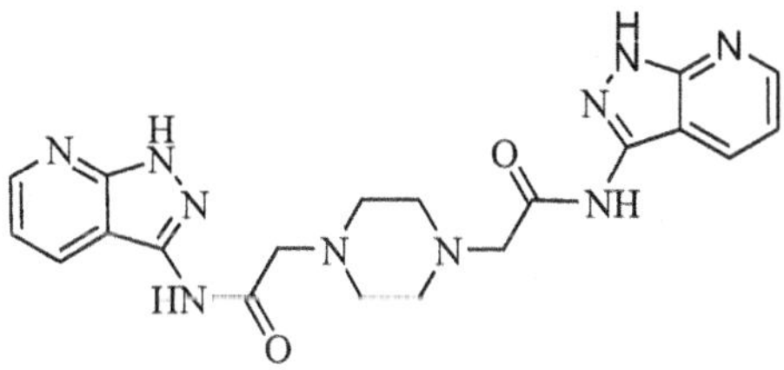

(23)

2-(Piperazin-1-yl)-*N*-(*1H*-pyrazolo[3,4-b]pyridin-3-yl)acetamide derivatives were reported by Tarana Umar et al. as potent AChE and amyloid-β aggregation inhibitors. Docking studies revealed strong interactions of the designed ligands with the amino acid residues of the active site of the protein which directly correlated with the in vitro results. The most potent compound **(23)** had an excellent AChE inhibitory activity having IC_{50} value of 4.8 nM and TEM analysis confirmed self-induced inhibition of $A\beta_{1-42}$ aggregation [56] to the extent of 81.65%. Self-induced $A\beta_{1-42}$ aggregation is linked to fibril generation that is controlled by formation of β-sheet pattern which could assist aggregation of $A\beta_{1-42}$. The change in the β-sheet pattern during fibrillo-genesis is mainly stabilized by Glu22/Asp23 and Lys28 residue salt bridges in $A\beta_{1-42}$ along with hydrophobic interactions. Compound (**23**) was found to be having interactions with amino acids Met35, Leu34, Ile32, Ile31, Gly29, Lys28, Val24, Asp23, Glu22, Phe20, Ala21, Leu17, Lys16, His13, and other residues through hydrophobic interactions in the vicinity of Aβ C-terminus, and a strong

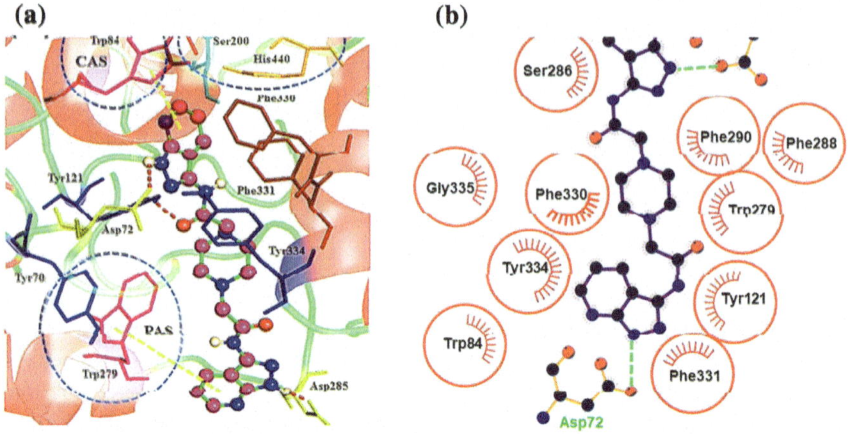

Fig. 3.10 Binding interactions of compound (**23**) with AChE. The inhibitor is displayed in ball and stick model while the key residues are displayed in stick model. Individual colors represent the residues like Ser200 (cyan), Glu199 (red), and His440 (orange), Tyr70, Tyr121, Tyr334 (blue), Trp84, Trp279 (hot-pink), Phe330, Phe331 (firebrick), and Asp72, Asp284 (yellow). Stick representation is used for other residues with carbons being shown in pink color; (**A**) depicts interactions of the active site key residues of AChE with compound (**23**) and plot (**B**) is Ligplot representation (reused with permission *source* Fig. 5, Umar et al. [56])

hydrogen bond was formed between carbonyl oxygen of carboxylate and N–H of Lys16 with a distance of 1.8 Å in the docking studies. Moreover, it was observed that intermolecular H-bonds existed between N–H of amide and oxygen of the hydroxyl of Glu22 (2.1 Å), N–H of pyrazole ring, and oxygen of hydroxyl of carboxylic acid group of Asp23 (2.6 Å), and finally another one between N–H of amide and Asp 23 carbonyl carboxylic acid (2.3 Å) (Fig. 3.10). Molecular modeling studies strongly predict excellent binding ability of compound **(23)** to the amino acids that interfere with β-sheets formation, leading to inhibition of $A\beta_{1-42}$ aggregation, and therefore, these interactions have a direct correlation with the experimental result of the lead compound offering activity in the nanomolar range.

3.2.4 Quinoline-Based Scaffolds as Anti-AD Agents

Zaib S. et al. disclosed a series of hybrid derivatives of quinoline-thiosemicarbazones as inhibitors of cholinesterase for the treatment of AD. Among the reported series, compound **(24)** proved to be a promising lead with 0.12 ± 0.02 μM IC_{50} value in the in vitro evaluation, which was five times more potent than the standard (i.e., galantamine, $IC_{50} = 0.62 \pm 0.01$ μM) used for the purpose of comparison. Molecular docking studies of the synthesized compound were performed using the PDB ID: 4BDT [57] structure of *h*AChE, determined by X-ray crystallography. The most potent compound **(24)** showed important interactions like T-shaped π–π interaction

(6.23 Å) by the quinoline ring with the amino acid residues of the active pocket, and thiosemicarbazone moiety interacted with Trp86 and formed π-sulfur bond (4.34 Å). H-bond was observed between Gly122 and the oxygen atom (2.82 Å). Tyr449 (5.86 Å) and Trp439 (4.68 Å) showed π-sulfur interactions with the sulfur atom of thiosemicarbazone group of compound (**24**). Oxygen of the methoxyl group interacted with Gly122 (2.82 Å) and formed H-bond, while non-bonding electron pair-π (2.80 Å) interaction was observed between chloro group and Tyr337. Furthermore, Glu452 displayed charge–charge attractive interaction (4.93 Å) with the morpholine nitrogen. It was clearly noted that the deep cleft of the catalytic site was not occupied by the methoxyl group. The most active compound (**24**) and Huprine W (Hup W) as a standard were used for molecular dynamics study. The study involved simulating the complexes of *h*AChE (PDB ID: 4BDT) with both the ligands, i.e., HupW and the selected inhibitor (**24**), taking the configuration of the docking poses of both molecules as the initial starting points in an aqueous environment for 30 ns.

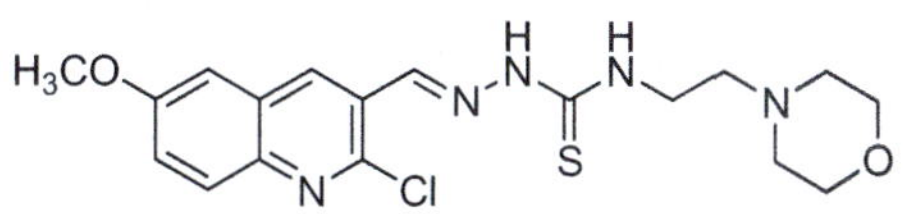

(24)

The results (Fig. 3.11) of the simulations were displayed as RMSD and RMSF values. The standard ligand (HupW) started showing stability from 5 ns onward, with only a small deviation observed during 2–4.8 ns. The chosen inhibitor (**24**) showed stability throughout the simulation period, with only a small deviation during the period 0.3–0.4 nm. The structure of the complex between the protein and compound (**24**) showed only a slight deviation between 14–17 ns and remained stable for the remaining period of simulation. Figure 3.11 reflects the root mean square fluctuations of both the inhibitors. A significant pattern of fluctuations was observed for the apo and holo forms of the protein during the simulation. The apo form of the protein started showing fluctuation between 0.1–0.3 nm with a rise to 0.4 nm, while the

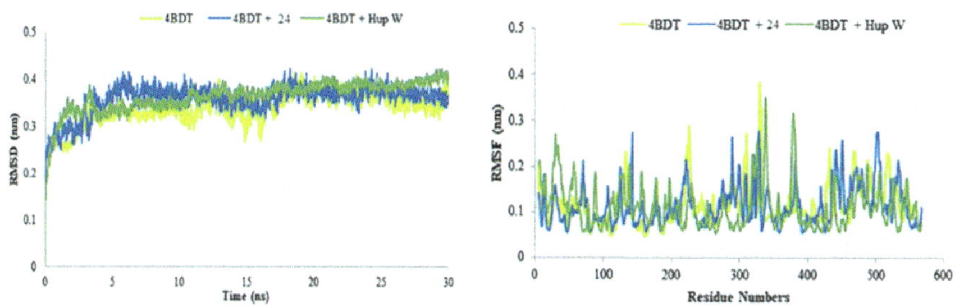

Fig. 3.11 RMSD and RMSF of the protein (4BDT) amino acid residues in the absence and presence of HupW and compound **(24)** during the simulation period of 30 ns (reused with permission *source* Fig. 7, Zaib et al. [57])

protein-inhibitor complex [*h*AChE-compound (**24**)], i.e., the holo form began with a fluctuation of 0.15 nm, varying between 0.1–0.28 nm with a minor increase. The holo protein, i.e., *h*AChE-HupW complex offered 0.10 and 0.30 nm fluctuations during the entire period of simulation of 30 ns. Both systems demonstrated high stability and limited level of fluctuations.

Elisabet et al. disclosed some potent BuChE and AChE inhibitors possessing additional antioxidant activity. Compound (**25**) from the series emerged as an interesting finding having significantly reduced oxidative stress and neuro-inflammation, improved basal synaptic efficacy, and much significantly reduced $A\beta_{42}$:$A\beta_{40}$ ratio in the hippocampus. The compound exhibited IC_{50} values of 1.1 and 7.3 nM against the enzymes *h*AChE and *h*BuChE, respectively [58].

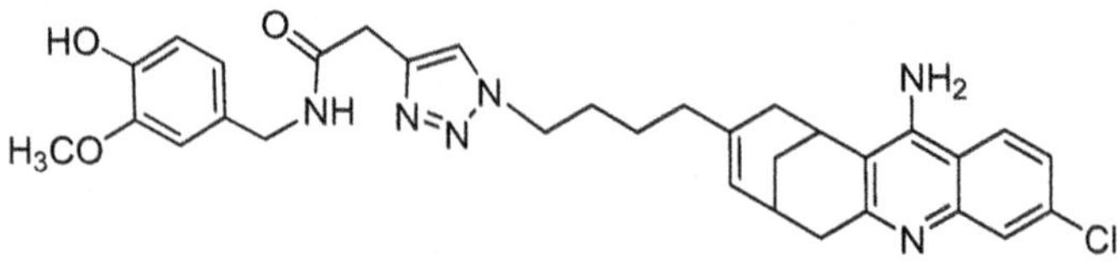

(25)

Molecular modeling studies revealed that the triazole ring showed π-stacking with the Phe290 side chain while the nitrogen of triazole ring formed H-bond with Phe288 and Phe295 in *h*AChE. Additionally, the nitrogen atom of amide group in compound (**25**) formed H-bonds with Ser286 as well as residual water present in the structure, which itself was H-bonded to Phe331. Ether oxygen of capsaicin fragment of compound (**25**) and hydroxyl group of Tyr121 makes a water bridge via water 78. While in case of BuChE, the chloro group of huprine was stacked in acyl-binding pocket, with Leu286 (O) and Trp231 at 4.0 and 4.1 Å, respectively. The aromatic system showed π-stacking with Phe329 and Trp231. Moreover, the amino group on 9-position got fitted between His438 and Ser198 (Fig. 3.12).

Novel cyclopentaquinoline derivatives were reported by Czarnecka et al. whereby they modified the tetrahydroacridine moiety and reported 6-chloronicotinamide derivative of cyclopentaquinoline (**26**) possessing *Eq*BuChE and *Ee*AChE inhibitory activities with IC_{50} values of 153 and 67 nM, respectively [59]. The compound (**26**) was also studied for its molecular interactions with the enzymes using docking studies on Gold Suite 5.1.

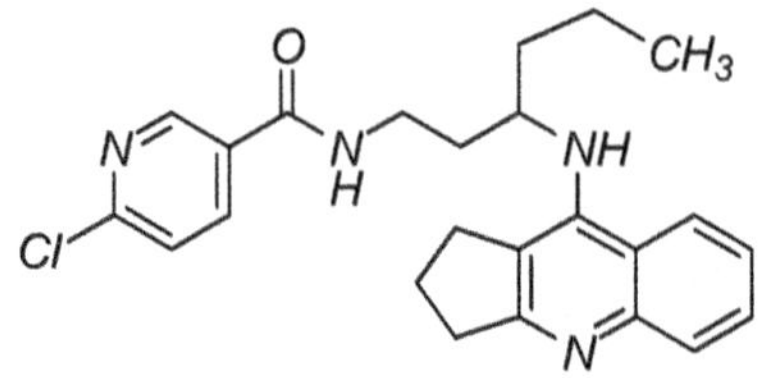

(26)

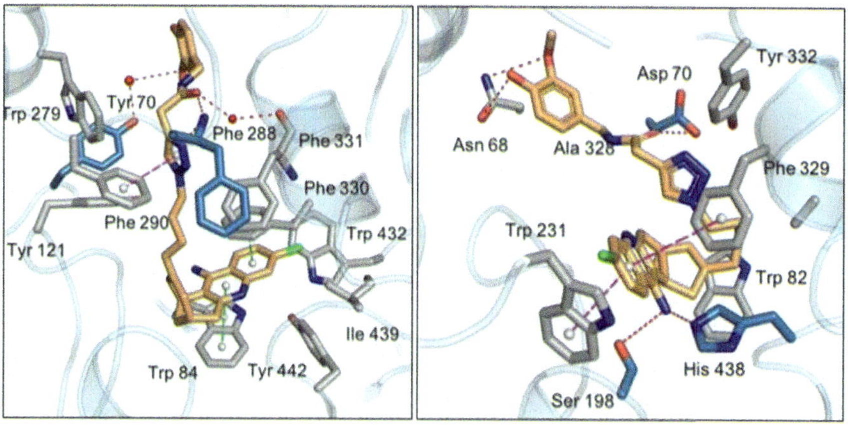

Fig. 3.12 Crystal structures of **25-***Tc*AChE and **25-***h*BuChE complexes (reused with permission *source* Fig. 5a, b, Viayna et al. [58])

The docking results clearly suggested that the modified tacrine fragment established interactions with the CAS site of the AChE enzyme, in which the cyclopentyl ring got sandwiched due to π–π stacking with Phe330 and Trp84 and cation-π interaction. His440 carbonyl carbon made hydrogen bonding with the protonated nitrogen while Trp279 and Tyr70 residues got into π–π stacking with chloropyridine ring and CH–π interactions with Tyr121 residue, in the AChE PAS site. Docking of compound (**26**) with BuChE revealed that the disposition of the cyclopentaquinoline ring was having similar pattern as was present with AChE, in which Trp82 established π–π stacking and hydrogen bonding with the main chain of His438. The 6-chloronicotinamide fragment of the compound (**26**) exhibited hydrophobic interactions with Ile356, Pro285, and Tyr282 near the PAS sub-site entrance and hydrogen bond formation with Tyr332 residue of the main chain.

Karolina et al. reported some novel cyclopentaquinoline-conjugated 9-acridine carboxylic acid derivatives as cholinesterase inhibitors. Compound (**27**) offered the highest inhibitory activity against both the enzymes AChE and BuChE yielding IC_{50} values of 273.33 and 03.73 nM, respectively [60]. In molecular modeling studies, it was revealed that the compound (**27**) possessed dual-binding modes. TRP86 and TRP286 in parallel disposition showed π-stacking with acridine and modified tacrine moieties. Oxygen of the amide group got stabilized by the H-bond formation with the TRY124. Complex of the compound with BuChE showed parallel disposition with the tacrine fragment, π-stacking with TRP82, and TRP231 and PHE329 residues were involved in the T-shaped π-stacking with acridine (Fig. 3.13).

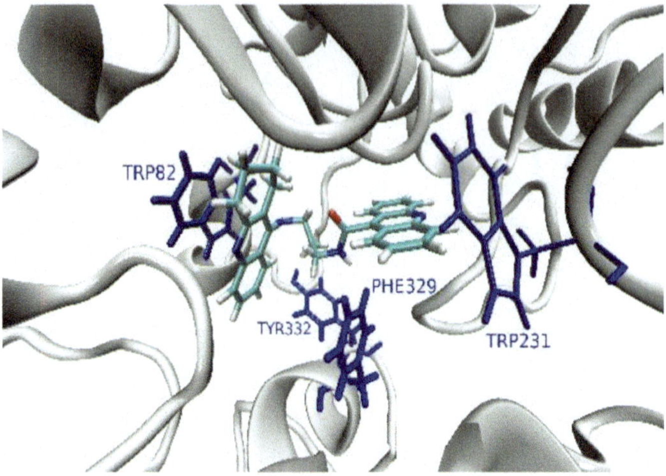

Fig. 3.13 Compound (**27**) in the binding pocket of BuChE (reused with permission *source* Fig. 6, Czarnecka et al. [59])

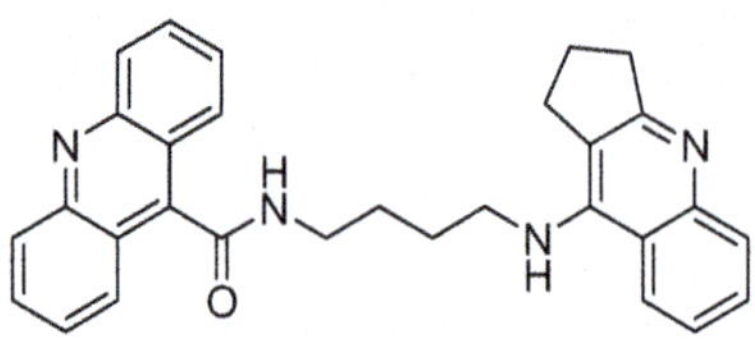

(27)

Safarizadeh et al. formulated a QSAR model for the inhibition of $A\beta_{1-42}$ peptide aggregation by the novel 2-arylethenylquinoline derivatives. Molecular modeling studies were also undertaken to look into the interactions of these novel molecules with the binding sites of the target [61]. AutoDock 4.2.5.1 was used for performing the docking studies on $A\beta_{1-42}$ (PDB 1IYT), and the peptides were analyzed in Discovery Studio, version 4.5. Results of the docking studies of the designed molecules 2-arylethenylquinoline with $A\beta_{1-42}$ peptide showed their binding with amino acid residues like LEU31, VAL24, MET35, LYS16, ALA21, LEU17, PHE20, and HIS13

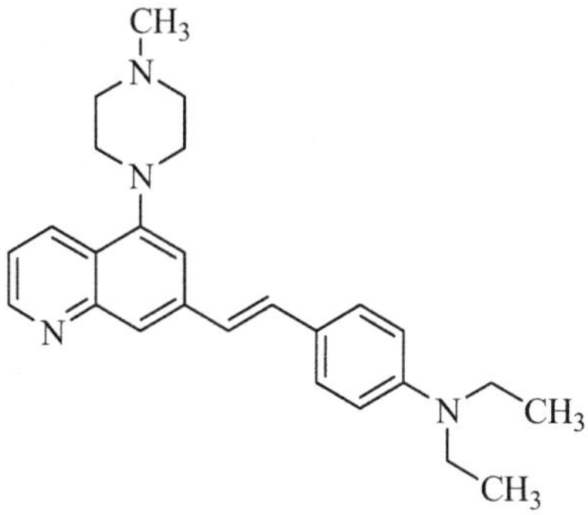

(28)

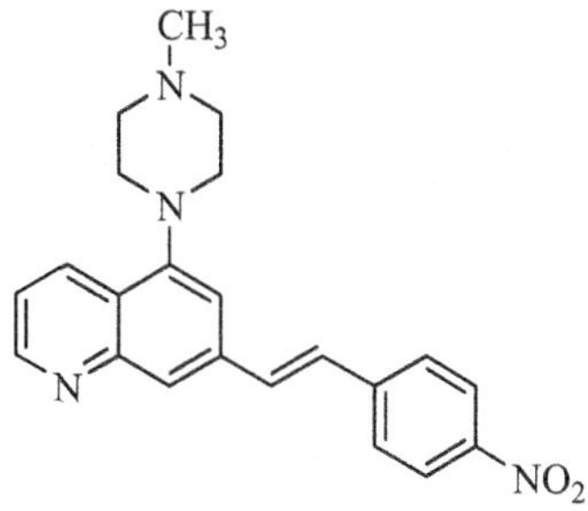

(29)

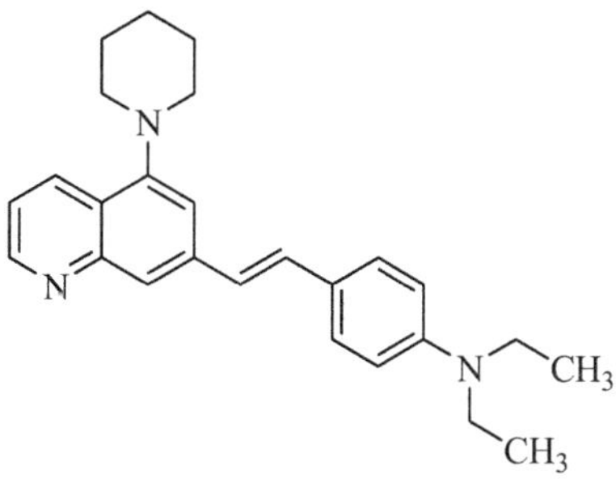

(30)

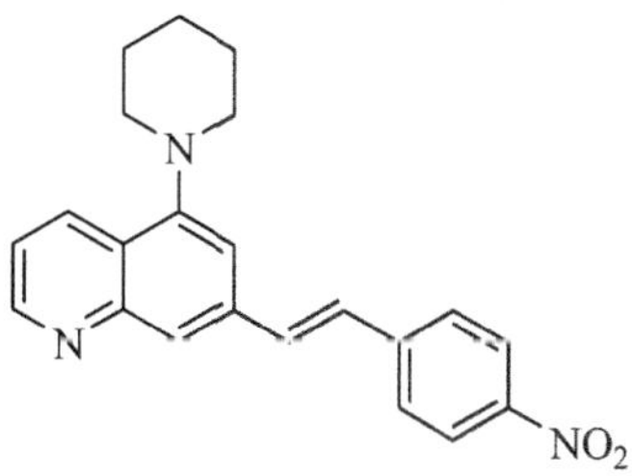

(31)

through hydrogen bonds along with hydrophobic interactions. The quinoline ring N-atom made hydrogen bonding specifically with ALA21 NH group along with another hydrogen bonding between the methyl group (donor) of diethylamino and PHE20 NH (acceptor). Moreover, significant level of hydrophobic interactions also occurred among the electron-donating groups on piperidyl ring with VAL24, ILE31, and LEU34, as well as of phenyl ring with LEU17, LYS16, and HIS13. Overall, the investigations revealed that with changes in substituents on the phenyl ring, substantial changes in the interactions were observed which could directly correlate with the activity and therefore, while designing novel quinoline compounds for the inhibition of $A\beta_{1-42}$ aggregation, the position of electron withdrawing and donating groups in the aryl ring should be given due importance in order to get more active agents.

Molecular dynamics was carried out for the most potent compounds (**28** and **29**) and the least potent ones (**30** and **31**) wherein RMSD was used to evaluate the stability of the $A\beta_{1-42}$ protein with respect to the defined molecules. The RMSD values were below 2 Å for $A\beta_{1-42}$ (RMSD 0.55 Å) with compounds (**28** RMSD 0.55 ± 0.09 Å and **29** RMSD0.47 ± 0.04 Å) which got equilibrated after 25 ns while the complexes of compounds (**30** and **31**) got stabilized at 32 and 25 ns respectively, which meant that the complexes were less stable and therefore weak in potency with RMSD (1.04 ± 0.03 Å and 1.01 ± 0.04 Å, respectively). The experimental results were in agreement with the inference obtained from the molecular dynamics studies, as evaluated by radius of gyration (Rg) plots.

Jiang et al. evaluated some novel PDE2 inhibitors with additional antioxidant activity [62]. Among the compounds of the designed series, compounds (**32** and **33**) exhibited potent PDE_{2A} inhibitory activity with IC_{50} values of 4.2 and 6.1 nM with significant level of

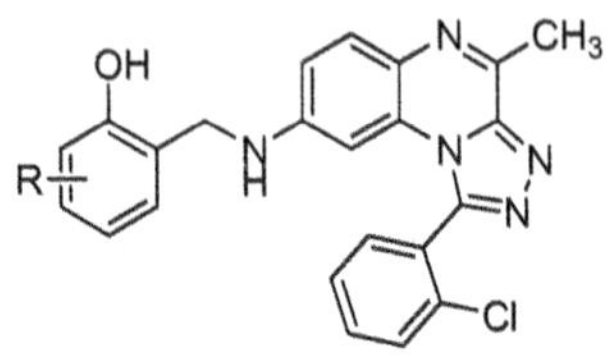

(**32**) R = 4-F; (**33**) R = 6-F

antioxidant activity [ORAC (Trolox) = 8.4 eq]. Compound (**32**) offered the least IC_{50} value among all of the designed compounds, against PDE_{2A}. In the docking pose of **32** with PDE2A, the triazoloquinoxaline scaffold showed indirect H-bond interactions with the Tyr655, Gln859, Gln812, and Tyr827 via three molecules of water. Direct π-σ and π–π interactions were observed with 6-F group of compound (**33**) with Ile826, Phe862, and Phe830 (Fig. 3.14). Hydrophobic interactions were observed among 2-chlorophenyl group and the neighboring residues Tyr655 and Leu770. Additionally, 6-F substituted phenolic group was suitably fitted in a pocket formed by four hydrophobic residues Leu770, Lue774, Phe862, and Ile866.

3.2.5 *Coumarin and Chromene-Based Scaffolds as Anti-AD Agents*

Along with cholinesterases, tau protein hyper-phosphorylation, and amyloid-β aggregation, AD is also linked with MAO-B enzyme activation. Changjun Zhang group reported a series of hybrid hydroxypyridinone-coumarin derivatives with potent inhibitory activity (IC_{50} value of 14.7 nM) against the enzyme MAO-B for which docking studies were also undertaken to discover the enzyme-ligand interactions and establish SAR for the same [63].

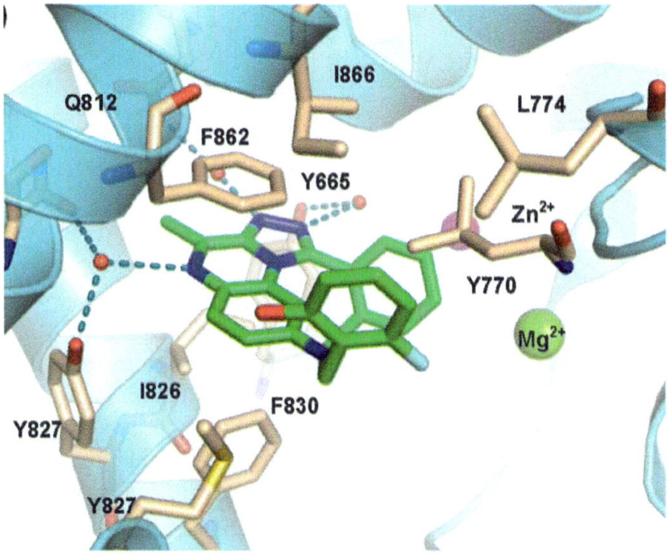

Fig. 3.14 Binding interactions of PDE2A (PDB ID: 4D09) with compound **(33)** (reused with permission *source* Fig. 2, Jiang et al. [62])

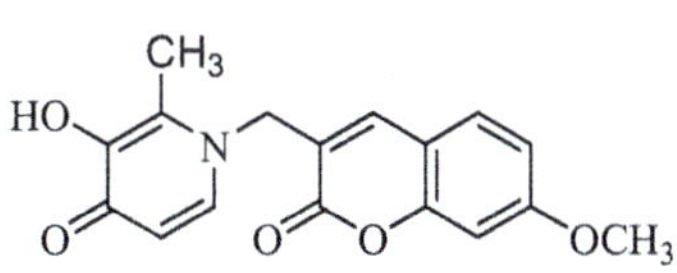

(34)

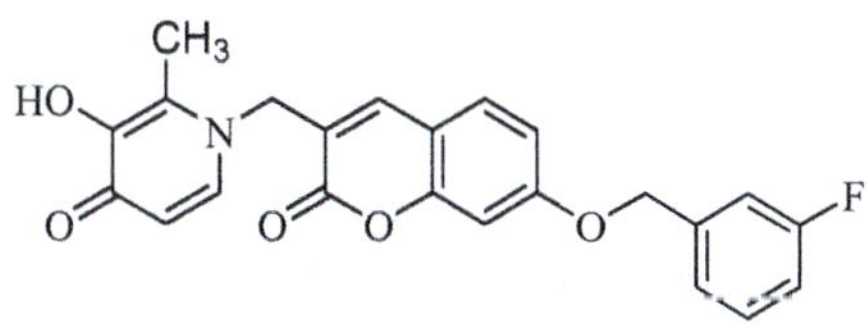

(35)

CDOCKER program was utilized to undertake molecular docking studies using the Discovery Studio 2016 software. On the basis of in vitro results, two compounds (**34** and **35**) were identified which revealed that the coumarin moiety showed lipophilic interactions along with hydroxypyridinone toward FAD cofactor and showed it clearly that hydroxypyridinone moiety acted as a key fragment as a metal chelator in binding to the MAO-B enzyme substrate. The pyran-carbonyl oxygen produced two hydrogen bonds with Gln206 and Tyr326 acting as the key component for imparting significant inhibitory activity to compound (**34**), while in case of compound (**35**), the benzyloxy fragment occupied the hydrophobic entrance and interacted with Phe103, Pro104, Leu164, Ile316, and Ile199. The hydroxyl of the pyridinone interacted by forming hydrogen bond with Gln206, while methyl part of pyridinone and Tyr435 offered π-alkyl interaction along with FAD which resulted in providing the highest activity to compound (**34**).

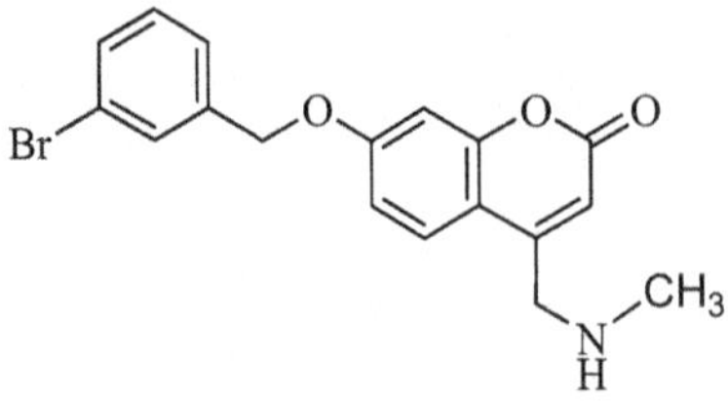

(36)

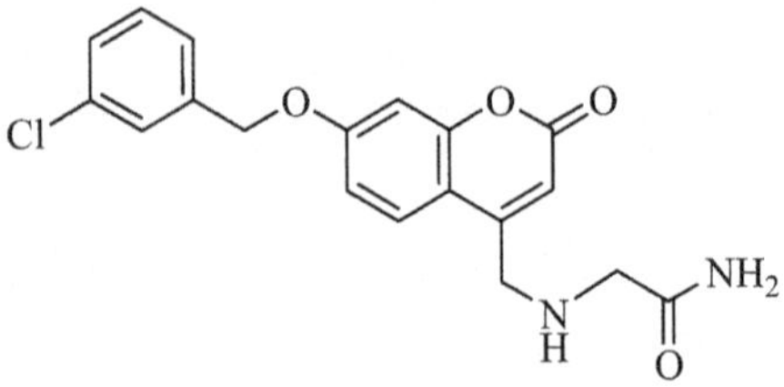

(37)

Another such series of coumarin derivatives was designed by Rullo et al. in which 7-benzyloxy-2*H*-chromen-2-one derivatives (**36, 37**) were reported possessing AChE, BuChE, and MAO-B (0.0022 ± 0.0001 μM) inhibitory activities [64]. The docking was performed on PDB ID: 6O4W using Maestro software package wherein the heterocyclic core was found to be binding to the AChE in the PAS site, just like the standard drug donepezil, even though the coumarin ring was not overlapping the Trp286 residue properly for making π–π interactions. The entire molecule was stabilized by π-cation and hydrogen bond interactions attaching the basic head to the entrance of the CAS gorge with Ser293 and Trp286 residues. Docking was repeated on another PDB ID: 6EY6, and the standard drug galantamine was docked for validation purpose, but the results were the same. Docking on PDB ID: 6F7Q of *h*BuChE showed possible binding of coumarin moiety of compound (**36**) at PAS Tyr332 and CAS Trp82 of BuChE, while compound (**37**) produced a twist which was not suited to interact with Trp82 resulting in comparatively lower anti-BuChE activity which was commensurate to the in vitro results.

Palareti et al. have disclosed a novel series of 4-isochromanones composed of *N*-benzylpyridinium moiety as potent acetylcholinestrase inhibitors. Compound (**38**) among the synthesized compounds has shown potent AChE/BuChE inhibitory activity with an IC_{50} value of 0.15 nM [65]. To understand the interactions of the reported compounds with the active site residues of AChE (PDB ID: 1EVE), molecular docking studies were undertaken. The docking studies revealed that compound (**38**) made hydrogen bond interactions between methoxyl groups and Trp286 residue similar to donepezil, wherein phenylpyridine moiety interacted with the CAS site and 4-isochromanone moiety interacted with the PAS site of AChE.

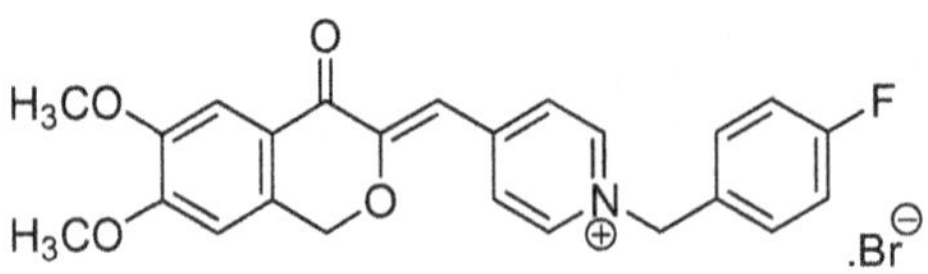

(38)

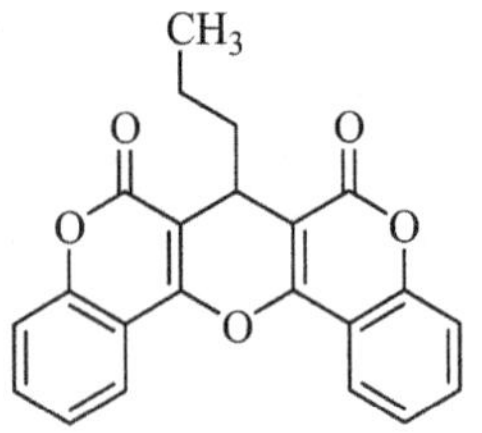

(39)

Maharajan et al. undertook investigations of 7-propyl-*6H*-pyranodichromene-6,8(*7H*)-dione **(39)**, a novel coumarin inhibitor of Alzheimer's as a potential drug acting at triple sites, using various in silico techniques including computational quantum calculation, molecular dynamics and molecular docking for the inhibition of β-secretase, glycogen synthase kinase, and acetylcholinesterase enzymes, which are potential targets for the treatment of AD [66].

Molecular docking studies were performed on the compound (**39**) taking the crystal structures of BACE1 (4GID), GSK3β (1Q41), and AChE (4EY6) from protein data bank. The proteins and ligands structures were optimized and docked in Schrodinger package-2014 (Maestro, 2014), and the interactions were identified in PyMol. The most potent compound (**39**) was docked within the binding sites of GSK3β, BACE1, and AChE using induced-fit docking offering the docking scores of −8.51, −8.62, and −10.48, respectively, while the Glide's free binding energies were −51.70, −51.83, and −52.71 kcal/M, respectively, which clearly predicted AChE-*biscoumarin* complex to be more stable with higher docking score as well as Glide free energy than the rest of the targets. Moreover, it also revealed that biscoumarin had strong binding affinity for the CAS site residues Ser203 and His447 of AChE with interaction distances of 2.4 and 3.2 Å, respectively. Molecular dynamics simulation was done with leap module and AMBERTOOLS14 with force field AMBER ff14SB for the newly synthesized coumarin derivatives up to 40 ns, wherein the RMSD showed similar deviations in the biscoumarin-AChE complex with maximum deviation of ~2.0 Å, making it highly stable, while biscoumarin-BACE1 had a sudden deviation around 22 and 33 ns due to absence of intermolecular interactions, although the deviation was ~2.8 Å which could be considered exhibiting high stability. The radius of gyration (Rg) values for GSK3β, BACE1, and AChE was 21.5, 21, and ~23 Å, respectively, indicating high compactness of the molecule within the proteins. The results after the MD stimulations of 40 ns showed strong hydrogen bonding between biscoumarin and the CAS site of AChE with a distance of 1.8 Å, while in case of GSK3β, some other neighboring interactions were seen, and the active site interactions disappeared due to the dynamic behavior of the compound. Free binding energies were determined to be −27.74, −21.19, and −26.43 kcal/M for AChE, BACE1, and GSK3β, respectively, by selecting 2000 frames from 0 to 40 ns (Fig. 3.15).

In 2020, Sepehri et al. reported synthesis, characterization, biological evaluation, and molecular docking of hybrid molecules consisting of coumarin-1,2,3-triazole-acetamides. They have reported some hybrid compounds containing coumarin scaffold and evaluated them against enzymes like human carbonic anhydrase I & II, α-glycosidase, α-amylase, acetylcholinesterase, and butyrylcholinesterase, offering Ki values ranging from 55.38 to 128.63 nM against α-Amy, 590.42–1104.36 nM against α-Gly, 27.17–1104.36 nM against BuChE, 24.85–132.85 nM against AChE, 508.55–1284.36 nM against hCA II, and 483.50–1243.04 nM against hCA I. Molecular modeling was undertaken to investigate the enzyme-ligand interactions. Compound (**40**) was found to be the most potent derivative against both the cholinesterase enzymes. It offered IC_{50} values of 33.12 and 29.35 nM against BuChE and AChE enzymes, respectively. The compound interacted with AChE amino acid residues

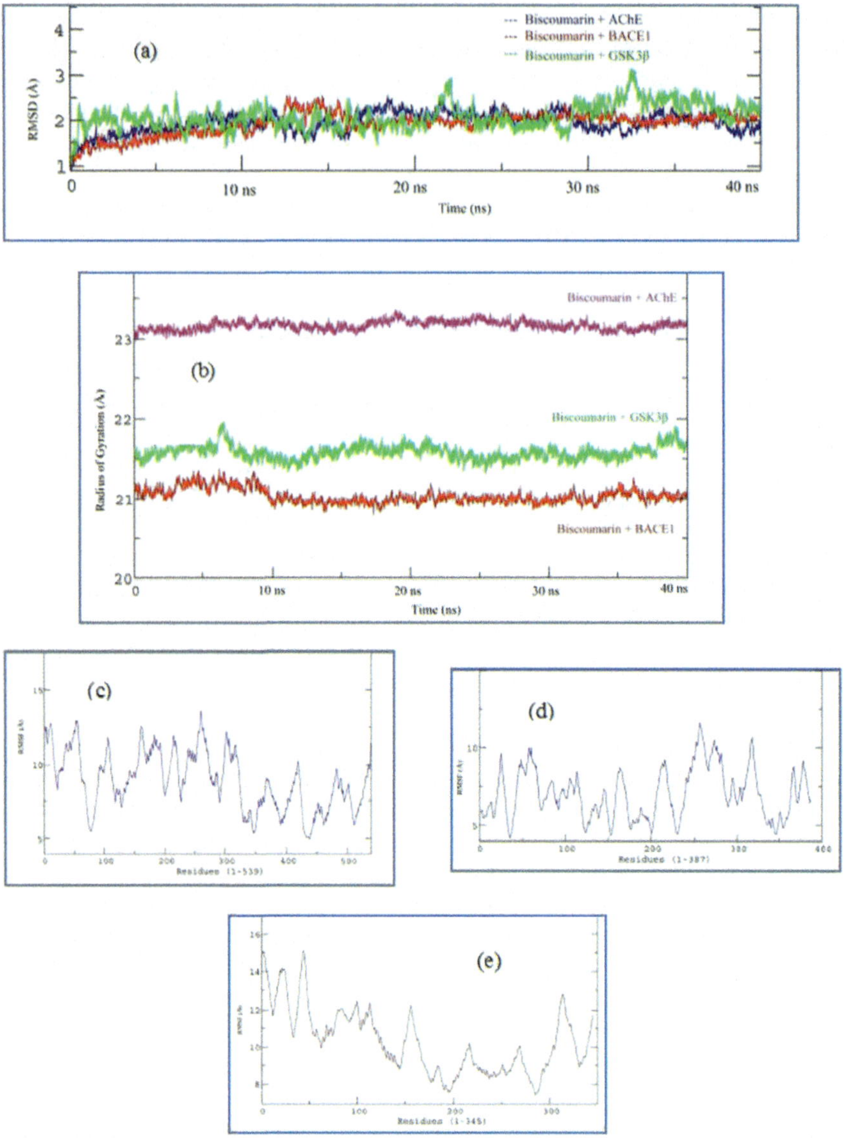

Fig. 3.15 MD simulation trajectories of biscoumarins with AChE, BACE1, and GSK3β for 40 ns. **a** RMSD, **b** radius of gyration, **c** RMSF of AChE, **d** RMSF of BACE1, and **e** RMSF of GSK3β (reused with permission *source* Fig. 5, Sivakumar et al. [66])

Trp84, Phe300, and Phe301 of the CAS region and with Trp279 and Tyr334 of the PAS sub-site [67].

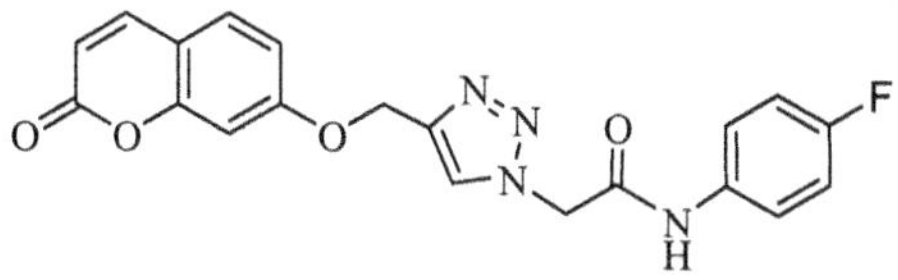

(40)

The potency of chromones as AChE antagonistic agents has been a key point in the investigations of anti-Alzheimer's drug discovery. One such study was carried out by Prayasee Baruah group wherein they designed new chromone analogs, mainly cyanochromone (CyC) (**41**) and aminomethylchromone (AMC) (**42**). After performing the docking studies, the compounds were evaluated for their in vitro activity [68].

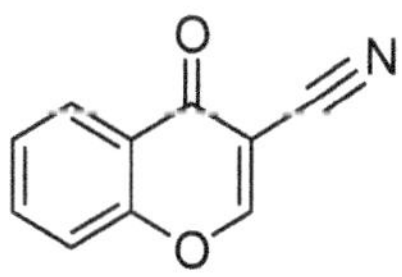

(41) CyC

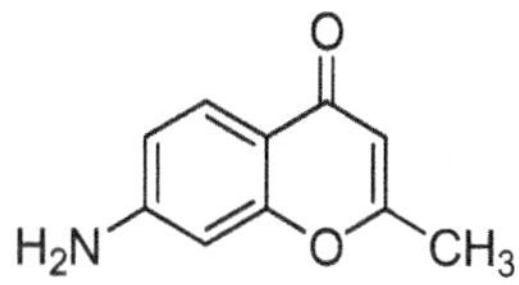

(42) AMC

Molecular docking was performed on HSA (PDB ID: 1AO6) and AChE (PDB ID: 1C2B) proteins taken from Protein Data Bank. The structures were docked using AutoDock Vina48, while the results were visualized in VMD1.9.352 for visualizing the molecular interactions of the proteins' residues with the ligands. With AChE, **CyC** was predicted to have a binding score of −35.02 kJ/mol making hydrogen bond interaction with residues Tyr12 and Tyr337. **CyC** made hydrophobic interactions with the amino acid residues Phe338, Gly124, Trp86, and Phe297. In comparison, **AMC** made hydrogen bond interactions with residues Tyr124 and Arg296 along with hydrophobic interactions with Tyr72, Trp286, Ser293, Phe297, and Ile294, offering a binding score of −33.37 kJ/mol in the modeling studies. Simulation in the PAS binding was performed in competition with Thioflavin-T that showed that **CyC (41)** and **AMC (42)** both replaced Thioflavin-T non-competitively at the PAS binding site, and this hypothesis was validated from the results of Achilles Blind Docking wherein it was seen that π–π stacking of **CyC** had taken place with the residues Trp 86 and Phe 295, which could be playing important role in the inhibitory activity of these chromones.

3.2.6 Pyrazole-Based Scaffolds as Anti-AD Agents

Taslimi et al. reported synthesis and carbonic anhydrase, α-glycosidase, and cholinesterase inhibition activities of some pyrazole-phthalazine derivatives [69]. The synthesized compounds were evaluated against α-glycosidase, BuChE, AChE, *h*CA I, and *h*CA II enzymes. To get an insight into the binding interactions of the synthesized compounds, modeling studies were performed with these enzymes. It was observed that the dioxo group was critically involved in deciding the potency of the inhibitory activities of the synthesized compounds. Compound (**43**) was observed to be exerting the highest inhibitory action against BuChE and AChE. Docking scores of compound (**43**) were found to be −9.91 and −12.31 against AChE and BuChE, respectively. Further, IC_{50} values for compound (**43**) were found to be 94.37 and 98.25 nM against AChE and BuChE enzymes, respectively.

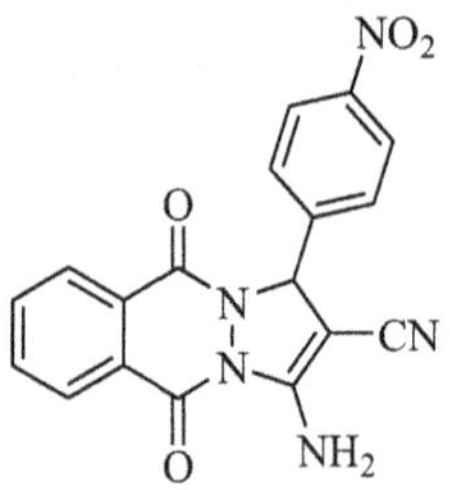

(43)

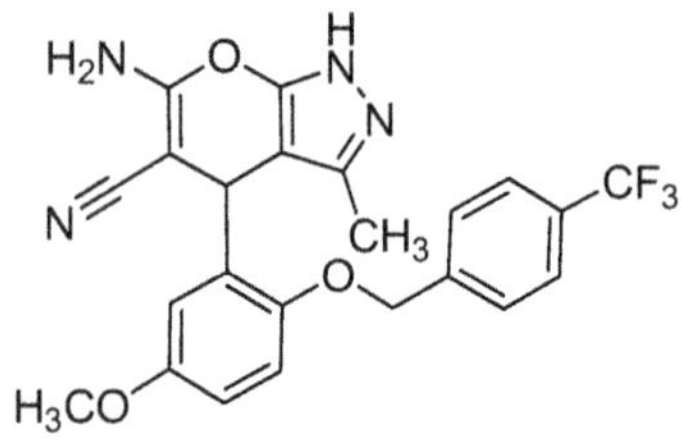

(44)

In 2021, Zhou et al. reported a series of dihydropyranopyrazole derivatives by modification of their previously reported compound **(*R*)-LZ77** by using structure-based designing approach [70]. After suitable modifications, one (**44**) of the resulting compounds showed an IC_{50} value of 41.5 nM for PDE2 inhibitory activity, while compound **(*R*)-LZ77** exhibited moderate activity against PDE2 with 261.3 nM as the IC_{50} value.

Based on the docking studies (Fig. 3.16), it was observed that the π–π stacking interactions were formed between the phenyl ring of compound **(*R*)-LZ77** and F862, and H-bonding between cyano group of thepyranopyrazole and Q859. Pyranopyrazole ring's oxygen atom established H-bonding with Y655 utilizing the water molecule as the bridge. A pocket (H-pocket) was formed by the amino acid residues L809, T805, I870, L770, H773, and F862. The benzyl side chain did not enter the hydrophobic pocket and remained outside the pocket. That makes it clear that if a side chain is introduced into the structure of **(*R*)-LZ77** that could occupy this pocket leading to improvement in the PDE2 inhibitory activity and selectivity in comparison with the other PDEs.

In the docking studies, it was observed that the oxygen atom of the 5-methoxyl of the enantiomer (+)-(**44**) established H-bond with Y827. This bonding improved the activity of this enantiomer over the derivative having a methyl group at this position. Position-2 benzyl side chain could enter the H-pocket of the enzyme and

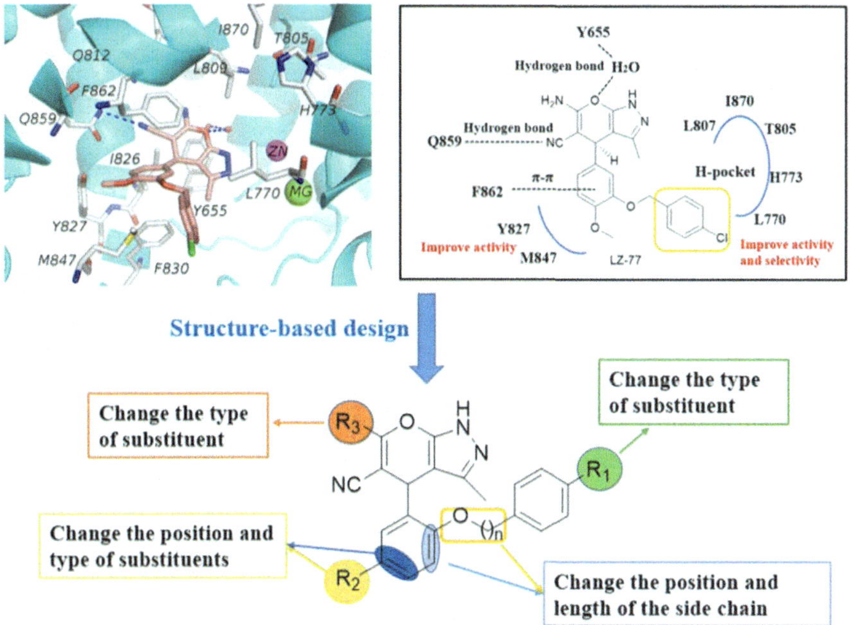

Fig. 3.16 Binding interactions of **(*R*)-LZ77** with PDE2 as seen in molecular docking studies (upper part), and the structure-based designing strategy for PDE2 inhibitors (lower structure) (reused with permission *source* Fig. 2, Zhou et al. [70])

establish strong hydrophobic interactions with amino acid residues L809, T805, I870, L770, H773, and F862, improving its inhibitory activity. H-pocket's size was large enough to accommodate the full side chain along with its 4-CF_3 substituent (Fig. 3.17a). It is interesting to note that in the case of the other enantiomer (−)-(**44**), the pyranopyrazole moiety shifted away from the H-pocket compromising its ability to make contacts with F862 and Q859 residues. So, the inhibitory activity of (−)-(**44**) against PDE2 got drastically reduced.

3.2.7 Benzimidazole and Benzodiazepine Derivatives as Anti-AD Agents

To study the acetylcholinesterase and butyrylcholinesterase inhibiting potential, Acar Cevik et al. designed compounds incorporating benzimidazole and triazole rings in the same chemical moiety and performed molecular modeling studies to find out the binding mechanism of these compounds with these proteins. Compounds (**45** and **46**) which contained 3,4-dihydroxyphenyl along with 5(6)-chloro substituent in the benzimidazole ring and alkyl substituent on the triazole ring were discovered to be

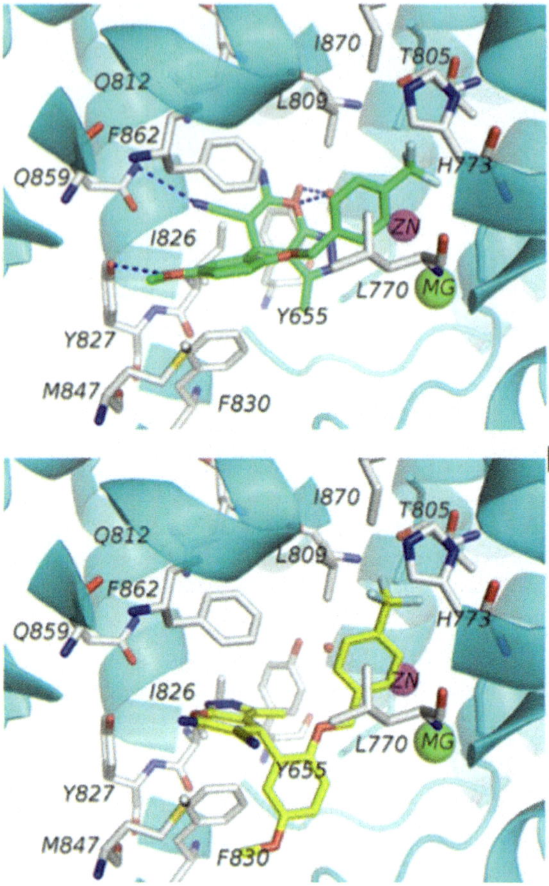

Fig. 3.17 Binding modes of both the enantiomers of compound **(44)** in the docking studies with PDE2. (**A**) (+)-isomer, (**B**) (−)-isomer (*source* Fig. 4, Zhou et al. [70])

effective AChE inhibitors. The Ki and IC_{50} values of compounds (**45** & **46**) were found out to be 26.2 nM and 31.9 ± 0.1 nM, respectively [71].

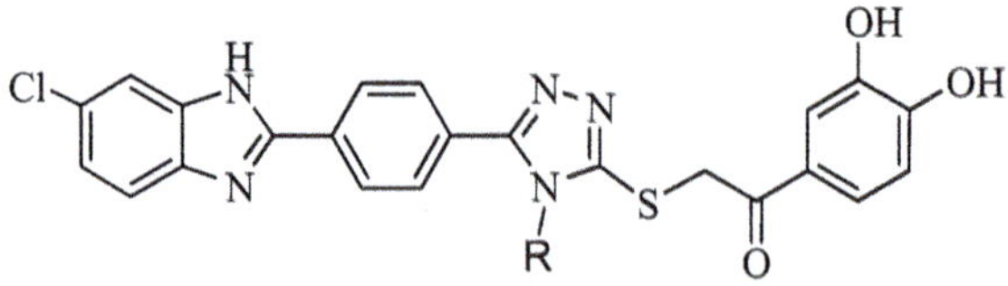

(45) R = CH_3; (46) R = C_2H_5

Docking studies of compounds (**45** & **46**) were undertaken using the 3D-structure (PDB ID: 4EY7) for AChE. It was found that the benzimidazole ring interacted with Trp286 of PAS through lipophilic interaction, while the polar dihydroxyphenyl-triazolylthioethanone side chain was bound to Trp86 residue of the CAS sub-site.

The results of **45** and **46** bound to the enzyme produced six mutual interactions, wherein 5(6)-chlorobenzimidazole ring and 3,4-dihdroxyphenyl ring offered π–π interactions with Trp286 and Trp86, respectively, and triazole N-atom interacted with Trp24 via H-hydrogen bonding. 3,4-dihdroxyphenyl ring was also found to be interacting with Gly120 and Tyr133 through polar interaction by acting as H-bond donor, and Tyr133 acting as acceptor. 3-Hydroxyl of phenyl ring also made hydrogen bonding with Glu202 residue through its carbonyl group.

An interesting series of benzodiazepine-1,2,3-triazole derivatives was reported possessing cholinesterase inhibitory activity in which the compounds inhibited butyrylcholinesterase (BuChE) enzyme but not acetylcholinesterase. Two compounds (**47** and **48**) were identified as sub-micromolar BuChE inhibitors offering IC_{50} values of 0.2 and 0.4 μM, respectively [72]. The docking studies were performed using AChE-donepezil complex (PDB ID: 4EY7) and BuChE-tacrine complex (PDB ID: 4BDS) on Autodock (ver.4.2.6). The docking results of compounds (**47** and **48**) vis-à-vis donepezil and tacrine indicated strong interactions with the ligands donepezil and tacrine. The standard ligands were re-docked into the active sites of the enzymes for the purpose of validation of the process. RMSD of the re-docked ligands in comparison with the co-crystallized structures was 1.12 and 0.062 Å, respectively, for tacrine and donepezil. The results obtained for compounds (**47** and **48**) clearly suggested that these structures could be ideal inhibitors of BuChE as they showed multiple interactions within its binding site (PAS). The dimethyl group of compound (**47**) produced π-alkyl interactions with Trp286 an aromatic residue present in PAS. Another π-anion interaction was seen between compound (**47**) phenyl ring and Asp74 residue, at the CAS site of AChE. π–π stacking interaction was seen between 3-fluorobenzyl group and Trp86, along with H-bond with Trp133 and Ala127. Additional hydrogen bonds were observed between the nitrogen of the triazole ring with both His447 and Ser203 residues. Docking studies performed on compound (**48**) indicated that the molecule was comfortably placed inside the active site of the enzyme with free binding energy of −8.6 kcal/M. The ether linkage oxygen atom formed H-bonds with Ser198 and His438 amino acid residues, while the phenyl group linker was involved with Glu197 in a π-anion interaction and with Trp82 in π–π interaction. The 4-methyl-subtituted benzyl ring produced another π–π stacking interaction with Tyr332 and Phe329.

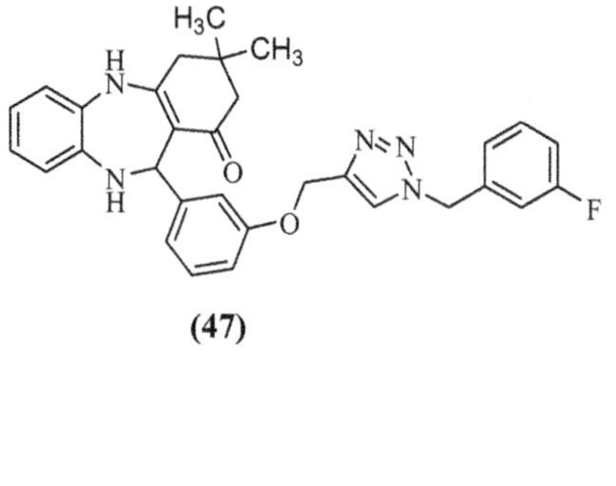

(47)

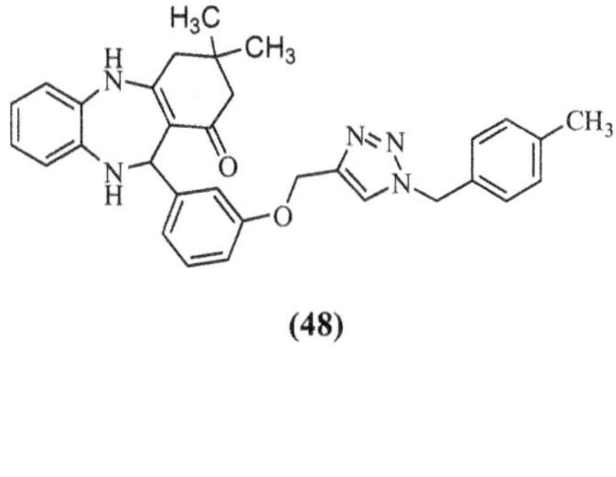

(48)

3.2.8 Thiazole Containing Compounds as Anti-AD Agents

Thiazole-piperazine hybrids as a novel class of compounds were developed by Osmaniye et al. to combat Alzheimer's disease. On cholinesterase enzymes, the inhibitory potential of compound **(49)** was found to be significant with IC_{50} value of 0.0317 μM against AChE, which was supported by the in silico studies [73].

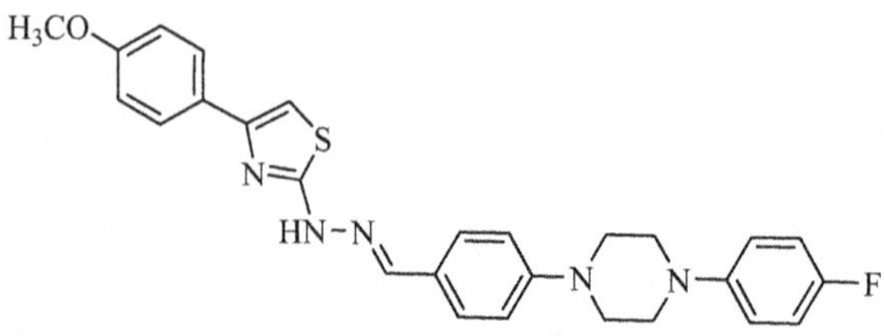

(49)

Molecular docking studies were performed on compound **(49)** and the standard drug donepezil using the crystal structure having PDB ID: 4EY7, wherein it was revealed that compound **(49)** was showing binding interactions to CAS and PAS sub-sites of AChE in a disposition similar to donepezil to Trp86 and Trp286, respectively, while the standard donepezil formed interactions with Trp286 and Phe295 of the PAS along with Trp86 of the CAS. Specifically, the thiazole ring of **(49)** formed π–π interaction with Trp286 indole ring system while the indole ring of Trp86 and 4-fluorophenyl ring offered π–π interactions. The piperidine phenyl ring and Tyr341 phenyl ring exhibited another π–π interaction. The nitrogen atoms of imine and piperidine ring jointly formed the basic centers wherein the imine group formed hydrogen bond with Tyr341, and the piperidine ring formed H-bond with hydroxyl group of Tyr337. Lastly, oxygen of 4-methoxyphenyl and carbonyl of Leu289 established hydrogen bond interactions resulting in manifestation of potent AChE inhibitory activity to the compound **(49)**.

3.2.9 Alkylamine Linked Derivatives as Anti-AD Agents

Synthesis and biological activity of phthalide tertiary amine derivatives have been reported by Luo L et al. as AChE inhibitors. The reported compounds demonstrated high selectivity and potency in AD [74]. Compound **(50)** offered the highest potency with an IC_{50} value of 2.66 nM against AChE. The compound **(50)** was bound to PAS as well as to CAS sub-sites as revealed in the docking studies. It was found to have excellent in vitro BBB permeation. Moreover, in the animal studies, the compound could exhibit reversal of scopolamine-induced memory deficit.

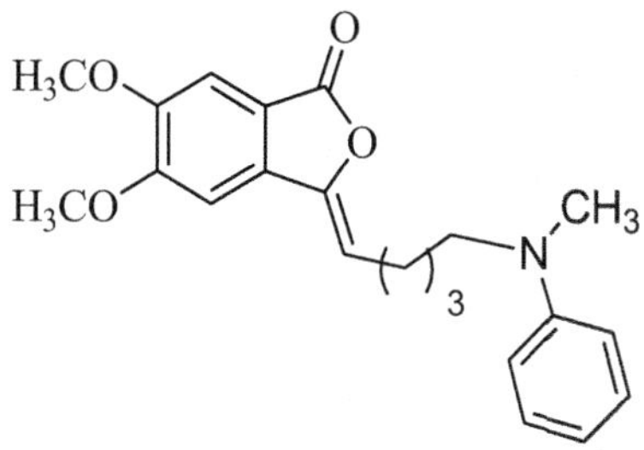

(50)

Some novel ferulic acid derivatives have been reported by Lan et al. for the treatment of Alzheimer's disease [75]. Most of the synthesized compounds exhibited potent AChE enzyme inhibition and self-induced β-amyloid (Aβ) aggregation activity with good antioxidative potential. Compound **(51)** offered IC_{50} values of 0.66 μM for BuChE and 9.7 nM for AChE, the lowest ones among all the compounds of the series. At a concentration of 20 nM, it offered 49.2% inhibition of Aβ aggregation. This was equivalent to 1.26 trolox antioxidant activity. Further, modeling and kinetic studies revealed that compound (**51**) interacted with CAS and PAS subsites of AChE simultaneously offering a mixed type of inhibition. It was found to show good BBB permeation potential in the PAMBA-BBB assay. All of the above observations proved that compound (**51**) could be a potential multifunctional lead for inhibiting ChE enzymes.

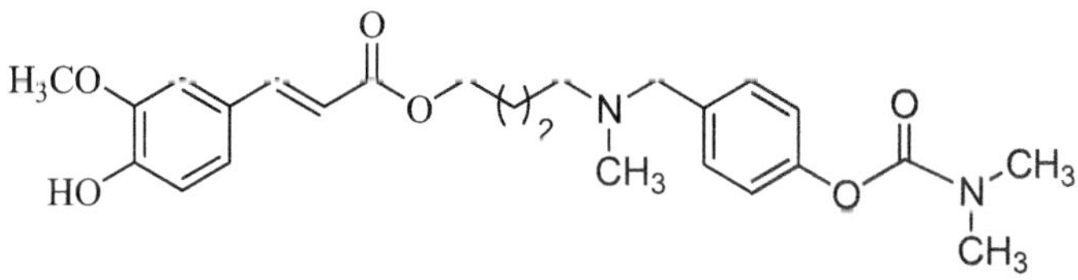

(51)

A combined study of molecular docking and QSAR and molecular dynamics of some AChE and BChE inhibitors have been reported by Daoud et al. [15]. The study was initially carried out on 36 compounds belonging to 4-[(diethylamino)methyl]phenyl ether class of derivatives. Compounds (**52** and **53**) have been identified as novel anti-Alzheimer's agents for further development on the basis of this study. Compound (**52**) offered an IC_{50} value of 0.084 μM against the enzyme AChE interacting with it via TYR34, TRP84, and TRP279 residues, and compound (**53**) with IC_{50} value 0.0091 μM was found to interact with BuChE.

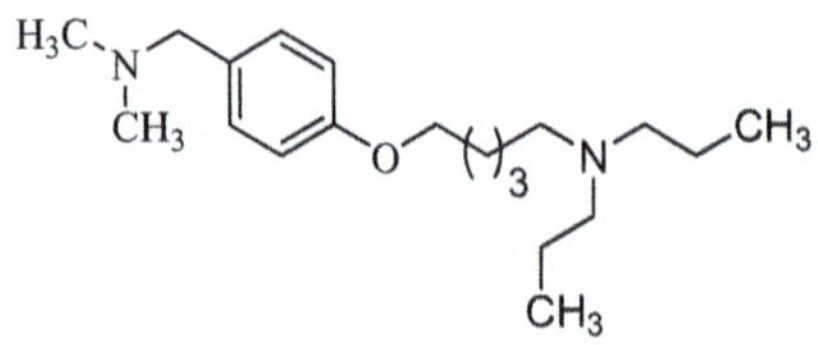

(52)

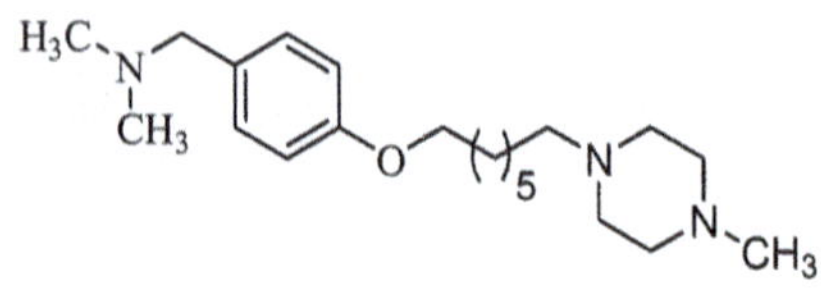

(53)

Begum et al. carried out in vitro as well as in silico studies of *N*-phthaloylglycine amide derivatives as BuChE inhibitors [76]. Schotten Baumann reaction was used to synthesize these amide derivatives. The compounds displayed potent BuChE inhibition activity in the in vitro evaluation. When compared with galantamine (IC_{50} value of 6.6 μM), compound **(54)** (IC_{50} value of 6.5 μM) was found to be equipotent to galantamine and the most active compound of the series. Molecular docking studies were undertaken for these compounds to unravel their high order of biological activity. A majority of the compounds of the series were fitting well in the BuChE enzyme binding sub-sites. The heterocyclic moiety of compound **(54)** established H-bonding interactions with TYR332 residue of the binding site. Further, it was found that π–π interactions were stabilizing the compound in the enzyme cavity (Fig. 3.18).

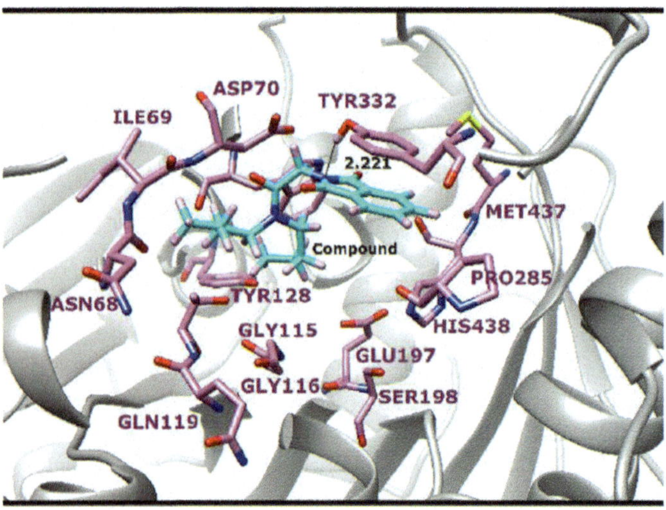

Fig. 3.18 Compound **(54)** docked within the binding gorge of BuChE (reused with permission *source* Fig. 3, Begum et al. [76])

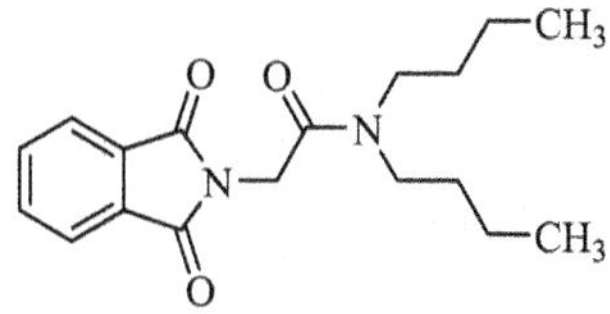

(54)

Xie et al. reported a series of hybrid molecules on the basis of clorgyline-rasagiline structure and assessed them for inhibition of amyloid-β aggregation and monoamine oxidase enzyme [77]. Among the synthesized compounds, a majority of them showed good *h*MAO-B inhibitory activity, but the chroman derivative **(55)** showed exceptionally good activity as *h*MAO-B inhibitor (IC_{50} value of 4 nM) and Aβ aggregation inhibitor (40.78% inhibition at 25 μM concentration).

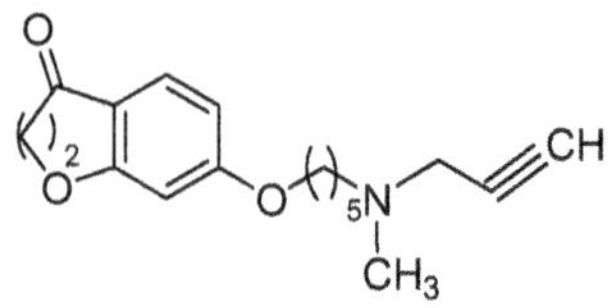

(55)

Docking studies were performed to understand the binding interactions occurring between compound **(55)** and *h*MAO-B (2V61) active site. It was observed that the ligand showed arene-H interactions with residue Tyr435 and van der Waals and hydrophobic interactions with leu717, Cys172, Ile198, Ile199, Ile316, Tyr326, Trp119, and Phe168 amino acid residues.

Gobec et al. have reported a series of tryptophan derivatives as selective BuChE inhibitors. Among the synthesized compounds of the series, compound **(56)** offered IC_{50} value of 2.8 nM against BuChE enzyme, the lowest among the series [78]. Molecular modeling studies were performed in the active site of the BuChE enzyme in order to assess the binding interactions of the compound. Modeling studies revealed that the compound was engaged into a π–π interaction with residue Trp231 at a distance of 4.3 Å at the acyl-binding pocket of the enzyme BuChE and also established H-bonding with Pro285 amino acid residue. Amino

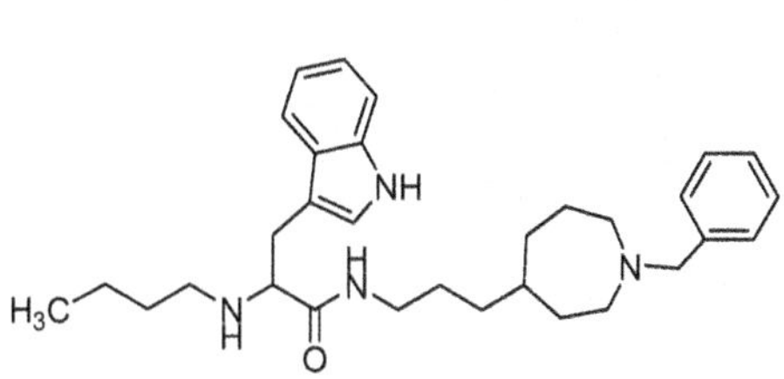

(56)

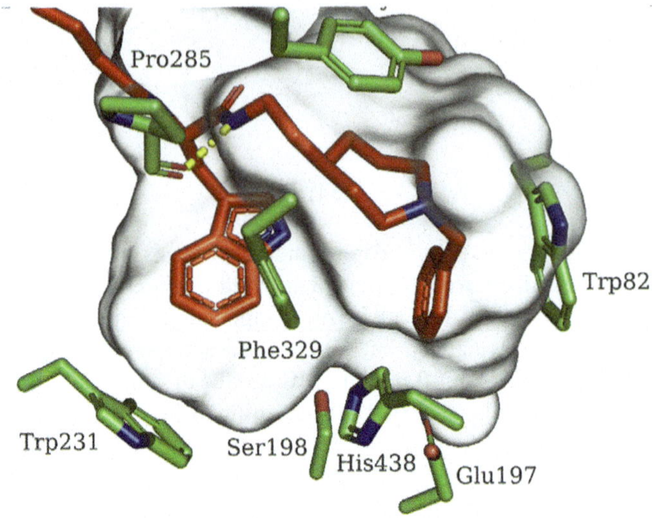

Fig. 3.19 Binding of compound (**56**) in the active binding site of BuChE (reused with permission *source* Fig. 6, Meden et al. [78])

acid residues Phe329, Gly116, and Gly117 were bound through van der Waals interactions with the compound (**56**), and residues Gln119, Ala277, Thr284, Ser287, and Asn289 made contacts through van der Waals interactions (Fig. 3.19).

3.2.9.1 Miscellaneous Compounds as Anti-AD Agents

Amide-based drugs have the potential to interact with various enzymes and receptors and generate a biological response. Hence, amide and carbamate scaffolds have become a popular topic in medicinal chemistry. Jiang and colleagues investigated the use of cannabidiol (CBD) and carbamate hybrids as selective inhibitors of ChE enzymes. They found that CBD could fit well into the groove of BuChE through several π–π interactions and hydrogen bonds. Earlier studies had shown that a hydroxyl group of CBD occupied enough space for interaction with Thr120. Docking studies indicated that the carbamate carbonyl group formed hydrogen bonds with the residues in the enzymes, and the nitrogen atom in the molecule was having crucial interactions with the binding site of ChEs. Based on these findings, the researchers designed 17 new compounds using a structural reassembly approach. Compounds **(57)** demonstrated high potency (IC_{50} = 5.3 nM) and selectivity for the inhibition of BuChE. The compound also showed good properties such as BBB penetration, pseudo-irreversible inhibition, antioxidant activity, safety, and neuroprotection. Another compound, **(58)**, also demonstrated nanomolar inhibition of *eq*BuChE (IC_{50} = 7.3 nM) and showed better *hBuChE* inhibition compared to rivastigmine. The molecular docking studies showed that phenyl halogen atom improved the lipid solubility and membrane permeability, increasing the absorption and transport of the drug. The substitution of chlorine or bromine atoms increased hydrophobic interactions and π–π interactions with Trp430 and Met437, enhancing the BuChE inhibitory activity further [79].

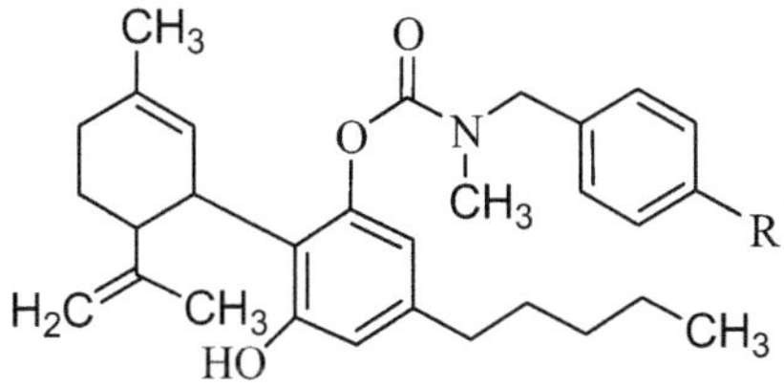

(57) R = Cl; (58) R = Br

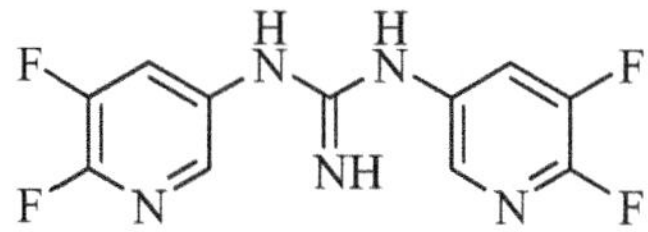

(59)

Sayyad et al. reported some biaryl guanidine derivatives and their biological evaluation for the treatment of Alzheimer's by inhibition of β-secretase enzyme [80]. The authors conducted virtual screening to predict compounds active against β-secretase. On the basis of virtual screening results, the authors synthesized 13 compounds and evaluated them for their biological activities through in vitro and in vivo experiments. Compound **(59)** was found to be the most active compound (IC_{50} value of 97 ± 0.91 nM) in the series. FRET assay for arresting β-secretase enzyme activity (99%) also indicated the activity potential of the compound. The compound contains fluorine atoms in its structure, which were responsible for improving its bioavailability in the enzyme's active domain of the flap region. In the Morris water maze novel object recognition test, the compound **(59)** was found to improve the scores significantly ($p < 0.05$).

Das et al. have reported dihyroactinidiolide (DA) **(60)** as a novel anti-Alzheimer's agent. Compound **(60)** has been synthesized from β ionone using oxidation method. From the in vitro studies, it has been revealed that this compound showed potent AChE inhibition with IC_{50} value of 34.03 nM [81]. Other than this, compound **(60)** was reported to possess DPPH and nitric oxide scavenging and metal chelating activities with IC_{50} values of 50 nM and about 270 nM, respectively. The compound showed no cytotoxicity toward N2a cells.

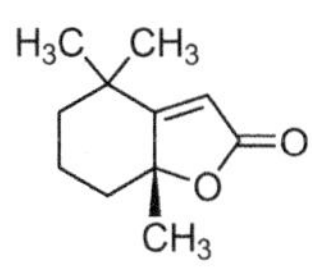

(60)

Molecular modeling studies also supported the analysis and showed that the compound **(60)** was involved in four hydrogen bond interactions through two O atoms of the lactone ring with residues GLY118, GLY119, and SER200 in the active site. To decode the **DA** structure–activity relationship, a standard compound γ-butyrolactone reported to increase acetylcholine levels in mice and rat, and γ-hydroxybutyrate levels in human brain were docked. From the docking studies, it was revealed that **DA** was also involved in four hydrogen bond interactions through its O and H groups with residues PHE288 and ARG289 in the AChE active site. At a concentration of 270 nM, **DA** was also reported to have de-aggregation and anti-aggregation potential on $A\beta_{25-34}$ fragment. In the docking studies, it offered a free binding energy of 3.62 kcal/mol, and it established hydrogen bonding with LYS28 residue of the enzyme.

Kumar et al. studied naphthofuran derivatives as potential inhibitors against GSK-3β and BACE1 through MM-PBSA binding energy analysis, molecular dynamics, and molecular docking studies. In comparison with the previously reported inhibitors, the two newly reported compounds NS7 (**61**) and NS9 (**62**) and exhibited much better binding affinities. Further, MM-PBSA analysis and molecular dynamics demonstrated that the binding interaction between these compounds and the enzymes was mainly due to hydrophobic interactions. This study also revealed that these naphthofuran derivatives have the potential to act as dual GSK-3β and BACE1 inhibitors by causing the enzymes to adopt a closed conformation and become inactive [82].

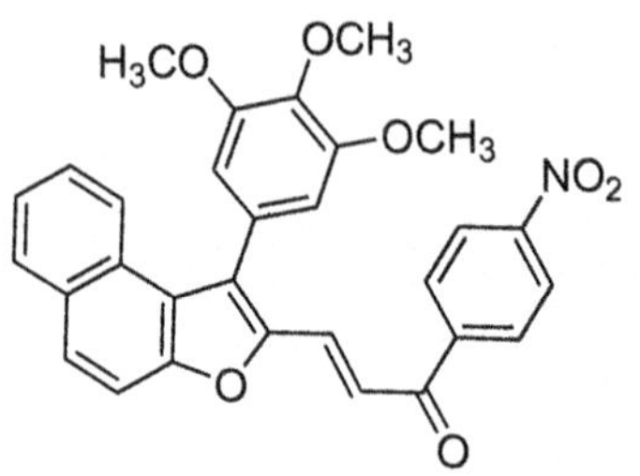

(61) NS7

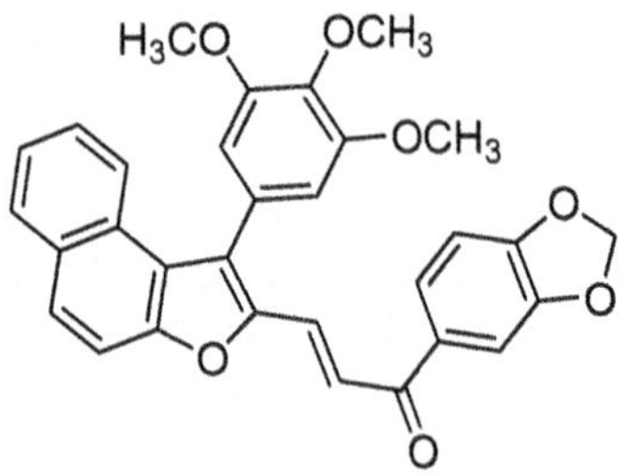

(62) NS9

Hu et al. reported dual AChE and PARP-1 inhibitors for the management of AD [83]. A group of 863 PARP-1 inhibitors possessing $IC_{50} \leq 10\ \mu M$ was selected by the authors in order to identify compounds with high affinity for AChE using virtual screening technique. Five crystal structures of AChE having some known inhibitors were utilized for virtual screening by applying same parameters on them. Using molecular dynamics approach, the complexes were also studied for their dynamic stability. Among the findings, compounds CID57390505 (**63**) and CID71605390 (**64**) attracted the spotlight by exhibiting high stability in the complex form with the enzyme AChE and high affinity for it in the dynamics and docking studies, respectively.

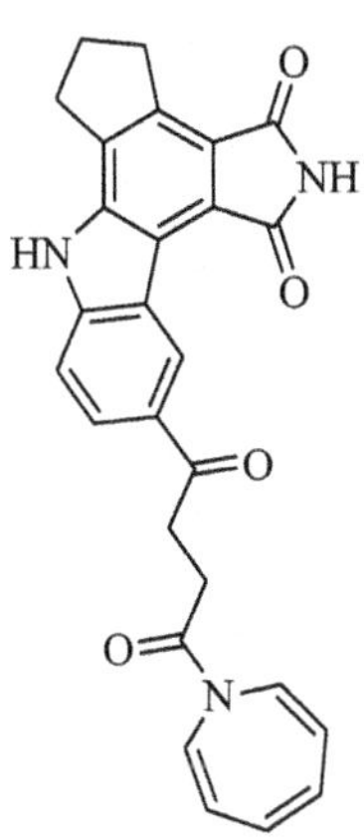

CID57390505

(63)

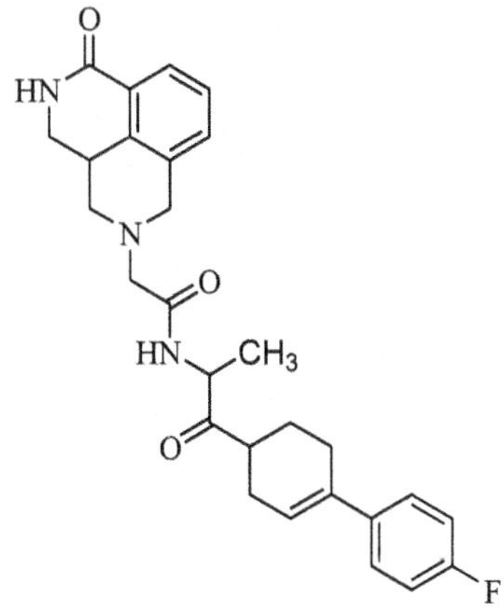

CID1605390

(64)

Hassan et al. used molecular docking and molecular dynamics to study the binding of Solanezumab and AZD3293 (**65**) to certain proteins like amyloid-β (Aβ) and β-secretase (BACE1) involved in the pathogenicity of AD [84]. The results indicated that hydrogen bonding was involved in the binding of AZD3293 (**65**) to BACE in its active site through the amino acid residues Lys107 and Asp32 having distances of 2.68 and 2.95 Å, respectively. Whereas, solanezumab interacted with Lys16 and Asp23 of Aβ peptide and formed hydrogen bond with heavy chains of Ser33 and Asp96, respectively. In case of binding of BACE to solanezumab, Asp96 formed H-bond with Lys16 with a distance of 2.82 Å, while Asp23 formed two hydrogen bonds with Ser33 having distances of 2.78 and 3.00 Å. AZD3293 offered better results in molecular dynamics offering RMSD fluctuation of 0.2 nm in comparison with 0.7 nm offered by solanezumab. All these results indicated that AZD3293 possessed an advantage over solanezumab for the management of Alzheimer's disease.

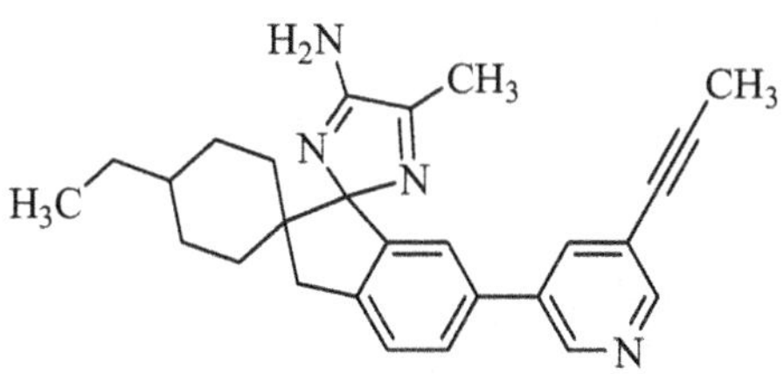

AZD3293

(65)

3.3 Conclusion

Among the main causes of deaths due to dementia, Alzheimer's is the most common one throughout the world. In the current scenario, around 50 million people worldwide suffer from dementia, and as per AD (Alzheimer's Dement) report by 2050, the number is bound to jump to more than 150 million. Many targets have been identified for the treatment of AD, but the exact causative factors responsible for the disease are not yet fully known, though sustained efforts are being made to unravel the mystery behind the disease. Computational techniques have proved of immense help to the researchers for discovering new drugs. These techniques are providing information about molecular interactions, ligand-enzyme interactions, stability of the molecular complexes formed after ligand-receptor interactions, and physicochemical properties of drugs. The present chapter provides a comprehensive information about the reports on the use of computational studies using MTDLs as hybrid molecules with a potential to be developed as future anti-AD drugs. Tacrine was the first approved drug which has been withdrawn from the market, but as discussed in this chapter, tacrine structure has been widely exploited for the discovery of novel AChE and BuChE inhibitors using molecular hybridization technique. It has been seen that heterocycles, fused or isolated, such as indole, quinoline, pyrimidine, and coumarin yielded potent AChE and BuChE inhibitors to be developed further as potential anti-AD drugs. Most of the hybrid molecules were binding to the active sites of AChE and BuChE enzymes and thus acted as inhibitors as indicated by these molecular modeling studies. Quinoline containing moieties were found to be good inhibitors of Aβ peptide which also exhibited good antioxidant properties. Some nitrogen containing heterocycles like pyrazole and pyrimidine showed additionally potent phosphodiesterase inhibitory activity.

In the changing paradigm, molecular modeling techniques are proving invaluable tools in the drug discovery process. In summary, this chapter has discussed some important findings involving molecular modeling techniques reported in literature for the discovery of MTDLs as anti-AD agents. The reported leads could pave the way for the discovery of new molecular entities useful for the treatment/cure of Alzheimer's disease.

References

1. Alzheimer's Disease Facts and Figures (2020) Alzheimer's Dement 16:391–460
2. Alzheimer's Disease Facts and Figures (2021) Alzheimer's Dement 17:327–406
3. Sperling RA, Aisen PS, Beckett LA, Bennett DA, Craft S, Fagan AM, Iwatsubo T, Jack CR Jr, Kaye J, Montine TJ, Park DC (2011) Toward defining the preclinical stages of Alzheimer's disease: recommendations from the National Institute on Aging-Alzheimer's Association workgroups on diagnostic guidelines for Alzheimer's disease. Alzheimers Dement 7(3):280–292
4. McKhann GM, Knopman DS, Chertkow H, Hyman BT, Jack CR Jr, Kawas CH, Klunk WE, Koroshetz WJ, Manly JJ, Mayeux R, Mohs RC (2011) The diagnosis of dementia due to

Alzheimer's disease: recommendations from the National Institute on Aging-Alzheimer's Association workgroups on diagnostic guidelines for Alzheimer's disease. Alzheimers Dement 7(3):263–269
5. Briggs R, Kennelly SP, O'Neill D (2016) Drug treatments in Alzheimer's disease. Clin Med 16(3):247
6. Reiss AB, Arain HA, Stecker MM, Siegart NM, Kasselman LJ (2018) Amyloid toxicity in Alzheimer's disease. Rev Neurosci 29(6):613–627
7. Forloni G, Balducci C (2018) Alzheimer's disease, oligomers, and inflammation. J Alzheimers Dis 62(3):1261–1276
8. Naseri NN, Wang H, Guo J, Sharma M, Luo W (2019) The complexity of tau in Alzheimer's disease. Neurosci Lett 13(705):183–194
9. Kabir MT, Uddin MS, Mamun AA, Jeandet P, Aleya L, Mansouri RA, Ashraf GM, Mathew B, Bin-Jumah MN, Abdel-Daim MM (2020) Combination drug therapy for the management of Alzheimer's disease. Int J Mol Sci 21(9):3272
10. Yang P, Sun F (2021) Aducanumab: the first targeted Alzheimer's therapy. Drug Discover Therapeut 15(3):166–168
11. Jeffrey C, Garam L, Travis M, Aaron R, Kate Z (2017) Alzheimer's disease drug development pipeline: 2017. Alzheimer's & Dement: Transl Res Clin Interv 3(3):367–384
12. Athar T, Al Balushi K, Khan SA (2021) Recent advances on drug development and emerging therapeutic agents for Alzheimer's disease. Mol Biol Rep 48(7):5629–5645
13. Oset-Gasque MJ, Marco-Contelles J (2018) Alzheimer's disease, the "one-molecule, one-target" paradigm, and the multitarget directed ligand approach. ACS Chem Neurosci 9(3):401–403
14. Murumkar PR, Le L, Truong TN, Yadav MR (2011) Determination of structural requirements of influenza neuraminidase type A inhibitors and binding interaction analysis with the active site of A/H1N1 by 3D-QSAR CoMFA and CoMSIA modeling. MedChemComm. 2(8):710–719
15. Daoud I, Melkemi N, Salah T, Ghalem S (2018) Combined QSAR, molecular docking and molecular dynamics study on new Acetylcholinesterase and Butyrylcholinesterase inhibitors. Comput Biol Chem 1(74):304–326
16. Murumkar PR, Sharma MK, Gupta P, Patel NM, Yadav MR (2022) Selection of suitable protein structure from Protein Data Bank: an important step in structure based drug design studies. Mini Rev Med Chem
17. Murumkar PR, Giridhar R, Yadav MR (2008) 3D-quantitative structure-activity relationship studies on benzothiadiazepine hydroxamates as inhibitors of tumor necrosis factor-α converting enzyme. Chem Biol Drug Des 71(4):363–373
18. DasGupta S, Murumkar PR, Giridhar R, Yadav MR (2009) Studies on novel 2-imidazolidinones and tetrahydropyrimidin-2 (1H)-ones as potential TACE inhibitors: design, synthesis, molecular modeling, and preliminary biological evaluation. Bioorg Med Chem 17(10):3604–3617
19. Sharma MK, Machhi J, Murumkar P, Yadav MR (2018) New role of phenothiazine derivatives as peripherally acting CB1 receptor antagonizing anti-obesity agents. Sci Rep 8(1):1–8
20. Murumkar PR, Gupta SD, Zambre VP, Giridhar R, Yadav MR (2009) Development of predictive 3D-QSAR CoMFA and CoMSIA models for β-aminohydroxamic acid-derived tumor necrosis factor-α converting enzyme inhibitors. Chem Biol Drug Des 73(1):97–107
21. Murumkar PR, Zambre VP, Yadav MR (2010) Development of predictive pharmacophore model for in silico screening, and 3D QSAR CoMFA and CoMSIA studies for lead optimization, for designing of potent tumor necrosis factor alpha converting enzyme inhibitors. J Comput Aided Mol Des 24:143–156
22. Sengupta P, Puri CS, Chokshi HA, Sheth CK, Midha AS, Chitturi TR, Thennati R, Murumkar PR, Yadav MR (2011) Synthesis, preliminary biological evaluation and molecular modeling of some new heterocyclic inhibitors of TACE. Eur J Med Chem 46(11):5549–5555
23. Murumkar PR, Sharma MK, Shinde AC, Bothara KG (2013) Three-dimensional quantitative structure–activity relationship CoMFA/CoMSIA on pyrrolidine-based tartrate diamides as TACE inhibitors. Med Chem Res 22:4192–4201

24. Murumkar PR, Sharma MK, Giridhar R, Yadav MR (2015) Virtual screening-based identification of lead molecules as selective TACE inhibitors. Med Chem Res 24:226–244
25. Sarkate AP, Murumkar PR, Lokwani DK, Kandhare AD, Bodhankar SL, Shinde DB, Bothara KG (2015) Design of selective TACE inhibitors using molecular docking studies: synthesis and preliminary evaluation of anti-inflammatory and TACE inhibitory activity. SAR QSAR Environ Res 26(11):905–923
26. Murumkar PR, Ghuge RB, Chauhan M, Barot RR, Sorathiya S, Choudhary KM, Joshi KD, Yadav MR (2020) Recent developments and strategies for the discovery of TACE inhibitors. Expert Opin Drug Discov 15(7):779–801
27. Sharma MK, Murumkar PR, Giridhar R, Yadav MR (2015) Exploring structural requirements for peripherally acting 1, 5-diaryl pyrazole-containing cannabinoid 1 receptor antagonists for the treatment of obesity. Mol Divers 19:871–893
28. Roy K (ed) (2018) Computational modeling of drugs against Alzheimer's disease. Springer, New York
29. Barmade M, Murumkar P, Kumar Sharma M, Shingala K, Giridhar R, Ram Yadav M (2015) Discovery of anti-malarial agents through application of in silico studies. Comb Chem High Throughput Screen 18(2):151–87
30. Yadav MR, Barmade MA, Chikhale RV, Murumkar PR (2018) Computational modelling of kinase inhibitors as anti-Alzheimer agents. Comput Model Drugs Against Alzheimer's Disease 347–417
31. Patel DV, Patel NR, Kanhed AM, Patel SP, Sinha A, Kansara DD, Mecwan AR, Patel SB, Upadhyay PN, Patel KB, Shah DB (2019) Novel multitarget directed triazinoindole derivatives as anti-Alzheimer agents. ACS Chem Neurosci 10(8):3635–3661
32. Khambete M, Murumkar P, Kumar A, Darreh-Shori T, De S, Yadav MR, Degani MS. Article details Pyrazoline containing molecules as multifunctional agents in Alzheimer's disease
33. Shidore M, Machhi J, Shingala K, Murumkar P, Sharma MK, Agrawal N, Tripathi A, Parikh Z, Pillai P, Yadav MR (2016) Benzylpiperidine-linked diarylthiazoles as potential anti-Alzheimer's agents: synthesis and biological evaluation. J Med Chem 59(12):5823–5846
34. Kanhed AM, Patel DV, Patel NR, Sinha A, Thakor PS, Patel KB, Prajapati NK, Patel KV, Yadav MR (2022) Indoloquinoxaline derivatives as promising multi-functional anti-Alzheimer agents. J Biomol Struct Dyn 40(6):2498–2515
35. Patel KB, Patel DV, Patel NR, Kanhed AM, Teli DM, Gandhi B, Shah BS, Chaudhary BN, Prajapati NK, Patel KV, Yadav MR (2022) Carbazole-based semicarbazones and hydrazones as multifunctional anti-Alzheimer agents. J Biomol Struct Dyn 40(20):10278–10299
36. Patel DV, Patel NR, Kanhed AM, Teli DM, Patel KB, Gandhi PM, Patel SP, Chaudhary BN, Shah DB, Prajapati NK, Patel KV (2020) Further studies on triazinoindoles as potential novel multitarget-directed anti-Alzheimer's agents. ACS Chem Neurosci 11(21):3557–3574
37. Machhi J, Sinha A, Patel P, Kanhed AM, Upadhyay P, Tripathi A, Parikh ZS, Chruvattil R, Pillai PP, Gupta S, Patel K (2016) Neuroprotective potential of novel multi-targeted isoalloxazine derivatives in rodent models of Alzheimer's disease through activation of canonical Wnt/β-catenin signalling pathway. Neurotox Res 29:495–513
38. Przybyłowska M, Dzierzbicka K, Kowalski S, Demkowicz S, Daśko M, Inkielewicz-Stepniak I (2022) Design, synthesis and biological evaluation of novel N-phosphorylated and O-phosphorylated tacrine derivatives as potential drugs against Alzheimer's disease. J Enzyme Inhib Med Chem 37(1):1012–1022
39. Yao H, Uras G, Zhang P, Xu S, Yin Y, Liu J, Qin S, Li X, Allen S, Bai R, Gong Q (2021) Discovery of novel tacrine-pyrimidone hybrids as potent dual AChE/GSK-3 inhibitors for the treatment of Alzheimer's disease. J Med Chem 64(11):7483–7506
40. Ozten O, Kurt BZ, Sonmez F, Dogan B, Durdagi S (2021) Synthesis, molecular docking and molecular dynamics studies of novel tacrine-carbamate derivatives as potent cholinesterase inhibitors. Bioorg Chem 1(115):105225
41. Chufarova N, Czarnecka K, Skibiński R, Cuchra M, Majsterek I, Szymański P (2018) New tacrine–acridine hybrids as promising multifunctional drugs for potential treatment of Alzheimer's disease. Arch Pharm 351(7):1800050

42. Li G, Hong G, Li X, Zhang Y, Xu Z, Mao L, Feng X, Liu T (2018) Synthesis and activity towards Alzheimer's disease in vitro: tacrine, phenolic acid and ligustrazine hybrids. Eur J Med Chem 25(148):238–254
43. Derabli C, Boulebd H, Abdelwahab AB, Boucheraine C, Zerrouki S, Bensouici C, Kirsch G, Boulcina R, Debache A (2020) Synthesis, biological evaluation and molecular docking studies of novel 2-alkylthiopyrimidino-tacrines as anticholinesterase agents and their DFT calculations. J Mol Struct 5(1209):127902
44. Zhu J, Yang H, Chen Y, Lin H, Li Q, Mo J, Bian Y, Pei Y, Sun H (2018) Synthesis, pharmacology and molecular docking on multifunctional tacrine-ferulic acid hybrids as cholinesterase inhibitors against Alzheimer's disease. J Enzyme Inhib Med Chem 33(1):496–506
45. Wieckowska A, Wichur T, Godyń J, Bucki A, Marcinkowska M, Siwek A, Wieckowski K, Zareba P, Knez D, Głuch-Lutwin M, Kazek G (2018) Novel multitarget-directed ligands aiming at symptoms and causes of Alzheimer's disease. ACS Chem Neurosci 9(5):1195–1214
46. Makhaeva GF, Kovaleva NV, Boltneva NP, Lushchekina SV, Rudakova EV, Stupina TS, Terentiev AA, Serkov IV, Proshin AN, Radchenko EV, Palyulin VA (2020) Conjugates of tacrine and 1, 2, 4-thiadiazole derivatives as new potential multifunctional agents for Alzheimer's disease treatment: Synthesis, quantum-chemical characterization, molecular docking, and biological evaluation. Bioorg Chem 1(94):103387
47. Shaikh S, Pavale G, Dhavan P, Singh P, Uparkar J, Vaidya SP, Jadhav BL, Ramana MM (2021) Design, synthesis and evaluation of dihydropyranoindole derivatives as potential cholinesterase inhibitors against Alzheimer's disease. Bioorg Chem 1(110):104770
48. He F, Chou CJ, Scheiner M, Poeta E, Yuan Chen N, Gunesch S, Hoffmann M, Sotriffer C, Monti B, Maurice T, Decker M (2021) Melatonin-and ferulic acid-based HDAC6 selective inhibitors exhibit pronounced immunomodulatory effects in vitro and neuroprotective effects in a pharmacological Alzheimer's disease mouse model. J Med Chem 64(7):3794–3812
49. Ghamari N, Dastmalchi S, Zarei O, Arias-Montaño JA, Reiner D, Ustun-Alkan F, Stark H, Hamzeh-Mivehroud M (2020) In silico and in vitro studies of two non-imidazole multiple targeting agents at histamine H3 receptors and cholinesterase enzymes. Chem Biol Drug Des 95(2):279–290
50. Lee HY, Fan SJ, Huang FI, Chao HY, Hsu KC, Lin TE, Yeh TK, Lai MJ, Li YH, Huang HL, Yang CR (2018) 5-Aroylindoles act as selective histone deacetylase 6 inhibitors ameliorating Alzheimer's disease phenotypes. J Med Chem 61(16):7087–7102
51. Lozinskaya NA, Babkov DA, Zaryanova EV, Bezsonova EN, Efremov AM, Tsymlyakov MD, Anikina LV, Zakharyascheva OY, Borisov AV, Perfilova VN, Tyurenkov IN (2019) Synthesis and biological evaluation of 3-substituted 2-oxindole derivatives as new glycogen synthase kinase 3β inhibitors. Bioorg Med Chem 27(9):1804–1817
52. Manzoor S, Prajapati SK, Majumdar S, Raza MK, Gabr MT, Kumar S, Pal K, Rashid H, Kumar S, Krishnamurthy S, Hoda N (2021) Discovery of new phenyl sulfonyl-pyrimidine carboxylate derivatives as the potential multi-target drugs with effective anti-Alzheimer's action: design, synthesis, crystal structure and in-vitro biological evaluation. Eur J Med Chem 5(215):113224
53. Zhang C, Zhou Q, Wu XN, Huang YD, Zhou J, Lai Z, Wu Y, Luo HB (2018) Discovery of novel PDE9A inhibitors with antioxidant activities for treatment of Alzheimer's disease. J Enzyme Inhib Med Chem 33(1):260–270
54. Ghobadian R, Nadri H, Moradi A, Bukhari SN, Mahdavi M, Asadi M, Akbarzadeh T, Khaleghzadeh-Ahangar H, Sharifzadeh M, Amini M (2018) Design, synthesis, and biological evaluation of selective and potent Carbazole-based butyrylcholinesterase inhibitors. Bioorg Med Chem 26(17):4952–4962
55. Kumar J, Gill A, Shaikh M, Singh A, Shandilya A, Jameel E, Sharma N, Mrinal N, Hoda N, Jayaram B (2018) Pyrimidine-triazolopyrimidine and pyrimidine-pyridine hybrids as potential acetylcholinesterase inhibitors for Alzheimer's disease. ChemistrySelect 3(2):736–747
56. Umar T, Shalini S, Raza MK, Gusain S, Kumar J, Seth P, Tiwari M, Hoda N (2019) A multifunctional therapeutic approach: synthesis, biological evaluation, crystal structure and molecular docking of diversified 1H-pyrazolo [3, 4-b] pyridine derivatives against Alzheimer's disease. Eur J Med Chem 1(175):2–19

57. Zaib S, Munir R, Younas MT, Kausar N, Ibrar A, Aqsa S, Shahid N, Asif TT, Alsaab HO, Khan I (2021) Hybrid quinoline-thiosemicarbazone therapeutics as a new treatment opportunity for Alzheimer's disease-synthesis, in vitro cholinesterase inhibitory potential and computational modeling analysis. Molecules 26(21):6573
58. Viayna E, Coquelle N, Cieslikiewicz-Bouet M, Cisternas P, Oliva CA, Sánchez-López E, Ettcheto M, Bartolini M, De Simone A, Ricchini M, Rendina M (2020) Discovery of a potent dual inhibitor of acetylcholinesterase and butyrylcholinesterase with antioxidant activity that alleviates Alzheimer-like pathology in old APP/PS1 mice. J Med Chem 64(1):812–839
59. Czarnecka K, Girek M, Kręcisz P, Skibiński R, Łątka K, Jończyk J, Bajda M, Kabziński J, Majsterek I, Szymczyk P, Szymański P (2019) Discovery of new cyclopentaquinoline analogues as multifunctional agents for the treatment of Alzheimer's disease. Int J Mol Sci 20(3):498
60. Maciejewska K, Czarnecka K, Kręcisz P, Niedziałek D, Wieczorek G, Skibiński R, Szymański P (2022) Novel cyclopentaquinoline and acridine analogs as multifunctional, potent drug candidates in Alzheimer's disease. Int J Mol Sci 23(11):5876
61. Safarizadeh H, Garkani-Nejad Z (2019) Molecular docking, molecular dynamics simulations and QSAR studies on some of 2-arylethenylquinoline derivatives for inhibition of Alzheimer's amyloid-β aggregation: insight into mechanism of interactions and parameters for design of new inhibitors. J Mol Graph Model 1(87):129–143
62. Jiang MY, Han C, Zhang C, Zhou Q, Zhang B, Le ML, Huang MX, Wu Y, Luo HB (2021) Discovery of effective phosphodiesterase 2 inhibitors with antioxidant activities for the treatment of Alzheimer's disease. Bioorg Med Chem Lett 1(41):128016
63. Zhang C, Yang K, Yu S, Su J, Yuan S, Han J, Chen Y, Gu J, Zhou T, Bai R, Xie Y (2019) Design, synthesis and biological evaluation of hydroxypyridinone-coumarin hybrids as multimodal monoamine oxidase B inhibitors and iron chelates against Alzheimer's disease. Eur J Med Chem 15(180):367–382
64. Rullo M, Catto M, Carrieri A, de Candia M, Altomare CD, Pisani L (2019) Chasing ChEs-MAO B multi-targeting 4-aminomethyl-7-benzyloxy-2H-chromen-2-ones. Molecules 24(24):4507
65. Palareti G, Legnani C, Cosmi B, Antonucci E, Erba N, Poli D, Testa S, Tosetto A (2016) DULCIS (D-dimer-ultrasonography in combination italian study) investigators (see appendix), De Micheli V, Ghirarduzzi A. Comparison between different D-D imer cutoff values to assess the individual risk of recurrent venous thromboembolism: analysis of results obtained in the DULCIS study. Int Lab Hematol 38(1):42–9
66. Sivakumar M, Saravanan K, Saravanan V, Sugarthi S, Kumar SM, Alhaji Isa M, Rajakumar P, Aravindhan S (2020) Discovery of new potential triplet acting inhibitor for Alzheimer's disease via X-ray crystallography, molecular docking and molecular dynamics. J Biomol Struct Dyn 38(7):1903–1917
67. Sepehri N, Mohammadi-Khanaposhtani M, Asemanipoor N, Hosseini S, Biglar M, Larijani B, Mahdavi M, Hamedifar H, Taslimi P, Sadeghian N, Gulcin I (2020) Synthesis, characterization, molecular docking, and biological activities of coumarin–1, 2, 3-triazole-acetamide hybrid derivatives. Arch Pharm 353(10):2000109
68. Baruah P, Rohman MA, Yesylevskyy SO, Mitra S (2019) Therapeutic potency of substituted chromones as Alzheimer's drug: Elucidation of acetylcholinesterase inhibitory activity through spectroscopic and molecular modelling investigation. BioImpacts 9(2):79
69. Taslimi P, Turhan K, Türkan F, Karaman HS, Turgut Z, Gulcin I (2020) Cholinesterases, α-glycosidase, and carbonic anhydrase inhibition properties of 1H-pyrazolo [1, 2-b] phthalazine-5, 10-dione derivatives: Synthetic analogues for the treatment of Alzheimer's disease and diabetes mellitus. Bioorg Chem 1(97):103647
70. Zhou Y, Li J, Yuan H, Su R, Huang Y, Huang Y, Li Z, Wu Y, Luo H, Zhang C, Huang L (2021) Design, synthesis, and evaluation of dihydropyranopyrazole derivatives as novel PDE2 inhibitors for the treatment of Alzheimer's disease. Molecules 26(10):3034
71. Acar Cevik U, Saglik BN, Levent S, Osmaniye D, Kaya Cavuşoglu B, Ozkay Y, Kaplancikli ZA (2019) Synthesis and AChE-inhibitory activity of new benzimidazole derivatives. Molecules 24(5):861

72. Mehrazar M, Hassankalhori M, Toolabi M, Goli F, Moghimi S, Nadri H, Bukhari SN, Firoozpour L, Foroumadi A (2020) Design and synthesis of benzodiazepine-1, 2, 3-triazole hybrid derivatives as selective butyrylcholinesterase inhibitors. Mol Diversity 24:997–1013
73. Osmaniye D, Sağlık BN, Acar Çevik U, Levent S, Kaya Çavuşoğlu B, Özkay Y, Kaplancıklı ZA, Turan G (2019) Synthesis and AChE inhibitory activity of novel thiazolylhydrazone derivatives. Molecules 24(13):2392
74. Luo L, Song Q, Li Y, Cao Z, Qiang X, Tan Z, Deng Y (2020) Design, synthesis and evaluation of phthalide alkyl tertiary amine derivatives as promising acetylcholinesterase inhibitors with high potency and selectivity against Alzheimer's disease. Bioorg Med Chem 28(8):115400
75. Lan JS, Zeng RF, Jiang XY, Hou JW, Liu Y, Hu ZH, Li HX, Li Y, Xie SS, Ding Y, Zhang T (2020) Design, synthesis and evaluation of novel ferulic acid derivatives as multi-target-directed ligands for the treatment of Alzheimer's disease. Bioorg Chem 1(94):103413
76. Begum S, Nizami SS, Mahmood U, Masood S, Iftikhar S, Saied S (2018) In-vitro evaluation and in-silico studies applied on newly synthesized amide derivatives of N-phthaloylglycine as Butyrylcholinesterase (BChE) inhibitors. Comput Biol Chem 1(74):212–217
77. Xie SS, Liu J, Tang C, Pang C, Li Q, Qin Y, Nong X, Zhang Z, Guo J, Cheng M, Tang W (2020) Design, synthesis and biological evaluation of rasagiline-clorgyline hybrids as novel dual inhibitors of monoamine oxidase-B and amyloid-β aggregation against Alzheimer's disease. Eur J Med Chem 15(202):112475
78. Meden A, Knez D, Malikowska-Racia N, Brazzolotto X, Nachon F, Svete J, Sałat K, Grošelj U, Gobec S (2020) Structure-activity relationship study of tryptophan-based butyrylcholinesterase inhibitors. Eur J Med Chem 15(208):112766
79. Jiang X, Zhang Z, Zuo J, Wu C, Zha L, Xu Y, Wang S, Shi J, Liu XH, Zhang J, Tang W (2021) Novel cannabidiol−carbamate hybrids as selective BuChE inhibitors: docking-based fragment reassembly for the development of potential therapeutic agents against Alzheimer's disease. Eur J Med Chem 5(223):113735
80. Ali S, Asad MH, Khan F, Murtaza G, Rizvanov AA, Iqbal J, Babak B, Hussain I (2020) Biological evaluation of newly synthesized biaryl guanidine derivatives to arrest β-secretase enzymatic activity involved in Alzheimer's disease. Biomed Res Int 11:2020
81. Das M, Prakash S, Nayak C, Thangavel N, Singh SK, Manisankar P, Devi KP (2018) Dihydroactinidiolide, a natural product against Aβ25-35 induced toxicity in Neuro2a cells: synthesis, in silico and in vitro studies. Bioorg Chem 1(81):340–349
82. Kumar A, Srivastava G, Negi AS, Sharma A (2019) Docking, molecular dynamics, binding energy-MM-PBSA studies of naphthofuran derivatives to identify potential dual inhibitors against BACE1 and GSK-3β. J Biomol Struct Dyn 37(2):275–290
83. Hu XM, Dong W, Cui ZW, Gao CZ, Yu ZJ, Yuan Q, Min ZL (2018) In silico identification of AChE and PARP-1 dual-targeted inhibitors of Alzheimer's disease. J Mol Model 24:1–9
84. Hassan M, Shahzadi S, Seo SY, Alashwal H, Zaki N, Moustafa AA (2018) Molecular docking and dynamic simulation of AZD3293 and solanezumab effects against BACE1 to treat Alzheimer's disease. Front Comput Neurosci 1(12):34

Chapter 4
Computational Modeling in the Development of Antiviral Agents

Priyank Purohit, Pobitra Borah, Sangeeta Hazarika, Gaurav Joshi, and Pran Kishore Deb

Abstract As a result of the damage that viruses have done over time, humans have developed a variety of defenses against viral illnesses, such as vaccines and antiviral drugs for treatment. Since the 1950s, new viral illnesses including AIDS, Hepatitis, and coronavirus infections like SARS, MERS, and COVID-19 have periodically emerged, posing a challenge to the development of antiviral drugs. The creation of computer models is an interactive, iterative process that blends empirical datasets with known facts and assumptions (knowledge-driven or data-driven approach). In order to allow system simulation, the generated models should ideally offer reusability, composability, and interoperability. We surmise that the development of computational and mathematical frameworks will not only assist the development of newer antivirals, but simulating viral infections will also help in incorporating progressive immunosenescence and finding host genetic factors to expand the knowledge of infectious disease to an unprecedented level of detail. In addition to the fundamental molecular aspects of viral infection, this chapter emphasizes the fundamentals of computer modeling and discusses the relationship between in silico experiments and viral infections.

Keywords Virus–receptor interactions · Virtual screening · Molecular docking · In silico · Antiviral drug development

P. Purohit · P. Borah (✉) · S. Hazarika · G. Joshi
School of Pharmacy, Graphic Era Hill University, Dehradun, Uttarakhand 248002, India
e-mail: pobitrab.phe15@gmail.com

S. Hazarika
Department of Pharmaceutical Engineering and Technology, Indian Institute of Technology (Banaras Hindu University), Varanasi, Uttar Pradesh 221005, India

P. K. Deb
Department of Pharmaceutical Sciences, Faculty of Pharmacy, Philadelphia University, PO Box 1, Amman 19392, Jordan

S. Kar and J. Leszczynski (eds.), *Current Trends in Computational Modeling for Drug Discovery*, Challenges and Advances in Computational Chemistry and Physics 35,
https://doi.org/10.1007/978-3-031-33871-7_4

4.1 Introduction

In today's scenario, it is still a matter of debate about how to define viruses. Although many different names have been given to viruses over the years, including "pathogenic entities", "microscopic parasites", "small-sized infectious agents", "collections of genes surrounded by a protein coat", and "organic particles", all of these definitions are conclusive of the fact: viruses are non-living beings as they are dependent on other cells for survival and replication. In spite of their shared need for host cells for replication, viruses display a remarkable degree of diversity, with four distinct types recognized based on their physical characteristics: endogenous viral components, temperate viruses, virus-like particles, and virus particles or virions [1]. Of these, virus particles cause human infection. Nucleic acid genome and an inner nucleic acid core containing genetic information like DNA or RNA are enclosed and protected by a protein casing called a capsid. There are viruses that have an additional coating of a lipid bilayer taken from host cell membranes. Most extracellular enveloped virions include additional viral particles including surface glycoproteins and matrix proteins [2]. Multiple cell and virion surface factors regulate host cell contact and intracellular viral tropism, facilitating effective viral propagation and replication. Two major modes of viral tropism have been identified: receptor-dependent tropism, in which invasion is initiated by the binding of virus to specific molecules of host cells, and receptor-independent tropism, where a vector is needed by virus to enter the host cell [3]. Most viruses employ receptor-dependent tropism, meaning that the exclusive expression of a given receptor on a particular cell type is predictive of that cell type's susceptibility to viral infection, cell entrance, and viral reproduction. Therefore, the viral structure can be divided based upon two main functions: (1) the infectious function leading to viral entrance into the host cells by causing identification of cellular receptors by surface proteins and (2) the protective function carried out by capsid proteins [4]. The host immune system exerts selection pressure on the viral surface proteins or protein domains. Antibody-binding areas, also known as epitopes, undergo frequent mutation as a result of this selection pressure resulting in the continual emergence of new viral entities or strains. Besides mutation via selection, some viruses may result in novel viral subtypes with genetically recombined regions of nucleic acid. Viruses with greater adaptability are often to blame for the widespread distribution of illness throughout vast areas or even globally, leading to pandemics.

By analyzing the biological interactions of medications with viral molecular components using available datasets, computational modeling offers a more effective way to speed up drug discovery or repurposing of antiviral treatments. Target-based and disease-based computational techniques are both used to identify drug molecules. While the latter makes use of pre-existing databases to identify novel indications for current medications [5] by comparing them with illness characteristics, the former generates data on drug–target interaction. A fully regulated in silico reconstruction of mechanisms and interaction processes is possible with computational modeling and simulation [6]. Following that, experiments are carried out computationally in a "simulation". Mathematical and computational modeling consequently addresses an increasing number of crucial aspects of infection dynamics [7]. Being able to determine previously unknown infection parameters makes this particularly intriguing. It is now beyond dispute that CADD plays a critical role in the drug design process and that, over time, it can both speed up the discovery of viable therapeutic candidates and lower the associated costs. CADD and its approaches can have an impact on different stages of the drug discovery pipeline by using knowledge and information about biological targets or knowledge about a ligand with demonstrated bioactivity. Investigations into the mechanism of drug action are made possible by CADD methods' capacity to illuminate drug–target interactions at the atomistic level, revealing atomistic insights that have an impact on drug development. Since CADD approaches also enable modeling of complicated biological processes that, up until now, seemed impossible to investigate using experiments, they strive to complement rather than replace established in vitro and in vivo experimental procedures. We speculated that the creation of mathematical frameworks for modeling viral infections will aid in integrating progressive immunosenescence and identifying host genetic influences to enhance the understanding of infectious disease to an unheard-of level of detail. The basics of computer modeling are highlighted in this chapter, and we go into greater detail on relating in silico experiments to viral infections.

4.2 Brief History and Structure of Viruses

Viruses are the smallest (~20–300 nm), non-cellular, obligate intracellular infectious agents that utilize the host cellular machinery for their survival and multiplication. Among 400 viruses known to be pathogenic to human, the zoonoses that spread from insects or animals to humans are particularly considered to be more dangerous for the human race [8, 9]. Transmission of the viruses occurs by various routes, including air or droplet infection (e.g., smallpox, influenza, measles, chicken pox, mumps, viral pneumonia, rubella, coronavirus disease-2019, etc.), bite of arthropods (e.g., Colorado tick fever and yellow fever); physical contact (e.g., genital herpes, common cold, rabies, cold sores, acquired immunodeficiency syndrome or AIDS, etc.); and water- or food-borne (e.g., hepatitis A, poliomyelitis, and viral gastroenteritis). Historically, viral infections led to millions of casualties, including epidemics or pandemics starting from the smallpox epidemic (AD 165–180 and AD 251–266) to Spanish flu (1918–1920), human immunodeficiency virus (HIV) infection (1981–present), severe acute respiratory syndrome or SARS (2002–2003), Middle East respiratory syndrome or MERS (2012–present), West African Ebola (2014–2016), and the COVID-19 (2019–present) pandemic [10, 11]. Accordingly, the humans are adopting different counter-measures over time to control the viral infections or their spreads, which includes development of antiviral agents and vaccines. As the viruses are dependent on host cell machinery, it is difficult to develop an effective and safer antiviral drug without potentially harming the host [12].

The viruses can be classified as DNA viruses or RNA viruses simply because they either contain DNA or RNA but not both. The majority of RNA viruses have a single-stranded RNA (ssRNA) molecule, although a few may contain a double-stranded RNA (dsRNA) molecule. If the RNA base sequence is identical to viral mRNA, it is referred to as the positive (+) strand, whereas if it is complementary, it is referred to as the negative (−) strand. In contrast, the majority of DNA viruses possess dsDNA, while only a few include ssDNA [6]. A comprehensive taxonomy of viruses is available at https://talk.ictvonline.org/. Based on their morphological architecture, viruses can have helical symmetry, icosahedral (cubic) symmetry, or a complex structure. For instance, viruses like influenza virus demonstrate standard helical symmetry with viral genome bound to protein subunits (nucleocapsid helix) encased in a lipid envelope, whereas human papillomavirus exhibits icosahedral symmetry as they have 20 equilateral triangular faces in the capsid with 12 vertices. Contrary to this, viruses (e.g., poxvirus) that lack helical or icosahedral symmetry may exhibit more complicated forms such as an ovoid, spherical, or brick-shaped pattern [13].

Apart from these, viruses can be further classified based on various considerations, such as the type of genome they contain (DNA versus RNA, single strand versus double strand, sense versus antisense, and linear versus circular), physicochemical properties, replication mechanism, and so on. The genetic blueprints that a virus uses to replicate and achieve host lethality may vary slightly between copies. This is especially true for retroviruses and some other RNA viruses [9]. In spite of destroying the host cell and its machinery, certain viruses alter the cellular physiology. Unchecked cellular proliferation is attributed to several infections that may later progress into malignancy. Hepatitis B and C are two viruses that can infect the liver and lead to chronic liver disease. For some people, the effects of chronic hepatitis can last for years or even decades. In certain populations, the mild forms of chronic hepatitis never progress to liver failure; however, some individuals may eventually develop cirrhosis and liver failure. Typically, viruses target a single cell type. For example, the common cold virus only infects cells in the upper respiratory tract. Though viruses specifically infect plants and animals, some of them are only transmitted by human beings [10].

4.3 Mechanism of Viral Infections

As an obligate intracellular parasite, viruses are absolutely dependent on the host resources for their growth and replication processes. In a nutshell, the replication cycle of a virus starts with the attachment phase, in which the virus attaches to the host cell receptor primarily through electrostatic adsorption in a fortuitous manner. For instance, HIV binds the CD4 receptor, rhinoviruses recognize the ICAM-1 receptor, and Epstein-Barr virus targets the CD21 receptor [14]. A study reported that certain mammalian cells inhibit viral entry in non-infected cells by producing proteins capable of restricting the binding of virus to the receptor [15]. Next, in the penetration phase, viruses invade the cell by various mechanisms, viz. envelope/ membrane fusion, receptor-mediated endocytosis, or through direct penetration of membrane. It is followed by the viral genome (DNA or RNA) release in the cell and exhibits an eclipse period, where the virus exploits host machinery to make necessary viral proteins and undergoes replication. Permissive cells promote viral replication to favor new virion production. Mostly, the host metabolic processes are directed toward the generation of viral products, which ultimately lead to host cell destruction. However, if the cell metabolism remains intact, the infected cell tends to survive. Abortive infections may be observed in cases of defective virus invasion or non-permissive cell infections. This results in latent infection, which can later cause malignancy [14, 16].

The replication processes vary depending on the type of virus; however, transcription followed by protein translation remains the eventual goal [17]. Typically, the viruses containing dsDNA (except for poxvirus) require the host enzyme DNA-dependent RNA polymerase for viral protein production, including essential enzymes like DNA-dependent DNA polymerase. On the other hand, the hepatitis B virus entails DNA polymerase to repair the DNA damages before transcription [18]. Contrariwise, an ssDNA virus initially needs to convert its genetic material into dsDNA (sometimes with the assistance of a helper virus) prior to the translation of mRNA. Notably, the RNA viruses have demonstrated evolution over time to obtain different reproductive strategies compared to the DNA viruses. To understand the ssRNA virus replication process, they may be grouped into three types. The first group (e.g., togaviruses, picornaviruses, and flaviviruses) contains (+)-sense-strand RNA, which is processed by the ribosome into polyproteins. Subsequent autocatalysis of polyproteins by the proteases yields the viral proteins. Nevertheless, a few viruses, such as togaviruses, possess only a segment of the RNA for translation. The template sense strand is used to yield the antisense strand by using RNA polymerase. Later, the yielded antisense strands are modified into infectious and sense-strand RNAs, which are packed inside virions for future transmission processes [14, 19]. On the other hand, another virus group, including paramyxoviruses, orthomyxoviruses, arenaviruses, and rhabdoviruses, contains antisense RNA that is required for transcription. These viruses utilize RNA transcriptase (usually found within the virion) to generate short sense-strand RNAs coding for essential viral enzymes required for replication. Subsequently, these enzymes are utilized to produce full-length sense RNAs and multiple antisense genome in the progeny [14, 20]. The third group includes retroviruses with ssRNA found in dimeric forms of both sense and antisense strands. The reverse transcriptase (RT), carried in the virion, converts the retroviral RNA to dsDNA. Transcription results in complementary DNA strand formation, while RNAse H and ribonucleases digest the original strand. Ensuring this dsDNA synthesis follows, which is later incorporated into the host DNA via integrase. The result of transcription is full-length RNA (progeny) that is packaged as virion materials and polyproteins that are required for the generation of essential proteins [21–23].

During replication, viral genome-encoded structural proteins such as the lipid envelope that contains antigenic glycoproteins are also formed. Cellular egression occurs either by destruction of the cell due to the assembly of corresponding virions (as in case of non-enveloped viruses including reoviruses, picornaviruses, etc.) or by budding and shedding extracellularly following virion insertion into the membrane (mostly seen in enveloped viruses such as herpesviruses, togaviruses, coronaviruses, retroviruses, hepadnaviruses, and antisense RNA viruses). Generally, the manifestation of symptoms (though they may sometimes remain asymptomatic) is considered a consequence of rapid viral replication leading to host cell injuries. Overall, viral infection and its pathogenesis remain dependent on viral entry, replication, transmission, cellular injury, immune responses, and viral shedding. A detailed insight into viral diseases is available in existing literature [refer to 14, 16, 24, 25].

4.4 Computational Modeling in Viral Infections

Mathematics and logic form the basis of computational modeling of biological phenomena. In case of viral infections, computational biology techniques allow close observation of the molecular interactions in the form of simulation of the biological membranes, protein folding, protein–ligand complexes, protein–protein complexes, and molecular structures [26, 27]. For instance, application of Statistical Mechanics and Molecular Dynamics methods allows simulation of different protein folding conformations. These methods can be used to simulate conformational changes of viral restriction factors (e.g., BST-2, IFITM, APOBEC3G, and SAMHDI) during the activation of innate immune systems associated with viral infections such as Zika, HIV, Hepatitis C, Western Nile, and Influenza [28]. Similarly, Molecular Modeling is utilized for the 3D representation of protein structures via conformational analysis. Several other computer programs allow construction and identification of small-molecule library by analyzing different protein databases (e.g., Protein Data Bank or PDB) [29]. The computational protein–protein interaction (PPI) prediction tools mainly utilize information such as protein structure, physiochemical properties, sequence information, semantic analysis, and earlier investigated interactions between cellular proteins and viral proteins [30]. In spite of high recommendation of classical machine learning tools in the prediction of PPI (given the availability of plenty of known interactions), newer models are demanded for the prediction of virus–host PPI. Instead of doing learning tasks on separate domains, multitask learning exploits relationships among various domains and learns the problem concurrently, which improves performance [31, 32].

Broadly, computational chemistry can be divided into Molecular Mechanics and Quantum Mechanics methods. The former method is based on Newtonian physics and can calculate very big molecules by viewing a molecule as a collection of spheres held together by springs. As long as the atoms' hybridization is maintained, the parameters for the molecule's atoms in this method are essentially constant across varied structures [33]. On the other hand, delineating a linear combination of atomic orbitals as molecular orbitals, the Quantum Mechanics approaches take into account the motion of electrons while assuming that the nucleus is fixed. They can be categorized into ab initio or semi-empirical techniques. The ab initio approaches warranting solutions for a broad array of molecules make approximations based on the laws of quantum mechanics for directly solving Schrodinger's equation, whereas empirical parameters or pre-calculated parameters from Schrodinger's equation form basis of the development of semi-empirical approaches [33]. Application of molecular modeling of the interaction between fusion protein of human Respiratory Syncytial Virus (RSV) and inhibitory small molecules is an example of use of computational modeling in viral infections. To predict novel host–virus protein interactions, many computational techniques have been developed. In innovative host–virus interaction predictions, various predictive models have been put forth depending on the availability of interaction information.

The five axes of classification for computational models in virology are: (1) discrete versus continuous; (2) hypothesis-driven versus data-driven; (3) stochastic versus deterministic; (4) spatiotemporal versus temporal; and (5) white-box versus black-box models. Atomistic simulations and individual-virion simulations modeling atoms and virions, respectively, as discrete entities are two examples of discrete models. In continuous models, no independent representation of the individual entities is made; rather, monitoring of their dispersion in time and/or space or density is done. How a virion becomes infectious is a crucial subject in virus assembly that has been addressed using continuous modeling approaches. In the case of HIV, whereby virion infectivity is essentially gained by proteolytic maturation, a functional fusion machinery in the infectious virions is proposed to be assembled by matrix (MA) domain cleaving off from the envelope (ENV) domain including the viral glycoprotein, and allowing ENV to loosen up with the spread of MA in the virion and clustering the trimers [34]. Spatiotemporal models directly depict the spatial localization or dispersion of viruses. As opposed to geographic localization, temporal models follow the dynamic evolution of an aggregated variable, such as the total virus load or infection multiplicity. In viral kinetics models, this is typically the case [35]. On the other hand, in stochastic models, some events occur probabilistically, making it impossible to anticipate how the infection state will develop over time, but it is possible to assess the likelihood of various developments. For instance, a "white box" model for the modeling of infection transmission in a tissue culture model has been constructed that takes the local virion concentration into account [36]. On the contrary to stochastic models, deterministic models assume that all outcomes are known in advance; therefore, this type of model often necessitates thorough understanding of the underlying molecular mechanics. The extrapolation of deterministic models to acto-myosin-based cell motility has been done for several features of actin polymerization [37, 38]. If the interactions between the proteins that make up the capsid are known in atomic detail, deterministic models might be used to simulate the disruption of a non-enveloped virion. Such models might prove to be better predictive tools; for instance, as reported in the case of adenovirus, whereby it can foretell if during virion drifting motions on the cell surface, the mechanical forces acting on virus particles are adequate for the partial breakage of the virion upon entry [39–41]. Models that are hypothesis-driven are created based on an anticipation or a hunch, even in the absence of initial data. A hypothesis is first formulated, after which it is formalized—for instance, using mathematical formulas, sets of rules, or chemical reactions. The generated model is then put through a simulation to investigate its behavior and test it against observations and known facts in an effort to disprove the hypothesis. Instead of formalizing a hypothesis, models can be learned from data. This is frequently helpful in the early exploratory stages of a project or when looking for higher-order patterns in data that are hidden to the human eye. Although statistical analytic techniques have long done this, current advances in artificial intelligence and machine learning have given data-driven modeling a new level of sophistication. The correlations must be learned using supervised machine learning, which needs a lot of training data from well-known situations. It is therefore not unexpected that some of the earliest applications in virology were finding connections between viral infection state and host cell gene expression levels, for instance in hepatitis B virus infections [42]. Likewise, white-box models only have one or a few unidentified parameters. The most concrete proof that a mechanism is sufficient is a white-box model. It is strong proof that the modeled mechanisms are adequate if, for instance, all binding affinities, infection probabilities, and diffusion constants are measured independently and the model accurately reproduces the data. Black box models, in contrast to white-box models, are completely identified by parameter fitting. As a result, once the fundamental mechanism is understood, they offer indirect methods of estimating not directly measurable or observable quantities. They do, however, always leave some room for uncertainty regarding the real mechanism because many processes may be able to reproduce the same data for various parameter values [43]. The ability to draw statistical conclusions from experimental data is a crucial function of simulations. The Bayesian framework or the maximum-likelihood framework is commonly used for inference. For both, an observation process "forward model" is necessary. The mechanism of virus propagation seemingly marked by plaque dynamics can be deduced using virus plaque formation in infected tissue simulations [36].

4.4.1 Virtual Screening (VS)

By using algorithms to sort through a library of tiny chemicals, VS, an in silico approach, can identify those that have the potential ability to bind a target molecule. In contrast to experimental approaches like High Throughput Screening, in which numbers of compounds are experimentally evaluated, it has allowed scientists to obtain lead compounds for binding a biological target from millions of molecules in a dataset within a short span of time while being inexpensive. This is made feasible by the ongoing development and improvement of more powerful computers and reliable algorithms that steadily reduce the amount of time needed for such analyses while enhancing the precision of the projected interactions [44, 45]. Two distinct computer-aided drug design or CADD methodologies are structure-based drug design (SBDD) and ligand-based drug design (LBDD) (Fig. 4.1). Using SBDD techniques, it is

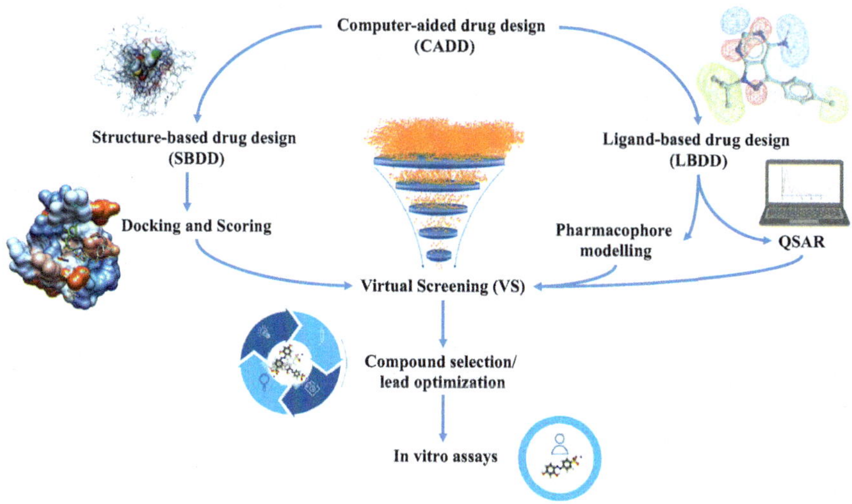

Fig. 4.1 Diagram depicting the workflow of virtual screening

possible to pinpoint key sites and interactions that are essential to each macromolecular target's specific biological activity. These macromolecular targets are typically proteins or RNA. Once the biological target has been identified, information on its structure must be gathered from a number of sources. The PDB is the most significant of these, as it contains macromolecular structures that have been solved experimentally by NMR, X-ray crystallography, or cryoelectron microscopy [33, 46]. The alternative, in the lack of a structure that can be solved experimentally, is to obtain a model that has previously been proposed or to build one based on homology using software like the resilient Modeler. The next step after obtaining the biological target's structure is to define the binding locations where the screening compounds will be docked. When ligands are present on repositories, the binding site may even be identified with structures if it is well-established. But research can be done using tools like MDpocket [47], which can simultaneously find transitional pockets and prospective target sites while investigating a new biological target or looking at additional potential allosteric sites in the structure. The next stage is the docking of the compounds in the biological target's binding site after it has been identified and clustered [48].

The structure–activity relationship (SAR)—a concept central to LBDD—is established by focusing on known compounds or ligands in order to determine a correlation between their physical–chemical characteristics and biological activity. The design of novel medications or drug optimization can be influenced by this knowledge [49]. Quantitative structure–activity relationship (QSAR), a method of calculation, allows for the quantification of molecular descriptors. Additionally, there is a chance for a connection with research into the compounds' tridimensional structure (3D-QSAR). Many methodologies go beyond tridimensional descriptors and explore ligands in a variety of other factors, such as solvation profile and various conformations. As an illustration, consider 4D-QSAR, which represents the many conformations of the same ligand as a new variable by investigating quantum mechanics (QM) or molecular mechanics (MM). The descriptors in any SAR methodology must be carefully picked depending on each inquiry, excluding highly correlated ones, in order to minimize overlapping data that can be found by multiple linear regression (MLR) analysis or even principal component analysis (PCA). [33]. Another analysis in LBDD suggests using a pharmacophore model to uncover more potent ligands. The minimum electrostatic energy and spatial distribution of some functional groups must be proposed as being required for effective activity against the targeted target when modeling. Functional groups in a known ligand or set of ligands, from hydrogen bond acceptors and donors to hydrophobic or aromatic groups, must be examined. Analyzing a pharmacophore model, which is closely related to the 3D-QSAR method, is a highly successful strategy to screen for new active chemicals as well as to improve the known ones. Software that can be utilized in this technique includes LigandScout, Discovery Studio Visualizer, and platforms for visualizing and analyzing structures.

The next steps are as follows after choosing the first strategy (either SBDD or LBDD) to go through virtual screening. The preparation of the chemicals database is the first and most important step. In particular, if the 3D geometry of such objects is not present in the selected database, it entails finding and getting it. A variety of programs, like Open Babel and LigPrep, can be used to transform 2D structures into 3D ones. The Cambridge Structure Database (CSD), which houses the crystal structures of compounds, can be used to obtain more reliable structures. Checking the tautomeric and protonation state is also necessary before moving on to the following steps to enable confidence and accurate docking interactions. Additionally, in terms of the target preparation, in addition to correcting the structure for a specific pH, an intriguing optional step can be to group the binding site among a collection of structures, either from various structures found in repositories or from MD simulations, in order to find a docking pose that is more precise. As a result, clustering can be done using MD software utilizing tools like GROMACS's clustering function or Amber's CPPTRAJ [44]. In most cases, compounds under investigation are docked in a grid assigned by one of the several docking programs (e.g., Glide and AutoDock4.2). In the subsequent phase, the best candidate ligands will be ranked and selected using the ratings based on electrostatic and Van der Waals interactions at the binding site that are provided in this stage. By evaluating and identifying the substances that can bind and those that do not interact in a stable manner at the biological binding site, this crucial phase aids in the reduction of the number of molecules that need to be studied [50]. The final stage before starting in vitro testing with the remaining collection of drugs is the examination of pharmacokinetics properties, which follows the energy of the complex protein–ligand [49]. One of the problems undermining the study is inaccuracies in the energy of the systems. Deep learning and machine learning are currently the most successful strategies created to address that. Machine learning can use a training dataset and be applied to the test's list of compounds, or it can use the test's list of compounds directly, in which case the algorithm will try to find a pattern without any prior knowledge. Examples of algorithms that can be used for this assessment include Naive Bayes (NB), K-Nearest Neighbor (KNN), and

Random Forest (RF). Contrarily, deep learning is regarded as a subset of machine learning techniques that utilizes neural networks. The issue is solved hierarchically using layers of different nonlinear layouts of processing units. Due to the inherent flexibility of its connections, it is a powerful technique to process many descriptors quickly, but reproducibility improvements are still needed [51].

4.4.2 Molecular Docking

One of the first steps in evaluating compounds as possible drug molecules is molecular docking, which works in tandem with VS and seeks to forecast the orientation and conformation of ligands at the target binding site (Fig. 4.2) [33]. The orientation and conformation of a ligand at the receptor-binding site are analyzed using a search approach in molecular docking, and the binding free energy, or affinity, of a receptor-ligand complex is predicted using a scoring function. Molecular docking requires the molecular structures of the macromolecule receptor and ligand. It is necessary to first define the three-dimensional structure of the receptor, which is frequently a protein. Ideally, this should be done using experimental data that may be available in databases like the Protein Data Bank (PDB). PDB contains the atom coordinates in individual molecules as well as complexes that have been resolved using experimental techniques like X-ray diffraction and nuclear magnetic resonance. Alternative computational methods can be employed for proteins that have not been solved yet. One of these methods is called comparative modeling, in which the amino acid sequence of the upcoming protein is compared to sequences of other proteins with existing structures [52]. Theoretical computation can also be used to undertake ab initio modeling for structure prediction [33]. The receptor-binding site that needs to be examined must be chosen manually and defined by providing the coordinates, or it must be chosen automatically using the coordinates of ligands that have already been bonded. Prediction algorithms can be employed to locate potential cavities in situations where the site is unknown. In order to reduce computing costs, a grid was created, which may be conceptualized as a box made up of a cubic network of points where the interaction energy is calculated. The grid delimits the binding site for the typical molecular docking calculation. Following the precalculation of the receptor potential energy for each grid point, the electrostatic and Lennard–Jones energies of the ligand's interaction with each grid point are computed [53]. Investigating the protonation states of the amino acids that interact at the receptor-binding site is also crucial. The structures of the ligands can be found in open databases like Zinc and PubChem. Dedicated molecular drawing applications like MarvinSketch, ChemDraw, and Avogadro can be used to create input files with three-dimensional coordinates [54]. When creating the ligand file, some factors need to be taken into consideration. Charges, stereochemical geometry, and the protonation states of chemical groups are some of these [55]. In addition to the pH, the interaction with the macromolecule binding site should be taken into account when determining the protonation states of the ligand groups. To utilize Molecular Docking with the receptor,

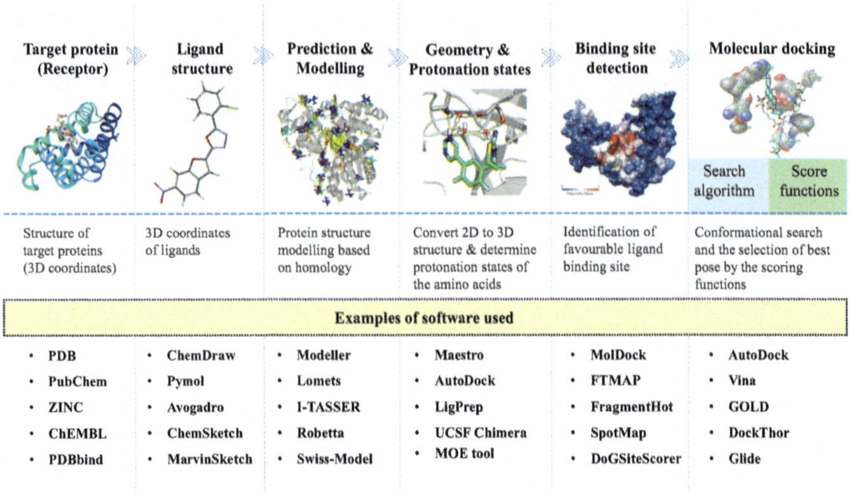

Fig. 4.2 Examples of software used in different stages of molecular docking

software like LigPrep can produce representative structures of these modifications [33]. The type of docking that needs to be computed must be specified because it has a direct impact on the outcomes. The computations can take either a rigid or flexible approach to the ligand and receptor-binding point. Due to the greater number of degrees of freedom that must be considered, flexible receptors and ligands have calculations that are more complex than those for rigid ones. Some docking software for flexible receptors and ligands is based on Koshland's induced-fit hypothesis, which states that both must tolerate slight changes in the binding that may take place [56].

4.4.3 Molecular Dynamics (MD)

The variation of atomic locations in a molecule as a function of time is described by Molecular Dynamics (MD), a computational method using Newton's laws. As a result, it differs from the docking technique, which relies mostly on stochastic processes. MD can be used to analyze traits including secondary structure content, side chain orientation, loop conformation, and the energy of long-term interactions with various molecules, like proteins and ligands. Because of this, it exhibits results comparable to those of experimental techniques like Nuclear Magnetic Resonance, but with the benefit of higher cost savings. Additionally, MD may be used to larger molecular systems, has better biological system reproduction, and takes less time [33]. MD simulation entails a number of processes, from getting the protein structure to parameterization, equilibration, minimization, simulation, and analysis. Molecular dynamics has several uses, including protein structure prediction, protein–lipid, protein–protein, and protein–ligand interaction analyses, viral studies, and drug creation for the widest range of disorders. The most challenging aspect of MD, in the opinion of many, is the analysis of MD simulations. The stability of variables like pressure, temperature, density, volume, and total energy during the course of the simulation needs to be examined first and foremost. You can proceed to the other analyses if these were stable. The root mean square deviation, which assesses how much the protein conformation changed throughout the course of the MD simulation in contrast to the initial crystallographic structure, is one of the most frequently performed analyses, particularly for proteins.

Apart from all these, numerous methods have been used in recent years to find novel antiviral agents with improved resistance profiles and new scaffolds, such as topology-matching design, targeted covalent inhibitors, proteolysis-targeting chimera (PROTAC), and ribonuclease targeting chimera (RIBOTAC) [57]. The "topology-matching design" theory for virus inhibitors was put to the test in 2020 by Nie et al. [58]. They created a nano-inhibitor that had the same nano-topological structure as the IAV virions and had hetero-multivalent inhibitory effects on HA and NA. The created nano-inhibitor was able to stop the viral particle's attachment to host cells and extracellular neutralization. The viral replication was significantly slowed down by six orders of magnitude and was inhibited to a level of more than 99.99% even 24 h after infection, indicating that a nano-inhibitor of this kind could be an effective anti-influenza agent. A spiky nano-inhibitor with a topology similar to IAV virions was also discovered by the researchers. The same group demonstrated topology-matched hetero-multivalent nanostructures as wide-spectrum IAV inhibitors [59]. The rational design of targeted covalent inhibitors has tremendously benefited from the advancement of structural biology and bioinformatics. Covalent inhibitors can engage with specific target proteins to produce covalent bonds that alter protein shape, impairing the protein's ability to function normally [60]. The process of creating a covalent bond with the target can be broken down into two related but distinct steps: (1) the inhibitor binds to the target reversibly, positioning the functional groups on weak electrophilic ligands next to particular nucleophilic residues on the protein; and (2) the ligand interacts with the functional groups involved in the protein to create the covalent bond [61]. PROTACs have emerged as a new paradigm for drug development to target proteins by encouraging and realizing the degradation of targets via the Ubiquitin–proteasome system [62]. PROTACs are hetero-bifunctional molecules made up of a linker, an E3 ubiquitin ligase recruiting ligand, and a ligand for the protein of interest (POI). To reduce the distance between them in vivo, bifunctional PROTAC molecules have one end that binds to the POI and the other end that attaches to an E3 ligase. The POI is subsequently ubiquitylated by an E2 enzyme through the E3 ligase, and the POI is then degraded by the proteasome [63, 64]. Antiviral agents have recently been steadily discovered using this technology. A novel method for RNA degradation is called Ribonuclease Targeting Chimera (RIBOTAC). In order to destroy the viral genome, RIBOTAC has an RNA-binding small molecule and a ribonuclease (RNase) L-recruiting module [65]. RNase L functions in innate immunity and is present in all cells in trace amounts as an inactive monomer. During viral infection, it becomes activated and dimerizes with innate substrate specificity [66].

To achieve the effect of selective cleavage, RIBOTACs locally recruit RNase L to the anticipated target.

4.5 Virus-Surface Proteins and Receptor Interaction

Enveloped viruses employ fusion glycoproteins as an adornment for their surface, and these proteins are crucial for the virus entry to the host cells. Fusion proteins are major components found on the viral surface, playing a crucial role in the infection process. As such, they are great targets for the development of antivirals and serve as the main immunogens in a variety of vaccine types. The refolding of fusion proteins from a highly strained and unstable prefusion conformation to a very stable postfusion conformation is assumed to supply the free energy that catalyzes the joining of host and virus membranes. The three main types of fusion proteins are Class I, Class II, and Class III fusion proteins [67]. Even though they all fuse in a similar way, these proteins have varied structures and respond to various stimuli. Class I proteins are tetramers with a higher α-helix content; in their postfusion conformation, they show a coiled-coil-helix encircled by three helices at the carboxy terminus (six α-helix bundles). Class II fusion proteins have a pre- and postfusion structural signature of β-sheets and are either homo- or heterodimers before fusion. Finally, class III fusion proteins have a mixed α-helical and β-sheet structure in both their prefusion and postfusion trimer stages. Class I and II fusion proteins' activation procedures need proteolytic processing of either the fusion protein or its partner protein; however, class III fusion proteins could not require processing to allow for cell entrance. Once fusion-competent, low pH is the main catalyst for class II fusion proteins, whereas class III fusion proteins are activated by interactions between a partner protein and a host cell receptor or by low pH. Class III fusion proteins may achieve an equilibrium between the prefusion and postfusion states, allowing the transition to be reversible, in contrast to the vast majority of viral fusion proteins that experience irreversible conformational changes upon activation (see Fig. 4.3) [68]. Class I fusion proteins are essential components of the fusion machinery of many different virus families, including the Orthomyxoviridae, Paramyxoviridae, Coronaviridae, Retroviridae, and Filoviridae. SARS-CoV-2 is only one example of a virus that has a class I fusion protein, but this kind of fusion protein has been extensively investigated since it is so common. In the context of the COVID-19, it is crucial to explore class I fusion proteins as targets of therapeutic importance. Class I fusion proteins, as an overarching example, are thought to be constructed from single-chain precursors that undergo proteolytic maturation to acquire fusion competence throughout their development [69]. A metastable structure that defines the prefusion state transitions through a number of intermediate phases after being activated to reach the postfusion state. The transition to a low pH environment and interactions with cell surface coreceptors (e.g., the envelope protein of HIV interacting with the CCR5, or interactions with cell surface receptors coupled with localized protease cleavage (e.g., the SARS-CoV-2 S protein binding to ACE2 receptor and is subsequently cleaved)

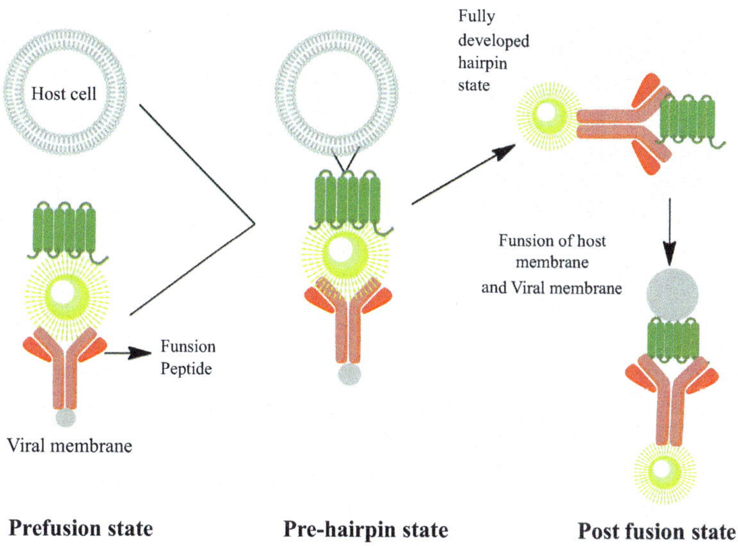

Fig. 4.3 General mechanism of virus–host cell interaction

are examples of triggers [70]. By stabilizing any of the conformations before the postfusion state is reached, the fusion process may be stopped. Inhibiting receptor binding is another option for avoiding receptor attachment altogether, and neutralizing antibodies are used for both of these purposes. For HA, for example, bnAbs target a conserved portion (the receptor-binding site pocket) in an otherwise variable region (the rest of the HA molecule). However, the antibodies' ability to neutralize a wide range of HA variants is hampered by the significant diversity around this site. Because of the high degree of conservation of the HA stalk, anti-stem antibodies tend to offer highly broad neutralizing potential because they target a particular epitope proximal to the fusion loop [71].

Numerous studies have examined the importance of certain cellular receptors as tropism-determining factors. The early studies on picornaviruses revealed a relationship between receptors on cells and the well-known in vivo tropism of poliovirus using organ minces and homogenates. Polioviruses could be absorbed by intestine and central nervous system (CNS) tissues from both humans and monkeys, but not by human lung, heart, or skin tissues. However, because receptors were also discovered on the monkey heart, human liver, and skeletal muscle, the connection was not strong [72]. In addition, poliovirus vaccine strains that do not cause cell damage have been demonstrated to attach to brain regions. Another piece of evidence linking cell surface receptors and pathogenicity is the similarity between the categorization of viruses into subgroups based on patterns of disease and the grouping of viruses by receptor specificities. Finally, picornaviruses' growth specificities have been shown using organ cultures. Some types of rhinoviruses, for instance, can only replicate in trachea organ cultures. Coxsackie viruses A1 and A5 can develop in primary, differentiating mouse muscle cultures but not in non-differentiating mouse cultures. Receptors for human enteroviruses, on the other hand, have been found in non-host species and on tissues unrelated to the virus' pathogenesis [73].

The virulence and cell tropism of the reovirus serotypes 1 and 3 have been genetically characterized to differ from one another. Reovirus serotype 3 and clone 1 have been shown to have different central nervous system cell tropisms depending on the presence of the S1 gene, which codes for the hemagglutinin. In newborn mice, HA3 (which contains nine genes from type I and the gene encoding the hemagglutinin from type 3), causes a fatal encephalitis with neuronal destruction but no ependymal cell damage. The differing cell-type selectivity of the serotypes can be explained by the viral hemagglutinin's apparent interaction with receptors on the surface of either neuronal or ependymal cells. Reovirus type 1 and clone 3.HA1 binding to isolated human cells served as an in vitro confirmation of these findings [74].

The M form of the encephalomyocarditis (EMC) virus induces a diabetes-like disease in some mouse strains by infecting and destroying pancreatic β cells. Since cultured pancreatic beta cells from resistant strains of mice are less able to absorb infectious EMC virus than beta cells from strains that are susceptible to EMC-induced diabetes, it appears that genetically determined variations in the surface viral receptors on these cells may be one of the factors regulating susceptibility to disease [75].

Lymphocytes may have a more targeted immune response to some viruses if they include viral receptors. Measles virus has a receptor on T cells and measles infection is linked to decreased tuberculin skin hypersensitivity and decreased helper cell activity. As with humans (T8 + cells), the Ly2,3 fraction of murine T cells is a primary target for reovirus type 3. The viral hemagglutinin is responsible for this binding. Additionally, when examined in vitro, suppressor T cells produced by reovirus type 3 may prevent Con A proliferation. It appears that reovirus type 3 produces functionally active suppressor T cells in vitro before engaging the viral hemagglutinin with a specific receptor on the Ly2,3 subset of mice lymphocyte [76].

4.6 Antivirals Targeting Viral Surface Proteins

By inhibiting viral replication in host cells, several small-molecule antiviral medicines decrease viral load, illness severity, or death from viral infections. Current antiviral medications, such as those used to combat HIV (abacavir; Ziagen), HSV (famciclovir; Famvir), and HBV (tenofovir; Rescriptor), are a prime example of this phenomenon (lamivudine; Epivir). For respiratory viruses like influenza, RSV, or coronaviruses, however, there are very limited treatment choices [77]. Oseltamivir, zanamivir, and peramivir, three antivirals, have an additional inhibitory mechanism that blocks the release of viral offspring from infected cells by inhibiting the activity of the neuraminidase protein. They do not involve the fusion proteins themselves but instead interfere with the molecular pathway by binding to a surface glycoprotein. But they are only useful in the early stages of an infection [78].

Supportive therapy (including acetaminophen for pyrexia and i.v. fluids for dehydration) and neutralizing antibodies (e.g., palivizumab) are often used to treat RSV infections. It is important to note that the cost of antibody-based therapy may add up rapidly. There are now four small-molecule antivirals licensed by the FDA for treatment against influenza viruses. Among them are medications that are effective against influenza A and B viruses and that target the influenza neuraminidase protein, such as oseltamivir phosphate, zanamivir, and peramivir. The issue is that these drugs frequently only function when used promptly after infection. Despite the fact that small-molecule drugs and/or their combinations have been demonstrated to be effective, the research and development of innovative small molecules against an emerging pandemic like COVID-19 has proven to be challenging [79]. Challenges include extensive development times, limited specificity, and typically poor affinities (given the tiny surface area of small compounds) that may lead to unwanted side effects or the requirement for repeated dosing to reach positive therapeutic objectives. Already licensed medications may target additional dangerous viruses, making drug repurposing a possible option to generating novel small molecules for treating viral infections. Remdesivir, Gilead Sciences, USA, was in this position; it had been designed for Ebolavirus and now has emergency FDA permission for COVID-19. Also being evaluated for treatment against SARS-CoV-2 infections is the favipiravir, which was first authorized for use against influenza in Japan. Both compounds target the viral RNA-dependent RNA polymerase in their triphosphate forms after being phosphorylated in vivo as prodrugs (RdRp). To prevent viral replication and spread, bioactive triphosphates compete with purine nucleosides for incorporation into viral RNA.

Alternative medicines for treating viral infections include those based on proteins. Higher specificity and affinity than small-molecule medicines are made possible by the vast protein–protein and/or protein-glycan interfaces. Monoclonal antibodies are the most widely used protein-based therapies, although they may be expensive to produce and may not be very stable (bacterial diseases, viral infections, arthritis, cancer, etc.). Antibody-based treatments have also proven an effective strategy in the ongoing COVID-19 epidemic. In particular, two treatment options have been approved for emergency use by the FDA to protect against SARS-CoV-2: the Regeneron cocktail (REGN-COVTM), casirivimab along with imdevimab, and bamlanivimab combined with etesevimab, which target the receptor-binding domain of the S protein at different sites. This development happened incredibly fast, with the pandemic starting in February 2020, antibodies being found in May 2020, clinical trials starting in November 2020, and FDA approval coming in February 2021, respectively [80].

Using tiny protein-based inhibitors is an alternate approach to developing big antibody molecules. Tiny, extremely stable, and easy-to-produce proteins may be developed to block functional locations in a target biomolecule, similar to naturally occurring inhibitors where small proteins attach to bigger proteins to occlude their action (say cystatin in trypsin). Recent research has shown that computational design may create very potent protein–protein inhibitors. The capacity of the influenza HA protein to facilitate fusion or attach to human cell receptors has been blocked, for instance, by inhibitors built from scratch to target-specific sections of the protein. Inhibitors of both classes were able to prevent and treat influenza in mice. Highly stable picomolar inhibitors were created lately by blocking the receptor-binding site of the S protein with small protein inhibitors [81]. These fine-tuned molecules prevented viral replication for three days and reversed infection for twenty-one days in Golden Syrian hamsters. The low molecular weight (<5 kDa) of this antibody compared to the typical 150 kDa of a monoclonal antibody means that far less material has to be generated to have the same efficiency, opening up a fantastic possibility to provide for a greater number of individuals. The use of computational design has several benefits over more conventional approaches, one of which is the robust binding of these tiny proteins [82]. Computational protein design, in contrast to the traditional method of identifying monoclonal antibodies from a pool, gives the designer full authority over the inhibitor's target area and allows for the optimization of the interaction of interest. Last but not least, tiny protein inhibitors are a potential method to combating viral infections since their great stability might circumvent storage and transportation hurdles that often impede the supply of medications to distant places [83].

4.7 Applications of Computational Modeling in Antiviral Drug Discovery

The study of viruses is an intriguing application of molecular dynamics. The satellite tobacco mosaic virus was the first fully atomized non-enveloped virus that Schulten and colleagues described in 2006 [84]. In terms of enveloped viruses, Ayton and Voth performed the first simulation of capsid of an immature HIV virus in 2010 [85]. The majority of investigations, however, focus on simulations of viral capsids and proteins since viral particles are too big for all-atom simulations. The study of more intricate mechanisms such replication, membrane attachment and fusion, viral maturation, and nuclear entry represents a constraint for this field. However, the study of viral proteins yields a wealth of crucial knowledge that also enables MD to develop drugs for viral diseases [86, 87]. Studying the connections between the Ebola virus' VP24 protein and the human protein KPNA5, Pappalardo and colleagues determined that this protein is essential for the virus' pathogenicity [88]. Molecular dynamics was used by Nasution and colleagues to undertake a drug design study that resulted in the identification of potential molecules [89]. Another example is Zhang and Zheng's study of the influenza B virus, which provided structural and dynamic details of the effect of serine triad on proton conduction in the tetrameric influenza B channel M2, essential to the virus life cycle [90]. To uncover possible inhibitors of the Zika NS2B-NS3 utilizing a collection of more than 7 million chemicals from the ZINC15 database, for instance, Bowen and colleagues used consensus-based docking, MD modeling, and binding energy estimates [91]. Although there is a long history of mathematical modeling of viral infections and immunity, the idea of quantifying the pathogen's within-host dynamics only became widely used in clinical practice after the publication of two seminal papers in 1995 [92, 93]. These papers used modeling to demonstrate that HIV infection was a highly dynamic process, in which the virus was rapidly replicating and being cleared within infected individuals. Due to the fact that HIV infection takes over ten years to progress to AIDS and because for the most of this time plasma HIV levels remain relatively stable, many people previously believed that HIV was a "dormant" virus comparable to other lentiviruses. When it was discovered that HIV replicated quickly (and erratically), it became clear that medication resistance was unavoidable if only one or two drugs were employed, for which the virus may develop resistance with one or two point mutations. As a result, it was determined that combination therapy involving at least three medications was required. The longevity of infected cells, the effectiveness of drug treatment, the modes of action of treatment, the dynamics of diverse populations of infected cells, etc., were all clarified by modeling HIV and then other infections.

To confirm the efficacy of drug combinations for epidemiological outbreaks in sizable populations, machine learning techniques can be used to create complex network models. In order to halt the spread of AIDS in the U.S. epidemic networks, Herrera and his colleagues proposed the first artificial neural network (ANN) model for the prediction of HAART cocktails in 2015. They did this by compiling datasets from ChEMBL (anti-HIV chemical compounds), the AIDSVu database (HIV surveillance reports), and the Census Bureau (socioeconomic data) [94]. In silico fragment-based drug design was carried out for drug–molecule interaction research, the first multitasking model for quantitative structure-biological effect relationships (mtk-QSBERs) was predicted, and its molecular components were virtually screened for the Hepatitis C virus infection [95]. Data on AIDS in the United States has been linked to ChEMBL data using ANNs. Network prediction models called ANNs are mostly employed in the creation of pharmaceuticals and medicinal chemistry. Using drug-drug similarity complex networks, Gonzalez-D'az et al. predicted novel antimicrobial medicines and targets using the MARkovian CHemicals IN SIlico DEsign (MARCH-INSIDE) method [96]. Novel stochastic moments are generated by computational analyses of structural stability connections [97]. In order to simulate their effects on the relationships between protein structure stability, Markovian Backbone Negentropies (MBNs) have been introduced. It has also been reported on an MBN for the computational investigation of structure/stability relationships based on a Markov chain model of electron delocalization along the protein backbone [98]. These methods primarily concentrate on cheminformatic methods for drug design, drug development, and drug–target interactions in in silico clinical fields, particularly with regard to viral illnesses.

The orthomyxoviridae family of RNA viruses is responsible for the annual seasonal pandemic of acute respiratory illness known as influenza. To combat influenza, pharmaceutical companies have focused on inhibiting neuraminidase (NA), the enzyme responsible for viral entry into the mucous membranes of the host. With this in mind, VS has been utilized effectively to discover new inhibitors of this target. They described a search of the National Cancer Institute (NCI) compound database using shape and chemical features to identify potential analogues of katsumadain A, a proven NA inhibitor. Within the micromolar range, five of the NA-inhibiting drugs tested (including four flavonoids) showed significant activity against three oseltamivir-susceptible H1N1 strains [99]. A strategy for forecasting influenza's dominant hemagglutinin gene, Plotkin et al. [100] proposed a virus and investigated the evolution of antigenic over the host genes. Clustering analysis was performed on the spatiotemporal viral swarm dispersal and the evolution of hemagglutinin structure. To cluster the viral sequences, a critical length scale in amino acid space was lastly used.

To combat HIV retrovirus, scientists have taken several different overarching strategies, including (1) preventing the retrovirus' genetic material from replicating or being inserted within the host cell, (2) preventing the virus from entering the cell, and (3) inhibiting the virus from fully assembling and maturing. In the first VS run, the known RT inhibitor dihydroxy benzoyl naphthyl hydrazone was used as the query molecule in a shape-based screening of the NCI database. The most active hits from this process were used for a second VS using a combination ligand-based technique that includes 2D and 3D similarity searches, and ligand-based pharmacophore screening. In micromolar concentrations, some compounds identified through new scaffolds exhibited strong anti-RT action, blocking both RT functions and RT-associated ribonuclease H [101]. Finally, the capsid protein has been identified as a promising target for the design of next-generation antiretroviral medicines because of its central involvement in HIV-1 assembly and maturation. The Curelli group performed a VS using docking and then looked for analogues. The C-terminal section of the HIV-1 capsid is the target of small-molecule inhibitors, and these have been discovered. In infected cells, the most effective chemicals significantly decreased infectivity [102]. Short linear motifs-based prediction methods for PPIs between HIV-1 and host proteins have been proposed by Becerra et al. [103]. They used three filtering techniques to find groups of linear motifs: (1) those conserved in viral proteins (C); (2) those found in disordered areas (D); and (3) those that are uncommon or sparse in a set of randomly generated viral sequences (R). To locate the disordered protein areas among the HIV-1 sequences and host sequences, these three sets have finally been used. Their research demonstrates that the majority of the virus's conserved linear motifs are found in disorganized areas. A group proposed a way to simulate the network of virus–human interactions using motif–domain interactions in [104]. A bi-clustering strategy was put forth by Mukhopadhyay et al. [105] to forecast HIV-1-infected proteins using interaction-based analysis. By using the bi-clustering technique, a set of association rules was extracted from the interaction of HIV-1 proteins. In order to forecast novel interactions between human proteins, a set of high-confidence rules was ultimately extracted. This same group [106] made further advancements to their work by adding type and direction (virus-to-host and host-to-virus)-based bi-clustering to already-known interactions to forecast novel host proteins.

An approach to anticipate the interactions between human proteins and HIV-1 based on protein structural similarity was proposed by Doolittle and group [107]. In this method, two crystal structures are compared to determine the structural similarity of host to that of pathogen proteins. It is assumed that targets will be identified for human proteins that share a lot of structural similarities with HIV proteins and their known interaction partners. They used a similar methodology to create a network of interactions between the host and the Dengue virus [108].

Several effective docking studies of the E-protein have been published. Two hit compounds demonstrated micromolar antiviral activity against dengue virus, while one molecule displayed antiviral activity against West Nile virus and yellow fever virus, which is an intriguing finding. Through docking of over a million in-house compounds, the authors found a molecule that suppresses dengue fusion in the low micromolar range. To combat HCV and other members of the Flaviviridae family, NS5 polymerase has emerged as a promising drug development target [109].

Although popular methods based on structure-based approaches, such as molecular docking, will always be helpful, they become computationally taxing when used to find pan-antivirals via multitarget inhibition against many viral proteins. One strategy focuses on models that combine perturbation theory and machine learning (PTML), which, by using Box-Jenkins operators, can combine in vitro screening data that includes various viral proteins from different viruses [110]. This makes it possible to examine FDA-approved medications and experimentally validate those that are projected to be versatile inhibitors of the viral proteins as prospective therapies for SARS-CoV-2 and other viral infections. In the second method, medicines are virtually screened using PTML models in inhibition/replication experiments against various viruses. Once more, in biological testing against SARS-CoV-2, the medications projected to be pan-antivirals will take precedence. Modeling anti-HIV action through multitarget inhibition [111], connecting in vitro activity with epidemiological consequences [96], and virtually screening antiretroviral drugs [112] are all successful uses of PTML models in antiviral research. PTML models could thus be simply modified and created for the two aforementioned scenarios. Alignment-free multitarget (AFMT) model development is the third strategy [113]. These computational methods are created by calculating molecular descriptors from the amino acid sequences of the proteins and the 2D structures of the ligands/inhibitors. The advantage of the AFMT models over their PTML equivalents is that the former can forecast the inhibitory potency of huge libraries of (either virtual or already synthesized) compounds against proteins other than those utilized to construct the AFMT models.

Drugs may be designed and engineered using computational approaches. Computational approaches are convenient for high throughput screening of existing medications due to their short time requirements, which may be used to find treatments for new ailments and to foresee their side effects. In a process known as drug repurposing, it is highly desirable to identify substances that have already been tested and approved for use on the market and that exhibit activity against the new issue in order to combat disease outbreaks or the ongoing emergence of drug-resistant diseases. The reason this is crucial is that when repurposing an already-commercial compound, several experimental steps can be skipped because information on their pharmacokinetic, pharmacodynamics, and toxicity profiles has been established, and they are, in some cases, safe for human administration [114]. Given that creating a new treatment is tedious and expensive before it can be made available to the public, this is particularly interesting and useful in the recent Zika, Ebola, and COVID-19 virus epidemics. Following the conventional steps in a VS process, the repurposing process starts evaluating databases of already-commercial chemicals like ZINC. Strong hardware and algorithms are consequently required for this fight since in an outbreak event urgency is needed to stop the disease's progress and the deaths that follow. The most recent instance of this is the COVID-19 pandemic, which tested scientists' abilities to the limit in order to find the quickest solutions to stop its spread, i.e., by developing, in a short period of time, the screening over thousand hits among 1.3 billion already regulated compounds against the virus main protease [114–116]. For instance, computational researches have identified Lopinavir and Ritonavir to be potential inhibitors against SARS-CoV-2 main protease (Mpro) [117, 118].

4.8 Conclusion

In virology, computational techniques have advanced significantly. They cover a wide range of machine learning techniques for automated model extraction, now-standard computational image analysis techniques, and identifying probable interactions between virions and host cell compartments. Since many of these advancements were initially motivated by virology-related applications, the project was truly interdisciplinary. In silico reconstitution is made possible by computer simulations of learned or postulated models, which can also demonstrate the sufficiency of a process rather than just a collection of constituent parts or molecules. Different resolutions of simulations are possible, ranging from continuum models to all-atom molecular dynamics simulations. In addition, we anticipate that a new generation of AI-driven protein modeling tools, e.g., AlphaFold, will offer even greater advancements in protein models for novel viruses. Nevertheless, when characterizing viral proteins, de novo modeling should be utilized with caution and supported by tests because their extremely wide structural repertoire might not be captured during training of an AI system. Furthermore, it is extremely difficult to characterize the structural properties of the macromolecular complexes generated by viral proteins. The next step is to create brand-new techniques for precise de novo protein complex characterization, similar to AI-driven techniques for protein structure prediction. We anticipate that computational modeling will continue to play a significant role in global viral disease response, treatment, and prevention given the historical contributions of mathematical and computer modeling to understanding the spread of infectious illnesses and their consequences in humans.

Conflict of Interest None to declare.

References

1. Narkhede YB, Gonzalez KJ, Strauch E-M (2021) Targeting viral surface proteins through structure-based design. Viruses 13:1320. https://doi.org/10.3390/v13071320
2. Lanjanian H, Nematzadeh S, Hosseini S et al (2021) High-throughput analysis of the interactions between viral proteins and host cell RNAs. Comput Biol Med 135:104611. https://doi.org/10.1016/j.compbiomed.2021.104611
3. Tellinghuisen TL, Rice CM (2002) Interaction between hepatitis C virus proteins and host cell factors. Curr Opin Microbiol 5:419–427. https://doi.org/10.1016/S1369-5274(02)00341-7
4. Carvalho FA, Carneiro FA, Martins IC et al (2012) Dengue virus capsid protein binding to hepatic lipid droplets (LD) is potassium ion dependent and is mediated by LD surface proteins. J Virol 86:2096–2108. https://doi.org/10.1128/JVI.06796-11
5. Kumar R, Harilal S, Al-Sehemi AG et al (2021) The chronicle of COVID-19 and possible strategies to curb the pandemic. Curr Med Chem 28:2852 2886. https://doi.org/10.2174/0929867327666200702151018
6. Ji Z, Yan K, Li W et al (2017) Mathematical and computational modeling in complex biological systems. Biomed Res Int 2017:e5958321. https://doi.org/10.1155/2017/5958321
7. Brodland GW (2015) How computational models can help unlock biological systems. Semin Cell Dev Biol 47–48:62–73. https://doi.org/10.1016/j.semcdb.2015.07.001
8. Woolhouse MEJ, Gowtage-Sequeria S (2005) Host range and emerging and reemerging pathogens. Emerg Infect Dis 11:1842–1847. https://doi.org/10.3201/eid1112.050997
9. Weiss RA, McMichael AJ (2004) Social and environmental risk factors in the emergence of infectious diseases. Nat Med 10:S70-76. https://doi.org/10.1038/nm1150
10. Huremović D (2019) Brief history of pandemics (pandemics throughout history). Psychiat Pandemics 7–35. https://doi.org/10.1007/978-3-030-15346-5_2
11. Piret J, Boivin G (2021) Pandemics throughout history. Front Microbiol 11. https://doi.org/10.3389/fmicb.2020.631736
12. He H (2013) Vaccines and antiviral agents. In: Current issues in molecular virology-viral genetics and biotechnological applications. IntechOpen
13. Fermin G (2018) Virion structure, genome organization, and taxonomy of viruses. viruses 17–54. https://doi.org/10.1016/B978-0-12-811257-1.00002-4
14. Lemke TL, Williams DA, Roche VF, Zito SW (2012) Foye's principles of medicinal chemistry, 7th edn. Wolters Kluwer Health Adis (ESP)
15. Ganser-Pornillos BK, Chandrasekaran V, Pornillos O et al (2011) Hexagonal assembly of a restricting TRIM5alpha protein. Proc Natl Acad Sci U S A 108:534–539. https://doi.org/10.1073/pnas.1013426108
16. Patrick GL (2013) An introduction to medicinal chemistry. Oxford University Press
17. Novoa RR, Calderita G, Arranz R et al (2005) Virus factories: associations of cell organelles for viral replication and morphogenesis. Biol Cell 97:147–172. https://doi.org/10.1042/BC20040058
18. Lu X, Block T (2004) Study of the early steps of the Hepatitis B Virus life cycle. Int J Med Sci 1:21–33
19. Mukhopadhyay S, Kuhn RJ, Rossmann MG (2005) A structural perspective of the flavivirus life cycle. Nat Rev Microbiol 3:13–22. https://doi.org/10.1038/nrmicro1067
20. Sedlmeier R, Neubert WJ (1998) The replicative complex of paramyxoviruses: structure and function. In: Maramorosch K, Murphy FA, Shatkin AJ (eds) Advances in virus research. Academic Press, pp 101–139
21. Rosenberg ZF, Fauci AS (1991) Immunopathogenesis of HIV infection. FASEB J 5:2382–2390. https://doi.org/10.1096/fasebj.5.10.1676689
22. Basavapathruni A, Anderson KS (2007) Reverse transcription of the HIV-1 pandemic. FASEB J 21:3795–3808. https://doi.org/10.1096/fj.07-8697rev
23. Goodman LS (2001) Goodman & Gilman's the pharmacological basis of therapeutics. McGraw-Hill, New York, pp 27–2141

24. De Clercq E, Li G (2016) Approved antiviral drugs over the past 50 years. Clin Microbiol Rev 29:695–747
25. De Clercq E (2004) Antivirals and antiviral strategies. Nat Rev Microbiol 2:704–720
26. Halder AK, Dutta P, Kundu M et al (2018) Review of computational methods for virus–host protein interaction prediction: a case study on novel Ebola–human interactions. Brief Funct Genomics 17:381–391. https://doi.org/10.1093/bfgp/elx026
27. Jenner AL, Aogo RA, Davis CL et al (2020) Leveraging computational modeling to understand infectious diseases. Curr Pathobiol Rep 8:149–161. https://doi.org/10.1007/s40139-020-00213-x
28. Silva T dos SC da (2019) Inibição da replicação do influenza através da modulação de fatores restritivos pelos ligantes dos receptores CCR5 e CXCR4
29. Hagen JB (2000) The origins of bioinformatics. Nat Rev Genet 1:231–236. https://doi.org/10.1038/35042090
30. Zheng L-L, Li C, Ping J et al (2014) The domain landscape of virus-host interactomes. Biomed Res Int 2014:867235. https://doi.org/10.1155/2014/867235
31. Kshirsagar M, Murugesan K, Carbonell JG, Klein-Seetharaman J (2017) Multitask matrix completion for learning protein interactions across diseases. J Comput Biol 24:501–514. https://doi.org/10.1089/cmb.2016.0201
32. Kshirsagar M, Carbonell J, Klein-Seetharaman J (2012) Techniques to cope with missing data in host–pathogen protein interaction prediction. Bioinformatics 28:i466–i472. https://doi.org/10.1093/bioinformatics/bts375
33. de Queiroz Simões RS, Ferreira MS, Dumas de Paula N et al (2020) Computational modeling in virus infections and virtual screening, docking, and molecular dynamics in drug design. Netw Syst Biol. https://doi.org/10.1007/978-3-030-51862-2_12
34. Chojnacki J, Staudt T, Glass B et al (2012) Maturation-dependent HIV-1 surface protein redistribution revealed by fluorescence nanoscopy. Science 338:524–528. https://doi.org/10.1126/science.1226359
35. Srivastava R, You L, Summers J, Yin J (2002) Stochastic versus deterministic modeling of intracellular viral kinetics. J Theor Biol 218:309–321. https://doi.org/10.1006/jtbi.2002.3078
36. Yakimovich A, Gumpert H, Burckhardt CJ et al (2012) Cell-free transmission of human adenovirus by passive mass transfer in cell culture simulated in a computer model. J Virol 86:10123–10137. https://doi.org/10.1128/JVI.01102-12
37. Labouesse C, Gabella C, Meister J-J et al (2016) Microsurgery-aided in-situ force probing reveals extensibility and viscoelastic properties of individual stress fibers. Sci Rep 6:23722. https://doi.org/10.1038/srep23722
38. Pollard TD, Cooper JA (2009) Actin, a central player in cell shape and movement. Science 326:1208–1212. https://doi.org/10.1126/science.1175862
39. Greber UF (2016) Virus and host mechanics support membrane penetration and cell entry. J Virol 90:3802–3805. https://doi.org/10.1128/JVI.02568-15
40. Nakano MY, Boucke K, Suomalainen M et al (2000) The first step of adenovirus type 2 disassembly occurs at the cell surface, independently of endocytosis and escape to the cytosol. J Virol 74:7085–7095
41. Burckhardt CJ, Suomalainen M, Schoenenberger P et al (2011) Drifting motions of the adenovirus receptor CAR and immobile integrins initiate virus uncoating and membrane lytic protein exposure. Cell Host Microbe 10:105–117. https://doi.org/10.1016/j.chom.2011.07.006
42. Iizuka N, Oka M, Yamada-Okabe H et al (2002) Comparison of gene expression profiles between hepatitis B virus- and hepatitis C virus-infected hepatocellular carcinoma by oligonucleotide microarray data on the basis of a supervised learning method. Cancer Res 62:3939–3944
43. Sbalzarini IF, Greber UF (2018) How computational models enable mechanistic insights into virus infection. Methods Mol Biol 1836:609–631. https://doi.org/10.1007/978-1-4939-8678-1_30

44. Sliwoski G, Kothiwale S, Meiler J, Lowe EW (2014) Computational methods in drug discovery. Pharmacol Rev 66:334–395. https://doi.org/10.1124/pr.112.007336
45. D'Souza S, Prema KV, Balaji S (2020) Machine learning models for drug–target interactions: current knowledge and future directions. Drug Discovery Today 25:748–756. https://doi.org/10.1016/j.drudis.2020.03.003
46. Muratov EN, Amaro R, Andrade CH et al (2021) A critical overview of computational approaches employed for COVID-19 drug discovery. Chem Soc Rev 50:9121–9151. https://doi.org/10.1039/D0CS01065K
47. Schmidtke P, Bidon-Chanal A, Luque FJ, Barril X (2011) MDpocket: open-source cavity detection and characterization on molecular dynamics trajectories. Bioinformatics 27:3276–3285. https://doi.org/10.1093/bioinformatics/btr550
48. Anderson AC (2003) The process of structure-based drug design. Chem Biol 10:787–797. https://doi.org/10.1016/j.chembiol.2003.09.002
49. Yu W, MacKerell AD (2017) Computer-aided drug design methods. Methods Mol Biol 1520:85–106. https://doi.org/10.1007/978-1-4939-6634-9_5
50. Kitchen DB, Decornez H, Furr JR, Bajorath J (2004) Docking and scoring in virtual screening for drug discovery: methods and applications. Nat Rev Drug Discov 3:935–949. https://doi.org/10.1038/nrd1549
51. Wang D, Cui C, Ding X, et al (2019) Improving the virtual screening ability of target-specific scoring functions using deep learning methods. Front Pharmacol 10
52. Webb B, Sali A (2016) Comparative protein structure modeling using MODELLER. Curr Protoc Protein Sci 86:2.9.1–2.9.37. https://doi.org/10.1002/cpps.20
53. Torres PHM, Sodero ACR, Jofily P, Silva-Jr FP (2019) Key topics in molecular docking for drug design. Int J Mol Sci 20:E4574. https://doi.org/10.3390/ijms20184574
54. Hanwell MD, Curtis DE, Lonie DC et al (2012) Avogadro: an advanced semantic chemical editor, visualization, and analysis platform. Journal of Cheminformatics 4:17. https://doi.org/10.1186/1758-2946-4-17
55. Lousa D, Baptista AM, Soares CM (2012) Analyzing the molecular basis of enzyme stability in ethanol/water mixtures using molecular dynamics simulations. J Chem Inf Model 52:465–473. https://doi.org/10.1021/ci200455z
56. Pagadala NS, Syed K, Tuszynski J (2017) Software for molecular docking: a review. Biophys Rev 9:91–102. https://doi.org/10.1007/s12551-016-0247-1
57. Xu S, Ding D, Zhang X, et al (2022) Newly emerging strategies in antiviral drug discovery: dedicated to Prof. Dr. Erik De Clercq on occasion of his 80th anniversary. Molecules 27:850. https://doi.org/10.3390/molecules27030850
58. Nie C, Parshad B, Bhatia S et al (2020) Topology-matching design of an influenza-neutralizing spiky nanoparticle-based inhibitor with a dual mode of action. Angew Chem Int Ed 59:15532–15536. https://doi.org/10.1002/anie.202004832
59. Nie C, Stadtmüller M, Parshad B, et al (2021) Heteromultivalent topology-matched nanostructures as potent and broad-spectrum influenza A virus inhibitors. Sci Adv 7:eabd3803. https://doi.org/10.1126/sciadv.abd3803
60. Bauer RA (2015) Covalent inhibitors in drug discovery: from accidental discoveries to avoided liabilities and designed therapies. Drug Discov Today 20:1061–1073. https://doi.org/10.1016/j.drudis.2015.05.005
61. Gehringer M, Laufer SA (2019) Emerging and re-emerging warheads for targeted covalent inhibitors: applications in medicinal chemistry and chemical biology. J Med Chem 62:5673–5724. https://doi.org/10.1021/acs.jmedchem.8b01153
62. Lai AC, Crews CM (2017) Induced protein degradation: an emerging drug discovery paradigm. Nat Rev Drug Discov 16:101–114. https://doi.org/10.1038/nrd.2016.211
63. Paiva S-L, Crews CM (2019) Targeted protein degradation: elements of PROTAC design. Curr Opin Chem Biol 50:111–119. https://doi.org/10.1016/j.cbpa.2019.02.022
64. Li X, Song Y (2020) Proteolysis-targeting chimera (PROTAC) for targeted protein degradation and cancer therapy. J Hematol Oncol 13:50. https://doi.org/10.1186/s13045-020-00885-3

65. Costales MG, Aikawa H, Li Y et al (2020) Small-molecule targeted recruitment of a nuclease to cleave an oncogenic RNA in a mouse model of metastatic cancer. Proc Natl Acad Sci U S A 117:2406–2411. https://doi.org/10.1073/pnas.1914286117
66. Costales MG, Matsumoto Y, Velagapudi SP, Disney MD (2018) Small molecule targeted recruitment of a nuclease to RNA. J Am Chem Soc 140:6741–6744. https://doi.org/10.1021/jacs.8b01233
67. Sanders RW, Moore JP (2021) Virus vaccines: proteins prefer prolines. Cell Host Microbe 29:327–333. https://doi.org/10.1016/j.chom.2021.02.002
68. Castaño N, Cordts SC, Kurosu Jalil M et al (2021) Fomite transmission, physicochemical origin of virus-surface interactions, and disinfection strategies for enveloped viruses with applications to SARS-CoV-2. ACS Omega 6:6509–6527. https://doi.org/10.1021/acsomega.0c06335
69. Churin Y, Roderfeld M, Roeb E (2015) Hepatitis B virus large surface protein: function and fame. Hepatobiliary Surg Nutr 4:1–10. https://doi.org/10.3978/j.issn.2304-3881.2014.12.08
70. Hinuma S, Fujita K, Kuroda S (2021) Binding of nanoparticles harboring recombinant large surface protein of hepatitis B virus to scavenger receptor class B type 1. Viruses 13:1334. https://doi.org/10.3390/v13071334
71. Taha BA, Al Mashhadany Y, Bachok NN et al (2021) Detection of COVID-19 virus on surfaces using photonics: challenges and perspectives. Diagnostics (Basel) 11:1119. https://doi.org/10.3390/diagnostics11061119
72. Brown NA, Schrevens S, van Dijck P, Goldman GH (2018) Fungal G-protein-coupled receptors: mediators of pathogenesis and targets for disease control. Nat Microbiol 3:402–414. https://doi.org/10.1038/s41564-018-0127-5
73. Haverkamp A-K, Lehmbecker A, Spitzbarth I et al (2018) Experimental infection of dromedaries with Middle East respiratory syndrome-Coronavirus is accompanied by massive ciliary loss and depletion of the cell surface receptor dipeptidyl peptidase 4. Sci Rep 8:9778. https://doi.org/10.1038/s41598-018-28109-2
74. Ram S, Gulati S, Lewis LA et al (2018) A novel sialylation site on neisseria gonorrhoeae lipooligosaccharide links heptose II lactose expression with pathogenicity. Infect Immun 86:e00285-e318. https://doi.org/10.1128/IAI.00285-18
75. Maginnis MS (2018) Virus-receptor interactions: the key to cellular invasion. J Mol Biol 430:2590–2611. https://doi.org/10.1016/j.jmb.2018.06.024
76. Simone D, Al Mossawi MH, Bowness P (2018) Progress in our understanding of the pathogenesis of ankylosing spondylitis. Rheumatology (Oxford) 57:vi4–vi9. https://doi.org/10.1093/rheumatology/key001
77. Mazzon M, Ortega-Prieto AM, Imrie D et al (2019) Identification of Broad-spectrum antiviral compounds by targeting viral entry. Viruses 11:E176. https://doi.org/10.3390/v11020176
78. Villalón-Letelier F, Brooks AG, Londrigan SL. Reading PC MARCH8 restricts influenza A virus infectivity but does not downregulate viral glycoprotein expression at the surface of infected cells. mBio 12:e01484–21. https://doi.org/10.1128/mBio.01484-21
79. Krug RM, Aramini JM (2009) Emerging antiviral targets for influenza A virus. Trends Pharmacol Sci 30:269–277. https://doi.org/10.1016/j.tips.2009.03.002
80. Vigant F, Santos NC, Lee B (2015) Broad-spectrum antivirals against viral fusion. Nat Rev Microbiol 13:426–437. https://doi.org/10.1038/nrmicro3475
81. Meagher JL, Takata M, Gonçalves-Carneiro D et al (2019) Structure of the zinc-finger antiviral protein in complex with RNA reveals a mechanism for selective targeting of CG-rich viral sequences. Proc Natl Acad Sci U S A 116:24303–24309. https://doi.org/10.1073/pnas.1913232116
82. Chakravarty M, Vora A (2021) Nanotechnology-based antiviral therapeutics. Drug Deliv Transl Res 11:748–787. https://doi.org/10.1007/s13346-020-00818-0
83. Du L, Yang Y, Zhou Y et al (2017) MERS-CoV spike protein: a key target for antivirals. Expert Opin Ther Targets 21:131–143. https://doi.org/10.1080/14728222.2017.1271415
84. Freddolino PL, Arkhipov AS, Larson SB et al (2006) Molecular dynamics simulations of the complete satellite tobacco mosaic virus. Structure 14:437–449. https://doi.org/10.1016/j.str.2005.11.014

85. Ayton GS, Voth GA (2010) Multiscale computer simulation of the immature HIV-1 virion. Biophys J 99:2757–2765. https://doi.org/10.1016/j.bpj.2010.08.018
86. Huber RG, Marzinek JK, Holdbrook DA, Bond PJ (2017) Multiscale molecular dynamics simulation approaches to the structure and dynamics of viruses. Prog Biophys Mol Biol 128:121–132. https://doi.org/10.1016/j.pbiomolbio.2016.09.010
87. Perilla JR, Goh BC, Cassidy CK et al (2015) Molecular dynamics simulations of large macromolecular complexes. Curr Opin Struct Biol 31:64–74. https://doi.org/10.1016/j.sbi.2015.03.007
88. Pappalardo M, Collu F, Macpherson J et al (2017) Investigating Ebola virus pathogenicity using molecular dynamics. BMC Genomics 18:566. https://doi.org/10.1186/s12864-017-3912-2
89. Nasution MAF, Toepak EP, Alkaff AH, Tambunan USF (2018) Flexible docking-based molecular dynamics simulation of natural product compounds and Ebola virus nucleocapsid (EBOV NP): a computational approach to discover new drug for combating Ebola. BMC Bioinf 19:419. https://doi.org/10.1186/s12859-018-2387-8
90. Zhang Y, Zheng Q-C (2019) What are the effects of the serine triad on proton conduction of an influenza B M2 channel? An investigation by molecular dynamics simulations. Phys Chem Chem Phys 21:8820–8826. https://doi.org/10.1039/C9CP00612E
91. Bowen LR, Li DJ, Nola DT et al (2019) Identification of potential Zika virus NS2B-NS3 protease inhibitors via docking, molecular dynamics and consensus scoring-based virtual screening. J Mol Model 25:194. https://doi.org/10.1007/s00894-019-4076-6
92. Wei X, Ghosh SK, Taylor ME et al (1995) Viral dynamics in human immunodeficiency virus type 1 infection. Nature 373:117–122. https://doi.org/10.1038/373117a0
93. Ho DD, Neumann AU, Perelson AS et al (1995) Rapid turnover of plasma virions and CD4 lymphocytes in HIV-1 infection. Nature 373:123–126. https://doi.org/10.1038/373123a0
94. Herrera-Ibatá DM, Pazos A, Orbegozo-Medina RA et al (2015) Mapping chemical structure-activity information of HAART-drug cocktails over complex networks of AIDS epidemiology and socioeconomic data of U.S. counties. Biosystems 132–133:20–34. https://doi.org/10.1016/j.biosystems.2015.04.007
95. Speck-Planche A, Dias Soeiro Cordeiro MN (2017) Speeding up early drug discovery in antiviral research: a fragment-based in silico approach for the design of virtual anti-hepatitis C leads. ACS Comb Sci 19:501–512. https://doi.org/10.1021/acscombsci.7b00039
96. González-Díaz H, Herrera-Ibatá DM, Duardo-Sánchez A et al (2014) ANN multiscale model of anti-HIV drugs activity versus AIDS prevalence in the US at county level based on information indices of molecular graphs and social networks. J Chem Inf Model 54:744–755. https://doi.org/10.1021/ci400716y
97. González-Díaz H, Prado-Prado F, Ubeira FM (2008) Predicting antimicrobial drugs and targets with the MARCH-INSIDE approach. Curr Top Med Chem 8:1676–1690. https://doi.org/10.2174/156802608786786543
98. Ramos de Armas R, González Díaz H, Molina R, Uriarte E (2004) Markovian Backbone Negentropies: molecular descriptors for protein research. I. Predicting protein stability in Arc repressor mutants. Proteins 56:715–723. https://doi.org/10.1002/prot.20159
99. Ghosh S, Nie A, An J, Huang Z (2006) Structure-based virtual screening of chemical libraries for drug discovery. Curr Opin Chem Biol 10:194–202. https://doi.org/10.1016/j.cbpa.2006.04.002
100. Plotkin JB, Dushoff J, Levin SA (2002) Hemagglutinin sequence clusters and the antigenic evolution of influenza A virus. Proc Natl Acad Sci U S A 99:6263–6268. https://doi.org/10.1073/pnas.082110799
101. Chowdhury KH, Chowdhury MR, Mahmud S et al (2020) Drug Repurposing approach against novel coronavirus disease (COVID-19) through virtual screening targeting SARS-CoV-2 main protease. Biology (Basel) 10:2. https://doi.org/10.3390/biology10010002
102. Jang WD, Jeon S, Kim S, Lee SY (2021) Drugs repurposed for COVID-19 by virtual screening of 6218 drugs and cell-based assay. Proc Natl Acad Sci 118:e2024302118. https://doi.org/10.1073/pnas.2024302118

103. Becerra A, Bucheli VA, Moreno PA (2017) Prediction of virus-host protein-protein interactions mediated by short linear motifs. BMC Bioinformatics 18:163. https://doi.org/10.1186/s12859-017-1570-7
104. Segura-Cabrera A, García-Pérez CA, Guo X, Rodríguez-Pérez MA (2013) A viral-human interactome based on structural motif-domain interactions captures the human infectome. PLoS ONE 8:e71526. https://doi.org/10.1371/journal.pone.0071526
105. Mukhopadhyay A, Maulik U, Bandyopadhyay S (2012) A novel biclustering approach to association rule mining for predicting HIV-1–human protein interactions. PLoS ONE 7:e32289. https://doi.org/10.1371/journal.pone.0032289
106. Mukhopadhyay A, Ray S, Maulik U (2014) Incorporating the type and direction information in predicting novel regulatory interactions between HIV-1 and human proteins using a biclustering approach. BMC Bioinformatics 15:26. https://doi.org/10.1186/1471-2105-15-26
107. Doolittle JM, Gomez SM (2010) Structural similarity-based predictions of protein interactions between HIV-1 and Homo sapiens. Virology Journal 7:82. https://doi.org/10.1186/1743-422X-7-82
108. Doolittle JM, Gomez SM (2011) Mapping protein interactions between dengue virus and its human and insect hosts. PLoS Negl Trop Dis 5:e954. https://doi.org/10.1371/journal.pntd.0000954
109. Santos FRS, Nunes DAF, Lima WG et al (2020) Identification of Zika virus NS2B-NS3 protease inhibitors by structure-based virtual screening and drug repurposing approaches. J Chem Inf Model 60:731–737. https://doi.org/10.1021/acs.jcim.9b00933
110. Nocedo-Mena D, Cornelio C, Camacho-Corona MDR et al (2019) Modeling antibacterial activity with machine learning and fusion of chemical structure information with microorganism metabolic networks. J Chem Inf Model 59:1109–1120. https://doi.org/10.1021/acs.jcim.9b00034
111. Speck-Planche A, Kleandrova VV, Luan F, Cordeiro MNDS (2012) A ligand-based approach for the in silico discovery of multi-target inhibitors for proteins associated with HIV infection. Mol BioSyst 8:2188–2196. https://doi.org/10.1039/C2MB25093D
112. Vásquez-Domínguez E, Armijos-Jaramillo VD, Tejera E, González-Díaz H (2019) Multi-output perturbation-theory machine learning (PTML) Model of ChEMBL data for antiretroviral compounds. Mol Pharm 16:4200–4212. https://doi.org/10.1021/acs.molpharmaceut.9b00538
113. Viña D, Uriarte E, Orallo F, González-Díaz H (2009) Alignment-free prediction of a drug-target complex network based on parameters of drug connectivity and protein sequence of receptors. Mol Pharm 6:825–835. https://doi.org/10.1021/mp800102c
114. Zheng W, Sun W, Simeonov A (2018) Drug repurposing screens and synergistic drug-combinations for infectious diseases. Br J Pharmacol 175:181–191. https://doi.org/10.1111/bph.13895
115. Schuler J, Hudson ML, Schwartz D, Samudrala R (2017) A systematic review of computational drug discovery, development, and repurposing for Ebola virus disease treatment. Molecules 22:E1777. https://doi.org/10.3390/molecules22101777
116. Ton A-T, Gentile F, Hsing M et al (2020) Rapid identification of potential inhibitors of SARS-CoV-2 main protease by deep docking of 1.3 billion compounds. Mol Inform 39:e2000028. https://doi.org/10.1002/minf.202000028
117. Awad IE, Abu-Saleh AA-AA, Sharma S et al. High-throughput virtual screening of drug databanks for potential inhibitors of SARS-CoV-2 spike glycoprotein. J Biomol Struct Dyn 1–14. https://doi.org/10.1080/07391102.2020.1835721
118. Mahdian S, Zarrabi M, Panahi Y, Dabbagh S (2021) Repurposing FDA-approved drugs to fight COVID-19 using in silico methods: Targeting SARS-CoV-2 RdRp enzyme and host cell receptors (ACE2, CD147) through virtual screening and molecular dynamic simulations. Inform Med Unlocked 23:100541. https://doi.org/10.1016/j.imu.2021.100541

Chapter 5
Targeted Computational Approaches to Identify Potential Inhibitors for Nipah Virus

Sakshi Gautam and Manoj Kumar

Abstract Nipah virus (NiV) is a bat-borne, highly pathogenic RNA virus belonging to the family *Paramyxoviridae*. It is known for causing lethal encephalitis in humans with a high fatality rate. With time, the world has faced numerous outbreaks in various regions such as Malaysia, Bangladesh, Philippines, and India. In this chapter, we have summarized experimentally tested antivirals and computational approaches to predict potential inhibitors against NiV. Various studies have been conducted in vitro and in vivo to find the potent novel molecules or repurposed drugs. This section describes the drug's primary function in case of repurposing, targets, screening method, inhibition efficiency, etc., from important studies. The computational section describes the approaches to identifying the potent inhibitors against NiV. These approaches include machine learning and QSAR-based prediction, molecular docking, molecular dynamics, integrated structure- and network-based approach, Drug–target–drug network-based approach, etc. In conclusion, this work will be helpful for the researchers in examining antivirals against NiV.

Keywords Nipah virus · Antivirals · Inhibitors · Computational · Prediction · Repurposed drugs

5.1 Introduction

Paramyxoviruses such as Nipah virus (NiV) and Hendra virus (HeV) are life-threatening human pathogens that cause a range of clinical manifestations from asymptomatic to symptomatic infections such as acute respiratory illness and severe

S. Gautam · M. Kumar
Virology Unit and Bioinformatics Centre, Institute of Microbial Technology, Council of Scientific and Industrial Research (CSIR), Sector 39A, Chandigarh 160036, India

S. Gautam · M. Kumar (✉)
Academy of Scientific and Innovative Research (AcSIR), Ghaziabad 201002, India
e-mail: manojk@imtech.res.in

S. Kar and J. Leszczynski (eds.), *Current Trends in Computational Modeling for Drug Discovery*, Challenges and Advances in Computational Chemistry and Physics 35,
https://doi.org/10.1007/978-3-031-33871-7_5

acute encephalitis [1]. The mortality rate of NiV infection is approximately 40%–75%, which varies depending on the clinical management and epidemiological surveillance of the outbreak location (https://www.who.int/news-room/fact-sheets/detail/nipah-virus). NiV infection in humans was first reported in September of the year 1998 in Malaysia [2]. NiV virus was isolated from the cerebrospinal fluid (CSF) of a patient in the Sungai Nipah village in 1999, therefore named this virus as "Nipah". From 2001 to 2013, various NiV outbreaks have occurred in Bangladesh during the winter season in the 20 districts. A small outbreak of NiV also took place in the Philippines in 2014. Various NiV outbreaks have occurred in India but not as frequently as in Bangladesh. A large outbreak has taken place in Siliguri, West Bengal (2001), and a small outbreak in Nadia district, West Bengal (2007) [3]. In May 2018, a NiV outbreak also occurred in Kozhikode and Malappuram districts, Kerala [4, 5]. Till now, the mortality rates of all the outbreaks are up to 91% including the outbreak in Kerala in 2018 [6]. The natural reservoir of the NiV is the fruit bats; transmission occurs from bats to humans, bats to pigs, pigs to humans, date palm sap to humans, and between humans, bats, pigs, etc. [7].

NiV genome is a non-segmented negative sense, single-stranded RNA (ssRNA) having a length of 18.2 kb, part of the genus *Henipavirus*, family *Paramyxoviridae*, subfamily *Paramyxovirinae*, and order *Mononegavirales* [8]. It is categorized as a Biosafety Level-4 (BSL-4) pathogen due to the high pathogenicity and lack of efficacious therapeutics or vaccines. It encodes six structural proteins, i.e., fusion protein (F), attachment glycoprotein (G), nucleoprotein (N), large protein or RNA polymerase protein (L), phosphoprotein (P), and matrix protein. P gene encodes four non-structural proteins, i.e., V, W, C, and P proteins [9, 10].

The incubation period for the NiV varies from 4 days to 2 weeks. Then, the primary clinical non-specific symptoms start appearing such as fever, headache, vomiting, dizziness, segmental myoclonus, hypertension and tachycardia, hypotonia and areflexia, and abnormal pupillary reflexes. In severe cases, patients start developing symptoms such as pulmonary syndrome, diffuse alveolar shadowing and acute respiratory distress syndrome (ARDS), motor deficits, reduced level of consciousness [11, 12]

NiV virus emerged in 1988, and various outbreaks have occurred with very high mortality rates over the last two decades. As it is a BSL-4 pathogen, highly pathogenic in nature, minimal studies are carried out with a live virus, and no licensed therapeutics or vaccines are available to treat NiV disease. Till now, no antiviral is under clinical trial for the treatment of NiV infection. So, there is an urgent need to develop strategies to deal with NiV outbreaks. Various studies have been carried out to identify the potent inhibitors in vitro and in vivo against NiV. In this chapter, we discussed the antiviral drug discovery for NiV. This chapter is majorly focused on two aspects, one is experimental novel or repurposed drugs and the other is the computational approaches to identify inhibitors for NiV.

5.2 Experimentally Tested Repurposed Drugs or Novel Molecules Against NiV

Drug repurposing is an approach used for identifying different uses of already existing licensed or investigational or approved medicines [13]. It provides several advantages like low risk of failure, reduced time frame, and low investment to develop repurposed drugs [14]. Therefore, several research groups made an attempt to discover potent drugs either using repurposed drugs or develop novel molecules against NiV. For example, Bitsch et al., in 2001, conducted an open trial in 194 patients; 140 out of 194 were treated with ribavirin, and 54 were treated as control. In the case of treated patients, 128 patients were given oral ribavirin as 2 gm on day 1, 1.2 gm on day 2–4, 1.2gm on day 5–6, and 0.6gm on day 7–10, and 12 patients were given intravenous ribavirin as 30 mg/kg, 16 mg/kg in every 6 h for 4 days, and then 8 mg/kg in every 8 h for 3 days. Forty-five deaths (~32%) were observed in the treated group, and 29 deaths (~54%) were observed in the control group. Some frequent side effects were also observed, such as teratogenic effects, anemia, and jaundice. Overall, treatment with ribavirin in patients with Nipah encephalitis reduced mortality by 36% [15].

Georges-Courbot et al., in 2006, tested the activity of ribavirin, 5-ethynyl analogue (EICAR) of ribavirin, 6-aza-uridine, pyrazofurin, and poly(I)-poly(C_{12}U) against NiV in vitro as well as in vivo. In in vitro testing, effective concentration of ribavirin, 5-ethynyl analogue (EICAR) of ribavirin, 6-aza-uridine, pyrazofurin, and poly(I)-poly(C_{12}U) were found to be 100 μg/ml, 1 μg/ml, 0.25 μg/ml, 0.12 μg/ml, and 6.25 μg/ml, respectively. Then, they performed an in vivo toxicity assay to test the toxicity of EICAR and 6-aza-uridine in hamsters. In the case of EICAR, all animals were found to be healthy with normal body weight gain, whereas, in the case of 6-aza-uridine, all animals lost weight and were found dead within 6 days. Due to the less availability of some of these compounds, further experiments could not be performed. In the first experiment of efficacy testing, they gave ribavirin dose of 50 mg/kg/day to six animals, 6-aza-uridine dose of 175 mg/kg/day to six animals, and phosphate-buffered saline as controls to six animals. The mean day of death was found to be 5.1 $\pm$ 0.7 days for controls, 6.1 $\pm$ 0.7 days for 6-aza-uridine, and 6.8 $\pm$ 0.7 days for ribavirin. As no therapeutic effect was observed for the ribavirin, they decided to use a lower concentration of ribavirin in the second experiment. They gave ribavirin dose of 50 mg/kg twice daily to six animals, poly(I)-poly(C_{12}U) dose of 3 mg/kg once daily to six animals, and placebo to six animals. The mean day of death was found to be 7.3 $\pm$ 2.94 days for controls, 8.6 $\pm$ 2.94 days for ribavirin, and all animals were alive till the 14th day for poly(I)-poly(C_{12}U). Seven samples (two controls, two ribavirin-treated animals, and three 6-aza-uridine treated animals) were collected after death for checking the presence of NiV. Seven samples (two controls, three ribavirin-treated animals, and two 6-aza-uridine treated animals) were collected within the initial 6 days of infection and examined for the presence of NiV. Only one sample of 6-aza-uridine treated was found positive for anti-NiV IgG in the samples collected within the initial 6 days of infection. Moreover, they checked the lungs, livers, spleens, kidneys, and brains of all dead animals for the presence of NiV by real-time RT-PCR and by measuring infectious viral titers. The virus was found to be present in all the dead animals by the real-time RT- PCR, and the viral titer was not detectable in three samples only. From the second experiment, one sample was collected from the ribavirin-treated animals, and six samples were collected from the poly(I)-poly(C_{12}U)-treated animals. One poly(I)-poly(C_{12}U)-treated animal was sick for some reason and therefore not euthanatized, and the other five poly(I)-poly(C_{12}U)-treated animals were euthanatized on the 30th day after infection. The virus was not detected in any of the five poly(I)-poly(C_{12}U)-treated animals, whereas viral RNA was found in the liver of three animals and the kidney of one animal [16].

Aljofan et al., in 2009, developed an assay of high throughput screening (HTS) for the identification of antivirals against live NiV and HeV. They single blindly screened a library containing 8040 low molecular weight compounds. For the screening purpose, they used 10 μM compounds in triplicates and 90% efficacy of compounds as the minimum criteria for lead compounds. They identified 28 molecules as lead compounds and further characterized their IC_{50} and CC_{50} values. Three compounds out of 28 lead compounds were pointed out as gentian violet (also called crystal violet), brilliant green (also called malachite green), and gliotoxin. All three compounds effectively blocked the NiV and HeV. IC_{50} values of gliotoxin and brilliant green were tenfolds lower than ribavirin, whereas IC_{50} values of gentian violet were fourfolds lower than ribavirin for NiV. IC_{50} values of gliotoxin and brilliant green were found to be threefolds lower than ribavirin. In contrast, gentian violet was found to be a bit less effective as compared to ribavirin for HeV. All three compounds were more cytotoxic than ribavirin in Vero cells. All three compounds were found to be more effective against NiV than HeV based on the therapeutic index (TI) value. Antivirals' potency was also checked against parent pseudotyped virus (pVSV), influenza H1N1 virus, human parainfluenza virus type 3(HPIV3), pseudotyped Nipah virus (pNiV), and pseudotyped Hendra virus (pHeV). Similar inhibition rates were observed for these viruses except for gliotoxin inhibition for influenza virus in a dose-dependent manner. Further, they performed the real-time PCR to check the response of these compounds on the inflammatory response profile, i.e., changes in the cytokines IL-8 and TNF-α levels. They observed that brilliant green increases the IL-8 and TNF-α levels by 15–20 folds. Gliotoxin and gentian violet induced IL-8 levels by twofold and decreased TNF-α levels. All three compounds are used in antibacterial and antifungal therapy. They are unsuitable for internal intake; therefore, they can be used as disinfectants, positive control for experimental studies, and topical antiviral applications [17].

Pallister et al. in 2009 evaluated chloroquine's effectiveness, a known antimalarial agent in preventing lethal NiV infection in vivo using ferrets. Earlier research showed that chloroquine could prevent NiV infection at a concentration of $\geq$1 μM in vitro. For *the* in vivo study, they give 25 mg/kg/day dose of chloroquine to three ferrets 24 h before viral exposure and give 25 mg/kg/day dose of chloroquine to three ferrets 10 h after viral exposure. One control ferret was given 20% sucrose 24 h before viral exposure, and one control ferret was given 20% sucrose 10 h after viral exposure. The virus isolate used in this study was NiV-Malaysia isolate EUKK 19817. They observed no change in the pathology, virus distribution, viral RNA levels, and replication process of the control and the treated animals. Chloroquine concentrations in serum were higher than their requirement for anti-NiV activity in vitro. On the whole, chloroquine was not an effective drug for treating NiV infection [18].

Freiberg et al., in 2010, tested the combined treatment of chloroquine and ribavirin in the golden hamster model infected with NiV and HeV. They first checked in vitro activity of chloroquine and ribavirin in HeLa cell lines. Effect of drug treatment was checked by measuring viral titers in the supernatants of cells infected with the virus. When chloroquine was given, IC_{50} was found to be 0.71 and 0.62 μM for HeV and NiV, and in the case of ribavirin, IC_{50} was found to be 4.96 and 4.18 μM for HeV and NiV. Then, they checked the in vivo activity of chloroquine and ribavirin in the golden hamster model infected with NiV and HeV. For this, five animals were given ribavirin (30 mg kg^{-1} in every 12 h), five animals were given chloroquine (50 mg kg^{-1} every day), and five animals were given a combination of both chloroquine and ribavirin. One control group is virus-infected animals having vehicle solution only, and the other control group has drugs only. Ribavirin-treated NiV-infected hamsters delay death by up to 5 days but cannot prevent death, whereas no positive impact was seen in ribavirin-treated HeV-infected hamsters. Chloroquine-treated NiV and HeV-infected hamsters neither delay nor prevent the death, whereas it reduces the survival by 3 days for NiV and 2 days for HeV. Chloroquine in combination with ribavirin cannot increase the mean survival times of the animals compared to the control group. There is no clear-cut effect observed in either case of treatment, i.e., individually or in combination, so they checked whether high doses of these drugs would be able to provide positive outcomes or not. They provided ribavirin and chloroquine in three different doses, i.e., 100, 150, and 200 mg kg^{-1} day^{-1} and 50, 100, or 150 mg kg^{-1} day^{-1}. In the case of ribavirin, 100 mg kg^{-1} day^{-1} was the most effective concentration for survival and delayed death by 3 days after infection. About 150 and 200 mg kg^{-1} day^{-1} were less effective as these concentrations only delayed death by 1 day. The maximum tolerable dose of ribavirin was between 150 and 200 mg kg^{-1} day^{-1}. In the case of chloroquine, 50 mg kg^{-1} day^{-1} was found to be the most effective concentration for survival and delayed death by 5 days after infection. About 100 and 150 mg kg^{-1} day^{-1} were found to be less tolerated and lead to death by day 2. Altogether, chloroquine and ribavirin individually or in combination could not protect virus-infected hamsters [19].

Mohr et al., in 2015, examined the 163 kinase inhibitors targeting the PI3K/Akt pathway, hypothesizing that cellular kinases are necessary for the replication process in various BSL-4 viruses. They initially tested all these inhibitors to check their ability to block Lassa virus (LASV). They found out that eight compounds including control exhibited anti-LASV activity. Then, these eight compounds were checked for their ability to inhibit replication against Marburg virus (MARV-GFP), Alkhurma hemorrhagic fever virus (AHFV), Nipah virus (NiV-luc recombinant reporter viruses), Ebola virus (EBOV-GFP), and Crimean Congo hemorrhagic fever virus (CCHFV). Two compounds-OSU-03012 (also called AR-12) and BIBX 1382 dihydrochloride out of eight compounds were effective against LASV as well as other viruses. OSU-03012 inhibited replication of LASV with an EC_{50} of 0.5 μM; CC_{50} in the range of 3.1–5.7 μM and SI lied between 6 and 11, respectively. It also inhibited EBOV-GFP,

NiV-luc, and MARV-GFP with an EC_{50} of 0.3, 0.4, and 0.3 μM, CC_{50} of 6.4, 8.2, and 7.1 μM, and SI of 23, 21, and 20, correspondingly. No inhibition effect of OSU-03012 was observed on AHFV and CCHFV. BIBX 1382 dihydrochloride inhibited replication of MARV-GFP, LASV, and EBOV-GFP with an EC_{50} of 1.8, 3.2, and 1.1 μM, CC_{50} of 29.1, 15.3, and 19.8 μM, and SI of 19, 6, and 18. No inhibition effect of BIBX 1382 dihydrochloride was observed on CCHFV, NiV-Luc, and AHFV. Further, they examined the effect of OSU-03012 and BIBX 1382 dihydrochloride on viral titers of wild-type EBOV and LASV. They observed that both were reducing wild-type EBOV and LASV titers by 2–3 logs in a dose-dependent manner. Then, to validate the antiviral effect of OSU-03012 and BIBX 1382, they measured the RNA of LASV in various cell lines using qRT-PCR. Both OSU-03012 and BIBX 1382 decreased RNA levels of LASV in a dose-dependent manner in all the cell lines. Moreover, they also studied the mechanism of action of OSU-03012 and BIBX 1382. In this, firstly they checked whether these compounds are inhibiting the entry step of EBOV and LASV with the help of HIV particles pseudotyped with the viral glycoproteins. OSU-03012 was not able to block the glycoprotein-mediated entry step of both the viruses, whereas BIBX 1382 was able to block the glycoprotein-mediated entry step of both the viruses. Secondly, they checked that OSU-03012 and BIBX 1382 affect the formation of virus-like particles (VLP) of LASV Z protein. Both OSU-03012 and BIBX did not inhibit the release of VLP of LASV Z protein [20].

Rashighi and Harris in 2017 checked the antiviral potency of 4′-Azidocytidine (R1479), known to inhibit dengue virus, hepatitis C virus (HCV), and respiratory syncytial virus (RSV) by blocking the RNA-dependent RNA polymerase (RdRP) function. Because of the RdRP binding domain structure conservation all-over multiple families of viruses, they explored the antiviral activity against NiV and HeV. Firstly, they checked the potential of R1479 to block the reporter activity of luciferase or recombinant GFP expressing NiVs. From this, EC_{50} values for NiV-Luc2AM and NiV-GFP2AM were found to be 1.12 and 1.64 μM. Then, they checked the potential of R1479 to block the cytopathic effect (CPE) induced by wild-type NiV and HeV infections in NCI-H358 cells at 72 h post-infection (hpi) using a cell viability reduction assay. From this, EC_{50} values for NiV and HeV were found to be 2.63 and 1.75 μM. They also checked the potential of R1479 to block other viruses of the Paramyxoviridae family such as measles virus, human parainfluenza virus, mumps virus. Secondly, they assessed the inhibition ability of R1479 to block henipaviruses using immunofluorescence assays. Lastly, they performed a virus titer reduction (VTR) assay to check the potential of R1479 to block NiV and HeV and found EC50 values of 1.53 and 2.41 μM, respectively. These multiple assays showed that R1479 could efficiently inhibit henipaviruses with lower EC_{50} values (μM) [21].

Dawes et al., in 2018, evaluated the potential of favipiravir against NiV and HeV in vitro and assessed its ability in vivo against NiV-M. Favipiravir (T-705, 6-fluoro-3-hydroxy-2-pyrazinecarboxamine) is a purine analogue, RdRp inhibitor, and is already approved for use against influenza strains in Japan. It also exhibited antiviral activity against various viruses such as bunyaviruses, filoviruses, flaviviruses, enteroviruses, arenaviruses, noroviruses, alphaviruses, and rhabdoviruses. It has been checked in vitro for paramyxoviruses such as measles virus, human parainfluenza virus

3, avian metapneumovirus, respiratory syncytial virus, human metapneumovirus (hMPV), and Newcastle disease virus and in vivo against human metapneumovirus (hMPV). In this study, they checked the ability of favipiravir to block the NiV-Bangladesh (NiV-B), HeV, NiV-Malaysia (NiV-M), recombinant NiV expressing Gaussia luciferase, and eGFP (rNiV-Gluc-eGFP) using virus yield reduction assay in vitro. From this assay, EC_{50} values for NiV-B, HeV, NiV-M, and rNiV-Gluc-eGFP were found to be 14.82, 11.71, 44.24, and 14.57 μM, respectively. Selectivity index (SI) for NiV-B, HeV, NiV-M, and rNiV-Gluc-eGFP was found to be >67.47, >85.39, > 22.60, and >66.63. EC_{90} values for NiV-B, HeV, NiV-M, and rNiV-Gluc-eGFP were found to be 15.87, 16.49, 123.8, and 16.25 μM. Then, they assessed the effect of delayed favipiravir treatment on NiV infection by measuring luciferase activity at different time periods. They observed that favipiravir is able to inhibit NiV infection effectively when it is added instantly. Further, they performed in vivo studies by giving favipiravir orally and subcutaneously in the Syrian hamster model. When favipiravir was given orally instantly after infection twice daily for 14 days, they observed high levels of viral P gene expression in controls compared to treated animals. Neutralizing antibody levels were measured in terms of $PRNT_{50}$, which were >80 and >1280 in two treated animals and <20 levels in three treated animals. These observations showed that orally favipiravir is highly effective in preventing NiV when given instantly after infection. When favipiravir was given subcutaneously once daily for 14 days, they observed similar results. Lastly, they evaluated the pathological changes associated with the favipiravir-treated NiV-M-infected hamster model. For this, they inspected the lung, spleen, and brain of animals taken during the study and the survivors at 42 days post-infection using H&E staining and IHC techniques targeting NiV nucleoproteins. Vehicle-treated animals showed various symptoms such as perivascular infiltration of inflammatory cells, presence of NiV antigens in endothelial cells, interstitial pneumonia with alveolar edema or hemorrhage in lungs, red pulp cord displayed necrotic areas scattered with mononuclear or reticular cells with NiV antigens in spleen and meningitis with infiltration of neutrophils and mononuclear cells in the brain. Favipiravir-treated animals did not show any critical disease-associated findings, and no NiV antigens were present in the lung, spleen, and brain of animals [22].

Lo et al., in 2019, checked the efficacy of remdesivir against NiV-Bangladesh (NiV-B) genotype in African green monkeys (AGM). Remdesivir is a nucleotide analog prodrug exhibiting broad-spectrum antiviral activity against filoviruses, paramyxoviruses, and coronavirus. In this study, the first groups with four AGMs (two males and two females) were infected with a lethal dose of NiV-B through the intratracheally and intranasally route and, after 24 h, treated with remdesivir once daily for 12 days. The second groups with four AGMs (two males and two females) were infected with a lethal dose of NiV-B through intratracheally as well as intranasally route and, after 24 h, treated with a vehicle only once daily for 12 days. Vehicle-treated animals initially developed mild respiratory symptoms and progress toward respiratory distress, increased respiration rate, and decreased oxygen saturation but no change in body weight and temperature. Remdesivir-treated animals showed reduced appetites and developed respiratory problems, but no change in body weight, temperature, respiration rate, and oxygen saturation levels. Only one remdesivir-treated animal was found to be positive for viral RNA by quantitative reverse transcription-polymerase chain reaction (qRT-PCR); all others were negative during the experiment. The virus was found in the blood of all vehicle-treated animals. For virological analysis, 35 tissues were taken from every animal, and high viral loads were observed for vehicle-treated animals on the 7th and 8th day post-infection (dpi). Viral RNA was found in 11 tissues taken from three remdesivir-treated animals and in the brain of one remdesivir-treated animal at 92dpi. Further, histological examination showed the slightest indication of focal perivascular inflammation and absence of NiV antigen checked by immunohistochemistry (IHC). The animal in which viral RNA is present in the brain showed mild-to-moderate mononuclear meningoencephalitis. Lately, they checked the virus-neutralizing titers in the serum of remdesivir-treated animals collected at different time points. They observed low virus-neutralizing titers in all the animals by 19 dpi [23].

Lo et al., in 2020, checked the in vitro activity of β-D-4′-chloromethyl-2′-deoxy-2′-fluorocytidine, known to treat pediatric and adult respiratory syncytial virus infection and are currently under phase I and phase II clinical trials. β-D-4′-chloromethyl-2′-deoxy-2′-fluorocytidine (ALS-8112) is a parent nucleoside of the methyl propionate derivative lumicitabine (ALS-008176, JNJ-64041575). Their blind screening against recombinant reporter NiV expressing ZsGreen1 fluorescent protein (rNiV-ZsG) led them to find ALS-8112 as a hit compound for testing. Firstly, they checked the ability of ALS-8112 to inhibit rNiV-ZsG based on a green fluorescence signal, and they observed that ALS-8112 was able to block rNiV-ZsG in a dose-dependent manner with CC_{50} of >100 μM. They also checked the potential of ALS-8112 to inhibit other human respiratory viruses such as GFP expressing recombinant RSV (rgRSV224), measles virus (rMVEZ-GFP), human parainfluenza virus 3 (hPIV3-GFP), and rNiV-ZsG. They used two human respiratory epithelial cell lines—NCI-H358 and HSAEC1-KT. They checked the ability of ALS-8112 to inhibit viruses based on green fluorescence signal and found that it inhibited rgRSV224 in both the cell lines with EC50 of 1.23 μM in NCI-H358 and 0.36 μM in HSAEC1-KT cells. It also inhibits rNiV-ZsG with EC_{50} of 0.56 μM in NCI-H358 and 0.84 μM in HSAEC1-KT cells, whereas very low potent for hPIV3-GFP and rMVEZ-GFP(3) in both the cell lines. Secondly, as it is known that NiV produces in vitro cytopathic effect (CPE), so they checked the ability of ALS-8112 to block the CPE produced by wild-type NiV-Malaysia (NiV-M) genotype, NiV-Bangladesh (NiV-B) genotype, and rNiV-ZsG and measured by the cellular ATP levels luminescence. They found that it inhibited NiV-produced CPE in both the cell lines with EC_{50} in the range of 0.89–3.08 μM, CC_{50} of above 50 μM, and also calculated SI using CC50 and EC50 values. Lastly, they checked the ability of ALS-8112 to reduce the infectious virus titers against NiV-B and rNiV-ZsG. From this, they observed a 6 or 7 times reduction in virus titers of NiV-B and rNiV-ZsG, respectively. Moreover, they also captured fluorescence micrographs of infected cells with rNiV-ZsG with different concentrations of ALS-8112 at 48 hpi. In this, they observed that infected cells were decreasing in a dose-dependent manner and completely ablated infected cells at the concentration of 12.5 μM. Similarly, they checked the ALS-8112 toxicity in different cell lines such as primary human peripheral blood mononuclear (PBM), human epithelial lung (A549), human lymphoblastoid (CEM), human hepatocellular carcinoma cells (HepG2), and Vero cells. There was no toxicity observed in HepG2, Vero cells, and A549 cells, whereas toxicity is observed in PBM and CEM cells with CC_{50} of 4.2 μM and 2.8 μM, respectively. In vivo testing of nucleoside analogs generally results in various side effects such as pancreatitis, anemia, lactic acidosis, neutropenia. Due to these side effects, they measured the mitochondrial toxicity, bone marrow toxicity, and lactic acid production, as these are stress markers of the cells [24].

Several groups also used other methods to find effective therapy such as monoclonal antibodies, peptides, conjugated peptides, RNA interference (RNAi). Like Zhu et al., in 2006, reported the potent neutralizing humanized monoclonal antibodies (hMAbs) against the viral envelope glycoprotein (G) of NiV and HeV. They used soluble, purified, and oligomeric HeV G as the antigen for screening an extensive naive phage display library to identify potent antibodies. After multiple rounds of the panning process, seven hMAbs were selected based on the binding activity to a soluble HeV G. They performed a cell fusion assay to assess the potential of these antibodies in inhibiting entry and membrane fusion. Fab m101 was found to be the most potent cell fusion inhibitory activity, whereas m102 and m106 were found to display cross-reactivity. Conversion of m101 to Immunoglobulin G1 (IgG1) was responsible for exceptionally high cell fusion inhibition activity. About 12.5 μg/ml of IgG1 m101 was required to neutralize 100% infectious HeV, and only 1.6 μg/ml of IgG1 m101 was required to neutralize 98% infectious HeV. m101, m102, and m103 antibodies were competing, thus suggesting that these antibodies exhibit overlapping epitopes. This study showed that these humanized antibodies are new immunotherapy for treating HeV and NiV [25].

Similarly, Dang et al., in 2019, reported the antibody that targets the fusion glycoprotein (F) to inhibit the NiV and HeV infection. They performed the cloning, sequencing, and generated humanized antibody of 5B3 (h5B3.1). They also performed a neutralization assay showing that 5B3 and h5B3.1 potently inhibited NiV and HeV infection. Further, they determined the structure of the 5B3 antibody complex with NiV-F trimer using cryogenic-electron microscopy (cryo-EM). Complex structural analysis showed that 5B3 antibodies recognize perfusion-specific different epitopes, which are conserved in both NiV and HeV F. Overall, 5B3 antibody could be an effective therapy against Henipaviruses (HNVs) [26].

Likewise, Dang et al., in 2021, reported the neutralizing mouse monoclonal antibodies, i.e., 12B2 and 1F5, against the NiV and HeV fusion glycoprotein (F). They showed that both 12B2 and 1F5 antibodies bind the F protein with strong affinity and neutralize NiV and HeV in the BSL-4 containment. They determined the structure of the 12B2 antibody complex with NiV-F and 1F5 antibody complex with HeV F at 2.9 and 2.8 Å resolutions using cryogenic-electron microscopy (cryo-EM). Complex structural analysis showed that both the antibodies recognize perfusion-specific different epitopes, which are conserved in both NiV and HeV F. Further, they performed membrane fusion assay, which showed that both 12B2 and 1F5 antibodies hold the F protein in the perfusion conformation and block the changes that are essential for the membrane fusion. They also generated humanized 12B2 (h12B2) and 1F5 (h1F5) antibodies and assessed their neutralization ability against NiV-Malaysia (NiV-M), NiV-Bangladesh (NiV-B), and HeV. For this, they calculated IC_{50} using plaque reduction assay to determine virus neutralization with the increasing concentration of antibodies. They found that IC50 is in the range of 0.4–3.6 μg/ml for 12B2–h12B2 and 0.2–1.3 μg/ml for 1F5–h1F5 against NiV-B, NiV-M, and HeV. This study showed that antibodies could be an effective therapy against henipaviruses (HNVs) [27].

Mathieu et al., in 2018, engineered the new antiviral lipopeptides and assessed their efficacy in vitro and in vivo against NiV. They used the known "VIKI" sequence to engineer new antiviral lipopeptides and developed various lipopeptides such as VIKI-dPEG4-Toco, VIKI-dPEG4-Chol, VIKI-dPEG4, VIKI-dPEG4-bisToco, and VIKI-dPEG4-bisChol. Then, they checked these lipopeptides' protease sensitivity and observed that the presence of cholesterol or tocopherol increases the resistance to degradation by proteases. Based on this observation, they used VIKI-dPEG4-Toco and VIKI-dPEG4-Chol for further experiments. They correlated the effectiveness of VIKI-dPEG4-Toco and VIKI-dPEG4-Chol in hamsters against NiV. For this, they infect the hamsters with 10^6 pfu NiV and then treat intranasally with 10 mg/kg peptide or vehicle on days—1, 0, and 1 of infection. They observed that treatment with peptides improves survival and 100% death in untreated animals. Based on biodistribution, VIKI-dPEG4-Toco was further used to check the effectiveness of VIKI-dPEG4-Toco in African green monkeys (AGMs). For this, they infect the hamsters intratracheally with 2×10^7 pfu of NiV, then treated with 10 mg/kg peptide intratracheally and 2 mg/kg peptide subcutaneously on days—1, 0, and daily for 5 days. They observed that treatment with peptides leads to the protection from lethal outcomes. This study showed that conjugated peptides could effectively treat lethal NiV infection [28].

Mungall et al., in 2008, designed eight (four large polymerase (L) and four nucleocapsid (N) gene-specific) siRNA molecules against NiV, two N gene-specific against HeV, and two siRNA molecules as a control. They tested the ability of these siRNA molecules to block a henipavirus minigenome replication system and live virus in vitro. Three out of four L gene-specific siRNA could inhibit replication using the minigenome system. N gene-specific siRNA inhibits only live virus replication, indicating that targeting early expressed gene transcripts is more effective than late expressed gene transcripts. siRNA molecules targeting NiV infection were only partially effective in inhibiting HeV infection. Overall, this study illustrated that inhibiting henipavirus by RNA interference approach could be an effective therapy [29].

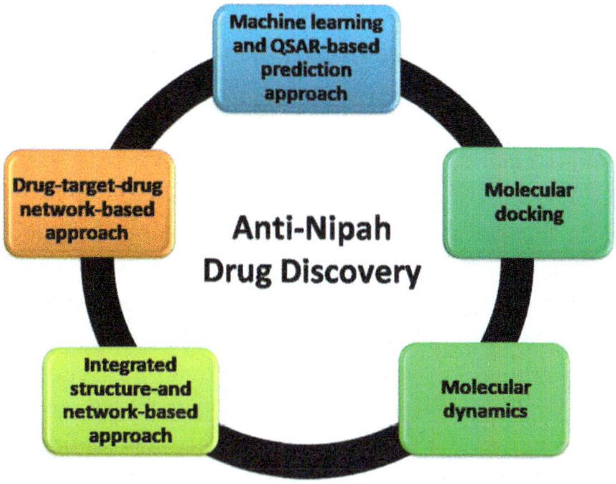

Fig. 5.1 Computational/in silico-based approaches to identify the potent drugs against NiV

5.3 Computational Approaches for the Identification of Antiviral Drugs for NiV

Many research groups made an effort to find the potential drugs using novel and repurposed drugs but still lack licensed drugs or vaccines, so there is a need to speed up the drug discovery process using various computational approaches. Several groups have used approaches as given in Fig. 5.1 to find the effective antivirals against NiV.

5.4 Machine Learning and QSAR-Based Prediction Approach

Machine learning is one of the important approaches in identifying inhibitors against various viruses. Several machine learning-based antiviral predictors using quantitative structure activity relationship (QSAR) information of molecules/peptides are available such as Anti-Ebola [30], anticorona [31], HIVprotI [32], anti-flavi [33], AVCpred [34], AVP-IC50 Pred [35], AVPpred [36]. Further, there are various antiviral databases which are available such as DrugRepV [37], AVPdb [38]. For NiV, Rajput et al., in 2019, developed a QSAR-based predictor and integrated it into a user-friendly web server "Anti-Nipah". Three hundred and thirteen experimentally tested chemicals were extracted from literature, and finally, 95 non-redundant chemicals with their IC_{50} values were converted into pIC_{50}. Simplified molecular-input line-entry systems (SMILES) of these chemicals were converted into a 3D-standard data format (3D-SDF). Further 3D-SDF was used to calculate 17,967 descriptors with the help of PaDel software. Then, these descriptors were used to fetch the most essential 42 features using "RemoveUseless" followed by "CfsSubsetEval". The model was developed using a support vector machine (SVM) through a tenfold cross-validation technique using these features. The model performance was evaluated using Pearson's correlation coefficient (PCC), Root mean absolute error (RMSE), Coefficient of determination (R2), and Mean absolute error (MAE). Training/testing and independent validation dataset showed the PCC of 0.82 and 0.92 during tenfold cross-validation. The applicability domain analysis by William's plot and scatter plot checked the robustness of the developed model. William's plot showed that all the points of training/testing and independent validation dataset lie in the range of threshold values of standard residues and leverage. A scatter plot between actual pIC_{50} and predicted pIC_{50} showed that all the points of training/testing and independent validation dataset lie near the trendline. Decoy set and chemical clustering using RApid DEcoy Retriever (RADER) software and compounds-specific bioactivity dendrogram (C-SPADE), respectively, also validated models robustness. Chemical clustering analysis displayed that these compounds are diverse in nature. Highly effective compounds with low IC_{50} formed clusters together and vice-versa. Some highly and less effective compounds were also clustered together. "Anti-Nipah" web server can be used for extracting information like inhibitors of the NiV given in the literature and patents, predicting the antiviral activity of a query molecule, and drawing the structure of query molecules [39].

5.5 Molecular Docking

Several research groups used molecular docking method to find the potent molecules against NiV. Molecular docking is a kind of modeling approach to understand the ligand and its target interactions [40]. Using this approach, Lipin et al., in 2021, examined the structure–property relationship of favipiravir, which is known to exhibit exemplary in vitro activity against NiV, and designed a series of 15 piperazine-substituted favipiravir derivatives and then computationally screened their interaction and ability to inhibit NiV-G protein. Density functional theory analysis was done to calculate all the derivatives' geometrical features and electronic properties. ADMET and toxicity analysis were done using the SWISSADME server and ProTox-II online tool, respectively. All the derivatives satisfied the Lipinski rule of five, high gastrointestinal absorption (GI), good solubility, and non-toxic, thus showing good oral bioavailability. PIC_{50} values were predicted using a web server, which showed that predicted PIC_{50} values for favipiravir derivatives were better than the parent favipiravir. Further molecular docking analysis and understanding of the binding mode and the affinity of favipiravir derivatives toward the NiV-G (PDB ID-3D11) were done using Maestro glide docking program and GLIDE—6.6. This study showed that piperazine-substituted favipiravir derivatives could be promising inhibitors against NiV [41].

Similarly, James et al., in 2021, prepared 22 derivatives of favipiravir having pyrazine and other heterocyclic groups as moiety Nipah glycoprotein, i.e., 3D11 was taken from protein data bank (PDB) for novel inhibitors against NiV using various computational approaches. Molecular docking studies were performed using all 22 derivatives and Nipah glycoprotein, i.e., 3D11, to determine the various conformations of these complexes. Further analysis showed that 13 derivatives have higher docking scores than the standard favipiravir; thus, it is proposed that these derivatives might have a perfect affinity for the NiV proteins. Docking scores for compound 5_Favipiravir, 4_Favipiravir, and 19_Favipiravir were found to be −6.16, −5.50, and −5.38 kcal/mol, respectively. These three compounds had pyrazole, imidazole, and pyrazinone as heterocyclic groups. Further physicochemical properties' analysis showed that all derivatives' properties lie in the expected value range, thus exhibiting good oral bioavailability. Lastly, in silico ADMET studies showed that derivatives have good scores for human oral absorption, human serum albumin binding, Caco-2 permeability, total solvent accessible surface area, etc. [42].

5.6 Molecular Dynamics

Some groups used molecular dynamics approach like Sen et al., in 2019, who used various in silico approaches like homology modeling or ab initio modeling, peptide designing, and molecular docking. Out of nine NiV proteins, four proteins, i.e., F, N, G, and P partial structures, were taken from the protein data bank (PDB), and using these structures, models were developed for the five proteins, i.e., M, W, V, L and F proteins with the help of homology modeling, ab initio modeling, and threading. They also contribute about 90% to the structural characterization of NiV proteins compared to the structural data available in the PDB of NiV proteins. With the help of these models, they designed four potent peptide inhibitors (one against F protein trimer, one against M protein dimer, and two against G-Protein-human ephrin-B2 receptor). Three independent molecular dynamics simulations were performed of 100 ns to check the stability of these four protein–peptide complexes. They also screened 22,685 compounds of the ZINC library against NiV proteins using AutoDock4 and Dock6.8 programs. Finally, predicted 146 small molecules as inhibitors that can bind G, F, P, N, and M NiV proteins and then cut down to 13 molecules. Three independent molecular dynamics simulations were performed of 50 ns to check the stability of these 13 protein–peptide complexes. They also determined the binding energies of these complexes with the help of MM/PBSA and found out that nine complexes had negative energy, one had positive energy, and three could not bind the proteins. Molecular docking studies also showed that few proposed inhibitors are already tested as repurposed drugs. For example—ZINC04829362 (Cyclopent-1-ene-1,2-dicarboxylic acid) is already known as an antiasthmatic and antipsoriatic drug and ZINC12362922 (Bicyclo[2.2.1]hepta-2,5-diene-2,3-dicarboxylic acid) known drug for depression and Parkinson's disease. Furthermore, they checked the effectiveness of recommended inhibitors against 15 (seven Malaysian, three Bangladeshi, and five Indian) strains of the NiV available in the NCBI Database. They checked the variations with the help of Multiple Sequence Alignment using MUSCLE and reduced it to only those variations that were in immediate contact with the inhibitors. Out of five residues changes (Lys236Arg, Asp188Glu, Gln211Arg, Asp252Gly, and Ile331Val), four were conservative substitutions, and one (Asp252Gly) was found to be a non-conservative change. Results showed that these recommended inhibitors could be potential antivirals against every single NiV strain [43].

Ropón-Palacios et al., in 2020, used virtual screening techniques like molecular docking and molecular dynamics to identify the potential novel antivirals against the NiV. One hundred and eighty-three ligands were taken from "The Pathogen Box Medicines for Malaria Venture (MMV, Geneva, Switzerland)" and allowed them to interact with NiV glycoprotein (NiV-G). NiV-G is involved in the entry of the virus by binding to the Ephirin-B2 (EFNB2) and Ephirin-B3 (EFNB3) receptors present on the surface of the host cells, therefore a promising target for inhibiting the virus infection. Of 183 ligands, three (MMV020537, MMV019838, and MMV688888) were potent inhibitors with binding energies of −11.8, −9.5, and −9.2 kcal/mol using a virtual screening approach. To refine the results of virtual screening, molecular docking was carried out with the Lamarkian hybrid genetic algorithm available in AutoDock. MMV020537 showed the binding energy of −14.29 kcal/mol (Kd = 0.03 nM), MMV019838 showed the binding energy of −10.23 kcal/mol (Kd = 31.61 nM), and MMV688888 showed the binding energy −11.82 kcal/mol (Kd = 2.18 nM). In both virtual screening and molecular docking, Ligand 1, i.e., MMV020537, had the lowest interaction energy. They also identified the two new residues, i.e., Cys240 and Arg236, present in the binding site and involved in the ligand recognition at 3.1 and 1.9 Å. Validation of molecular docking was done using X-Score as well as PLANTS software. Further molecular dynamics simulation studies showed that a complex formed between first ligand and NiV-G protein is found to be stable during production time (40 ns) [44].

Kalbhor et al., in 2021, took three chemical library databases—Asinex-Antiviral Library (8722), Enamine-Antiviral Library (3700), and ChemDiV-Antiviral Library (67,470) containing a total of 79,892 chemicals and NiV-G protein complex bound with cell surface receptor ephrin-B2 (PDB ID-2VSM) available in the Protein data bank (PDB) which were taken for analysis. Multi-step molecular docking analysis was performed to get those compounds docked with NiV-G protein using Glide-HTVS, Glide-SP, and Glide-XP. For further analysis, 299 compounds were selected based on the XP dock score and MM-GBSA score. Further pharmacokinetic analysis like ADME and synthetic accessibility properties have been carried out using 299 compounds. About 207 out of 299 compounds were found to be good with drug-likeness parameters like absorption, distribution, metabolism, and excretion. Then, toxicity-based analysis was performed using TOPKAT and found that 14 compounds are non-toxic in nature. Moreover, their molecular binding modes and intermolecular interactions were checked to cut down the compounds, and the final five compounds were obtained as NiV-G protein modulators. Then, they deeply explored the molecular binding interaction analysis of NiV-G protein and proposed inhibitors using the XP-docking method and protein–ligand interaction profiler (PLIP) tool. Molecular binding interaction analysis showed that Tyr581 residue of NiV-G protein was found to be a common residue involved in the H-bonding with various compounds, and Lys560, Gln559, Tyr581, Val507, Glu579, and Ile588 residues of NiV-G protein were found to be common residues involved in the hydrophobic interactions with G1, G2, G3, and G5 compounds. Moreover, they checked the ADME parameters, which showed that all five compounds exhibited drug-like properties like molecular weight of less than 500 g/mol, moderate-to-high soluble nature, orally active, good synthetic accessibility score, etc. Toxicity profiling using TOPKAT tool suggested that all five compounds are non-carcinogenic, non-toxic, and non-mutagenic. They also built an EGG-BOILED model to analyze two more essential parameters, HIA (Human Intestinal Absorption) and BBB (Blood–Brain Barrier). Further, they performed an MD simulation analysis of 100 ns to evaluate the stability, as well as the dynamic behavior of these NiV-G protein complexes using various parameters like root mean squared deviation (RMSD), radius of gyration (RoG), and root mean square fluctuation (RMSF). They also determined Molecular Mechanics Poisson–Boltzmann Surface Area (MM-PBSA) to determine binding free energies (ΔG) from all the MD simulations to deduce the energy contribution of recommended inhibitors in stabilizing the NiV-G protein complexes. Potential proposed inhibitors exhibit high negative ΔG values in the range of -166.246 to -226.652 kJ/mol and showed strong affinity toward the NiV-G protein complex [45].

Ahmed Bhuiyan et al., in 2022, took 92 compounds from Ambinter, but two compounds were duplicates, so the final 90 compounds and NiV-G protein (PDB ID-2VSM) from the RCSB protein data bank were used in this study. They predicted the active site of NiV-G with the help of the Computed Atlas for Surface Topography of Proteins (CASTp) server. Then, they performed molecular docking analysis using AutoDock Vina in which the top five compounds (CID: 11,096,158, CID: 11,861,102, Amb35795905, CID: 102,601,745) were selected, including control (CID: 24,139) based on binding affinity scores for further analysis. Pharmacokinetic properties and toxicity profiling of all five compounds were determined utilizing the SwissADME server and pkCSM server, which showed that all five compounds are suitable and non-toxic. Further molecular dynamics (MD) studies showed that all protein–ligand complexes are stable, but Amb33921182, i.e., 2- acetamido-2-deoxy-D-gluco-hexopyranose compound is the most potent drug candidate. They also calculated root mean square fluctuation (RMSF), root mean square deviation (RMSD), ligand properties, and protein–ligand contacts (P-L contact) [46].

Vinay Randhawa et al. in 2022 performed multi-target molecule screening using molecular docking and molecular dynamics approach to find the potential anti-NiV drugs targeting NiV-F, NiV-N, and NiV-G proteins. Potent known NiV inhibitors such as drugs, phytochemicals, and small molecules were extracted from the literature search using PubMed. Three-dimensional structures of drugs and small molecules were drawn manually using MarvinSketch v5.10.0 software (chemaxon.com), and 3D structures of phytochemicals were taken from Serpentina database. Three-dimensional structures of target proteins, i.e., NiV-F (PDB ID-5EVM), NiV-G (PDB ID-2VSM), and NiV-N (PDB ID-4CO6), were downloaded from RCSB protein data bank (PDB). Molecular docking studies were performed using QuickVina v 2.0 software, and then eight molecules (two chemicals and six phytochemicals) were selected for further analyses based on binding energy threshold values. ADME and pharmacokinetic properties were computed using ADMETlab web server, and three molecules were selected, i.e., two phytochemicals' molecules—CARS0358 and RASE0125—and one chemical-ND_nw_193.2 based on the z-score. Further molecular dynamics simulations were carried out with target only and with protein–ligand complexes for 5 ns. As a whole, two phytochemicals, i.e., CARS0358 (NA) and RASE0125 (17-O-Acetyl-nortetraphyllicine) were found to inhibit all the three targets of NiV, whereas one chemical-ND_nw_193 (RSV604) was found to inhibit NiV-N and NiV-G. CARS0358 (NA) and RASE0125 (17-O-Acetyl-nortetraphyllicine) are indole alkaloids that inhibit Zika and dengue virus infection. ND_nw_193 (RSV604) is a chemical drug that inhibits the human respiratory syncytial virus (RSV) [47]

5.7 Integrated Structure- and Network-Based Approach

Few researchers used integrated structure- and network-based approach like Pathania et al., in 2020, who used an integrated structure- and network-based drug discovery approach to identify the potential entry inhibitors for the NiV. For molecular docking, NiV-G crystal structure (2VSM) was downloaded from the protein data bank (PDB), and a small molecule library was prepared using 2327 Food and Drug Administration-approved drugs (FDA-approved drugs) taken from the DrugBank database. Then, structural optimization was performed with the help of the MMFF94 force field available in the OpenBabel v 2.4.0 program. Moreover, applying reasonable charges and hydrogen was added using AutoDock tools. For validating the molecular docking approach, CCDC/ASTEX datasets containing 305 protein–chemical complexes were refined to 265 complexes. Docking simulations utilized rigid receptors, flexible ligands, and a grid box around the binding pocket. Successful solutions were determined based on the best ligand mode with RMSD ≤ 2.0 Å compared to the experimental conformation. Then, four known molecules were taken from the CHEMBL database and docked with NiV-G to assess their docking protocol accuracy. Then, structure-optimized FDA-approved drugs were taken for screening against the NiV attachment glycoprotein (NiV-G) utilizing the molecular docking method. Seventeen drugs were found to be potent inhibitors against Nipah virus and then narrowed down to three novel inhibitors—nilotinib, acetyldigitoxin, and deslanoside following topological analysis of chemical–protein interaction network, formed by integrating drug–target network, human protein–protein interaction network, and NiV–human interaction network. Both acetyldigitoxin and deslanoside were previously known to be in the category of NiV inhibitors. In contrast, nilotinib is a part of benzanoids class, which was previously not identified as NiV inhibitors [48].

5.8 Drug–Target–Drug Network-Based Approach

Few researchers also used drug–target–drug network-based approach like Rajput et al., in 2020, who identified the repurposed drugs against 14 epidemic/pandemic causing viruses, including NiV, through drug-target-drug network analysis. In this, they manually extracted out drugs and their targets, which are already experimentally validated either in vitro or in vivo. Then, these extracted drug targets were used for fetching out new potent repurposed drugs. Then, prioritize the identified repurposed drugs based on confidence score, i.e., the number of drug targets mapped in the repurposed drugs divided by the total number of targets mapped to experimentally validated drugs. Sixteen repurposed drugs were found shared between NiV and HeV. Further, they performed the pathway analysis using the KEGGREST package in R/Bioconductor, which showed that most drug targets participated in cancer signaling pathways. Lastly, they performed molecular docking to validate and prioritize the identified repurposed drugs [49].

In conclusion, we have recapitulated the experimentally tested antivirals studies as well as the in silico approaches studies, which will be helpful for the researchers in antiviral drug discovery against NiV.

References

1. Eaton BT, Broder CC, Middleton D, Wang LF (2006) Hendra and Nipah viruses: different and dangerous. Nat Rev Microbiol 4(1):23–35. https://doi.org/10.1038/nrmicro1323
2. Pillai, V. S., Krishna, G., & Veettil, M. V. (2020). Nipah virus: past outbreaks and future containment. In: Viruses, vol 12, Issue 4. MDPI AG. https://doi.org/10.3390/v12040465
3. Banerjee S, Gupta N, Kodan P, Mittal A, Ray Y, Nischal N, Soneja M, Biswas A, Wig N (2019) Nipah virus disease: a rare and intractable disease. In: Intractable and rare diseases research, vol 8, Issue 1, pp 1–8. International Advancement Center for Medicine and Health Research. https://doi.org/10.5582/irdr.2018.01130
4. Aditi, Shariff M (2019) Nipah virus infection: a review. In: Epidemiology and infection, vol 147. Cambridge University Press. https://doi.org/10.1017/S0950268819000086
5. Skowron K, Bauza-Kaszewska J, Grudlewska-Buda K, Wiktorczyk-Kapischke N, Zacharski M, Bernaciak Z, Gospodarek-Komkowska E (2022) Nipah virus–Another threat from the world of zoonotic viruses. In: Frontiers in microbiology, vol 12. Frontiers Media S.A. https://doi.org/10.3389/fmicb.2021.811157
6. Arunkumar G, Chandni R, Mourya DT, Singh SK, Sadanandan R, Sudan P, Bhargava B (2019) Outbreak investigation of Nipah virus disease in Kerala, India, 2018. J Infect Dis 219(12):1867–1878. https://doi.org/10.1093/infdis/jiy612
7. Singh RK, Dhama K, Chakraborty S, Tiwari R, Natesan S, Khandia R, Munjal A, Vora KS, Latheef SK, Karthik K, Singh Malik Y, Singh R, Chaicumpa W, Mourya DT (2019) Nipah virus: epidemiology, pathology, immunobiology and advances in diagnosis, vaccine designing and control strategies—A comprehensive review. Veterinary Quarterly 39(1):26–55. https://doi.org/10.1080/01652176.2019.1580827
8. Ksiazek TG, Rota PA, Rollin PE (2011) A review of Nipah and Hendra viruses with an historical aside. In: Virus research, vol 162, issues 1–2, pp 173–183. https://doi.org/10.1016/j.virusres.2011.09.026

9. Harcourt BH, Tamin A, Ksiazek TG, Rollin PE, Anderson LJ, Bellini WJ, Rota PA (2000) Molecular characterization of Nipah virus, a newly emergent paramyxovirus. Virology 271(2):334–349. https://doi.org/10.1006/viro.2000.0340
10. Sun B, Jia L, Liang B, Chen Q, Liu D (2018) Phylogeography, transmission, and viral proteins of Nipah virus. Virologica Sinica 33(5):385–393. https://doi.org/10.1007/s12250-018-0050-1
11. Ochani RK, Batra S, Shaikh A, Asad A (2019) Nipah virus the rising epidemic: a review. Infezioni in Medicina 27(2):117–127
12. Lo MK, Rota PA (2008) The emergence of Nipah virus, a highly pathogenic paramyxovirus. J Clin Virol 43(4):396–400. https://doi.org/10.1016/j.jcv.2008.08.007
13. Ashburn TT, Thor KB (2004) Drug repositioning: identifying and developing new uses for existing drugs. Nat Rev Drug Discovery 3(8):673–683. https://doi.org/10.1038/nrd1468
14. Pushpakom S, Iorio F, Eyers PA, Escott KJ, Hopper S, Wells A, Doig A, Guilliams T, Latimer J, McNamee C, Norris A, Sanseau P, Cavalla D, Pirmohamed M (2018) Drug repurposing: progress, challenges and recommendations. In: Nature reviews drug discovery, Vol. 18, issue 1. Nature Publishing Group, pp 41–58. https://doi.org/10.1038/nrd.2018.168
15. Chong HT, Kamarulzaman A, Tan CT, Goh KJ, Thayaparan T, Kunjapan SR, Chew NK, Chua KB, Lam SK (2001) Treatment of acute Nipah encephalitis with ribavirin. Ann Neurol 49(6):810–813. https://doi.org/10.1002/ana.1062
16. Georges-Courbot MC, Contamin H, Faure C, Loth P, Baize S, Leyssen P, Neyts J, Deubel V (2006) Poly(I)-poly(C12U) but not ribavirin prevents death in a hamster model of Nipah virus infection. Antimicrob Agents Chemother 50(5):1768–1772. https://doi.org/10.1128/AAC.50.5.1768-1772.2006
17. Aljofan M, Sganga ML, Lo MK, Rootes CL, Porotto M, Meyer AG, Saubern S, Moscona A, Mungall BA (2009) Antiviral activity of gliotoxin, gentian violet and brilliant green against Nipah and Hendra virus in vitro. Virol J 6:1–13. https://doi.org/10.1186/1743-422X-6-187
18. Pallister J, Middleton D, Crameri G, Yamada M, Klein R, Hancock TJ, Foord A, Shiell B, Michalski W, Broder CC, Wang L-F (2009) Chloroquine administration does not prevent Nipah virus infection and disease in ferrets. J Virol 83(22):11979–11982. https://doi.org/10.1128/jvi.01847-09
19. Freiberg AN, Worthy MN, Lee B, Holbrook MR (2010) Combined chloroquine and ribavirin treatment does not prevent death in a hamster model of Nipah and Hendra virus infection. J Gen Virol 91(3):765–772. https://doi.org/10.1099/vir.0.017269-0
20. Mohr EL, McMullan LK, Lo MK, Spengler JR, Bergeron É, Albariño CG, Shrivastava-Ranjan P, Chiang CF, Nichol ST, Spiropoulou CF, Flint M (2015) Inhibitors of cellular kinases with broad-spectrum antiviral activity for hemorrhagic fever viruses. Antiviral Res 120:40–47. https://doi.org/10.1016/j.antiviral.2015.05.003
21. Hotard AL, He B, Nichol ST, Spiropoulou CF, Lo MK (2017) 4′-Azidocytidine (R1479) inhibits henipaviruses and other paramyxoviruses with high potency. Antiviral Res 144:147–152. https://doi.org/10.1016/j.antiviral.2017.06.011
22. Dawes BE, Kalveram B, Ikegami T, Juelich T, Smith JK, Zhang L, Park A, Lee B, Komeno T, Furuta Y, Freiberg AN (2018). Favipiravir (T-705) protects against Nipah virus infection in the hamster model /631/326/22/1295 /631/326/596/1296 /13/106 /14/35 /38/77 /82/51 /96/63 article. Sci Rep 8(1). https://doi.org/10.1038/s41598-018-25780-3
23. Lo MK, Feldmann F, Gary JM, Jordan R, Bannister R, Cronin J, Patel NR, Klena JD, Nichol ST, Cihlar T, Zaki SR, Feldmann H, Spiropoulou CF, De Wit E (2019) Remdesivir (GS-5734) protects African green monkeys from Nipah virus challenge. Sci Transl Med 11(494). https://doi.org/10.1126/scitranslmed.aau9242
24. Lo MK, Amblard F, Flint M, Chatterjee P, Kasthuri M, Li C, Russell O, Verma K, Bassit L, Schinazi RF, Nichol ST, Spiropoulou CF (2020) Potent in vitro activity of β-D-4′-chloromethyl-2′-deoxy-2′-fluorocytidine against Nipah virus. Antiviral Res 175:104712. https://doi.org/10.1016/j.antiviral.2020.104712
25. Zhu Z, Dimitrov AS, Bossart KN, Crameri G, Bishop KA, Choudhry V, Mungall BA, Feng Y-R, Choudhary A, Zhang M-Y, Feng Y, Wang L-F, Xiao X, Eaton BT, Broder CC, Dimitrov DS (2006) Potent neutralization of Hendra and Nipah viruses by human monoclonal antibodies. J Virol 80(2):891–899. https://doi.org/10.1128/jvi.80.2.891-899.2006

26. Dang HV, Chan YP, Park YJ, Snijder J, Da Silva SC, Vu B, Yan L, Feng YR, Rockx B, Geisbert TW, Mire CE, Broder CC, Veesler D (2019) An antibody against the F glycoprotein inhibits Nipah and Hendra virus infections. Nat Struct Mol Biol 26(10):980–987. https://doi.org/10.1038/s41594-019-0308-9
27. Dang HV, Cross RW, Borisevich V, Bornholdt ZA, West BR, Chan YP, Mire CE, Da Silva SC, Dimitrov AS, Yan L, Amaya M, Navaratnarajah CK, Zeitlin L, Geisbert TW, Broder CC, Veesler D (2021) Broadly neutralizing antibody cocktails targeting Nipah virus and Hendra virus fusion glycoproteins. Nat Struct Mol Biol 28(5):426–434. https://doi.org/10.1038/s41594-021-00584-8
28. Mathieu C, Porotto M, Figueira TN, Horvat B, Moscona A (2018) Fusion inhibitory lipopeptides engineered for prophylaxis of Nipah virus in primates. J Infect Dis 218(2):218–227. https://doi.org/10.1093/infdis/jiy152
29. Mungall BA, Schopman NCT, Lambeth LS, Doran TJ (2008) Inhibition of Henipavirus infection by RNA interference. Antiviral Res 80(3):324–331. https://doi.org/10.1016/j.antiviral.2008.07.004
30. Rajput A, Kumar M (2022) Anti-Ebola: an initiative to predict Ebola virus inhibitors through machine learning. Mol Diversity 26(3):1635–1644. https://doi.org/10.1007/s11030-021-10291-7
31. Rajput A, Thakur A, Mukhopadhyay A, Kamboj S, Rastogi A, Gautam S, Jassal H, Kumar M (2021) Prediction of repurposed drugs for Coronaviruses using artificial intelligence and machine learning. Comput Struct Biotechnol J 19:3133–3148. https://doi.org/10.1016/j.csbj.2021.05.037
32. Qureshi A, Rajput A, Kaur G, Kumar M (2018) HIVprotI: an integrated web based platform for prediction and design of HIV proteins inhibitors. J Cheminf 10(1). https://doi.org/10.1186/s13321-018-0266-y
33. Rajput A, Kumar M (2018) Anti-flavi: a web platform to predict inhibitors of flaviviruses using QSAR and peptidomimetic approaches. Front Microbiol 9:3121. https://doi.org/10.3389/fmicb.2018.03121
34. Qureshi A, Kaur G, Kumar M (2017) AVCpred: an integrated web server for prediction and design of antiviral compounds. Chem Biol Drug Des 89(1):74–83. https://doi.org/10.1111/cbdd.12834
35. Qureshi A, Tandon H, Kumar M (2015) AVP-IC50Pred: multiple machine learning techniques-based prediction of peptide antiviral activity in terms of half maximal inhibitory concentration (IC50). Biopolymers 104(6):753–763. https://doi.org/10.1002/bip.22703
36. Thakur N, Qureshi A, Kumar M (2012) AVPpred: collection and prediction of highly effective antiviral peptides. Nucl Acids Res 40(W1). https://doi.org/10.1093/nar/gks450
37. Rajput A, Kumar A, Megha K, Thakur A, Kumar M (2021) DrugRepV: a compendium of repurposed drugs and chemicals targeting epidemic and pandemic viruses. Brief Bioinform 22(2):1076–1084. https://doi.org/10.1093/bib/bbaa421
38. Qureshi A, Thakur N, Tandon H, Kumar M (2014) AVPdb: a database of experimentally validated antiviral peptides targeting medically important viruses. Nucl Acids Res 42(D1). https://doi.org/10.1093/nar/gkt1191
39. Rajput A, Kumar A, Kumar M (2019) Computational identification of inhibitors using QSAR approach against Nipah virus. Front Pharmacol 10(FEB). https://doi.org/10.3389/fphar.2019.00071
40. Dar AM, Mir S (2017) Molecular docking: approaches, types, applications and basic challenges. J Anal Bioanal Tech 08(02):8–10. https://doi.org/10.4172/2155-9872.1000356
41. Lipin R, Dhanabalan AK, Gunasekaran K, Solomon RV (2021) Piperazine-substituted derivatives of favipiravir for Nipah virus inhibition: What do in silico studies unravel? SN Appl Sci 3(1). https://doi.org/10.1007/s42452-020-04051-9
42. James JP, Apoorva, Monteiro SR., Sukesh KB, Varun A (2021) Design and identification of lead compounds targeting Nipah G attachment glycoprotein by in silico approaches. J Pharm Res Int 156–169. https://doi.org/10.9734/jpri/2021/v33i40a32232

43. Sen N, Kanitkar TR, Roy AA, Soni N, Amritkar K, Supekar S, Nair S, Singh G, Madhusudhan MS (2019) Predicting and designing therapeutics against the Nipah virus. PLoS Negl Trop Dis 13(12):e0007419. https://doi.org/10.1371/journal.pntd.0007419
44. Ropón-Palacios G, Chenet-Zuta ME, Olivos-Ramirez GE, Otazu K, Acurio-Saavedra J, Camps I (2020) Potential novel inhibitors against emerging zoonotic pathogen Nipah virus: a virtual screening and molecular dynamics approach. J Biomol Struct Dyn 38(11):3225–3234. https://doi.org/10.1080/07391102.2019.1655480
45. Kalbhor MS, Bhowmick S, Alanazi AM, Patil PC, Islam MA (2021) Multi-step molecular docking and dynamics simulation-based screening of large antiviral specific chemical libraries for identification of Nipah virus glycoprotein inhibitors. Biophys Chem 270:106537. https://doi.org/10.1016/j.bpc.2020.106537
46. Ahmed Bhuiyan M, Atia Keya N, Susan Mou F, Rahman Imon R, Alam R, Ahammad F (2020) Discovery of potential compounds against nipah virus: a molecular docking and dynamics simulation approaches. March. https://doi.org/10.21203/rs.3.rs-1398424/v1
47. Randhawa V, Pathania S, Kumar M (2022) Computational identification of potential multitarget inhibitors of Nipah virus by molecular docking and molecular dynamics. Microorganisms 10(6):1181. https://doi.org/10.3390/microorganisms10061181
48. Pathania S, Randhawa V, Kumar M (2020) Identifying potential entry inhibitors for emerging Nipah virus by molecular docking and chemical-protein interaction network. J Biomol Struct Dyn 38(17):5108–5125. https://doi.org/10.1080/07391102.2019.1696705
49. Rajput A, Thakur A, Rastogi A, Choudhury S, Kumar M (2021) Computational identification of repurposed drugs against viruses causing epidemics and pandemics via drug-target network analysis. Comput Biol Med 136:104677. https://doi.org/10.1016/j.compbiomed.2021.104677

Chapter 6
Role of Computational Modelling in Drug Discovery for HIV

Anish Gomatam, Afreen Khan, Kavita Raikuvar, Merwyn D'costa, and Evans Coutinho

Abstract With over 36 million people currently living with HIV, HIV/AIDS continues to have devastating effects on human health worldwide. Viral resistance to anti-HIV drugs remains a major cause of concern, necessitating a regimen of highly active antiretroviral therapy (HAART), which consists of a combination of multiple drugs for long-term clinical benefit. Clearly, the rapid development of novel molecules that can help change the present regimen to new drug combinations is critical for tackling the resistance problem. In this regard, computational methods have emerged as a valuable tool in HIV research, contributing greatly to our understanding of HIV biology and aiding in the design of potent anti-HIV compounds. This chapter gives an overview of the various computational strategies reported in the discovery of drugs for the treatment of HIV. A comprehensive overview of several structure-based and ligand-based computational methods is presented first; this is followed by some notable applications of these methods in the discovery of novel anti-HIV compounds. Finally, we discuss the emergence of powerful machine learning algorithms which have proven useful both in the design of new compounds and in the development of theoretical models that can predict resistance to antiretroviral therapy.

Keywords AIDS · Computational · HAART · HIV · Modelling

6.1 Background

Despite significant endeavours and treatment advancements since the Pasteur Institute in France isolated and identified the human immunodeficiency virus-1 (HIV-1), HIV has been a serious worldwide health threat [1]. Globally in 2020, around 37.7 million people were living with HIV, this number comprises 36 million adults and 1.7 million children under the age of 15. There were also 1.5 million new infections in 2020 with over 680,000 fatalities [2]. HIV belongs to the genus lentivirus

A. Gomatam · A. Khan · K. Raikuvar · M. D'costa · E. Coutinho (✉)
Department of Pharmaceutical Chemistry, Bombay College of Pharmacy, Mumbai, India
e-mail: evans.coutinho@bcp.edu.in

S. Kar and J. Leszczynski (eds.), *Current Trends in Computational Modeling for Drug Discovery*, Challenges and Advances in Computational Chemistry and Physics 35,
https://doi.org/10.1007/978-3-031-33871-7_6

within the retrovirus family [3]. Unlike some retroviruses, HIV is not capable of self-replication. HIV targets the body's immune system by attacking the CD4 T-cells over time in order to multiply and proliferate throughout the body; this directly impacts the immunity of the host, rendering the host more prone to opportunistic infections. In many cases, HIV progresses to its most advanced form, acquired immune deficiency syndrome (AIDS) if left untreated within 10 years, during which deadly infections and malignancies are common [4]. HIV usually spreads through sexual contact or by the transmission of blood, pre-ejaculate, sperm, and vaginal secretions. A pregnant HIV-positive woman may pass her infection to her unborn child through breast milk, blood, or vaginal discharge. An infected person may not exhibit any symptoms or might go through a brief period of influenza-like sickness. As the infection affects the immune system, physical weakness and enlargement of the neck lymph nodes are commonly observed, and in cases where adequate treatment is not given, life threatening conditions such as tuberculosis, severe bacterial infections, and even cancers such as lymphomas may occur [2, 5].

6.2 HIV Replication Cycle

The HIV-1 life cycle is complex and may be divided into two phases: early replication and late replication. The virion's adhesion to the T-cell surface and the proviral DNA's incorporation into the host genome are the first two stages of the early phase [6]. Proviral transcription begins at the late phase of replication and continues until fully infectious offspring virions are produced [7]. The various stages in the lifecycle of HIV and drug targets are depicted in Fig. 6.1 and are briefly discussed below [7–9].

1. **Binding**: The first step of viral attack is characterized by attachment of the virus to the T-cell surface through CD4 or CXCR4 or CCR5 receptors. This causes the binding of gp120, a monomeric virion associated protein, to form a gp120-CD4 complex.
2. **Fusion**: The viral envelope undergoes structural changes after adhering to the CD4 cell, causing the virus to fuse with the cell membrane resulting in bursting of the viral envelope. After entering the T-cell, the virus releases its RNA along with the enzymes—reverse transcriptase and integrase.
3. **Reverse transcription**: The enzyme reverse transcriptase converts HIV single strand RNA (ssRNA) to HIV double strand DNA (dsDNA), which allows it to enter the nucleus and combine with the cell's genetic material, resulting in virulent activity.
4. **Integration**: At this stage of the HIV life cycle, the enzyme integrase inserts newly transcribed viral DNA into the host DNA, causing the cell to become virulent.

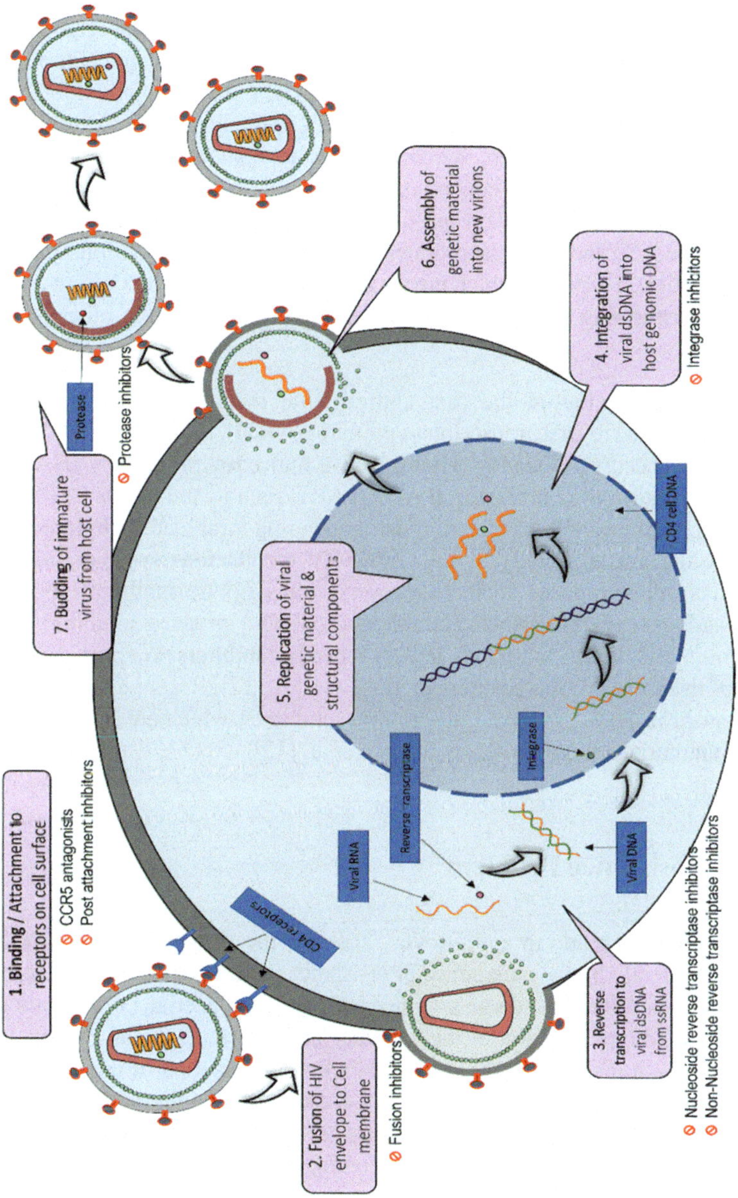

Fig. 6.1 Illustration of the HIV life cycle

5. **Replication**: After the viral DNA is integrated with the host DNA by integrase, the reproduction process of the virus begins. This process occurs when the virus begins to replicate or produce long chain HIV proteins using the machinery of the host cell.
6. **Assembly**: The sixth stage of the HIV lifecycle is the most crucial stage since it is here that the virus begins assembly of the components after manufacturing the essential components in the fifth stage. The new HIV RNA along with essential HIV precursor proteins generated by the host CD4 cells are ferried to the cell surface during this step, where the components are combined into a structurally complete but immature non-infectious virus.
7. **Budding**: As HIV pushes itself out of the host cell, it remains non-infectious. The HIV lifecycle ends with the production of mature infectious virions resulting from the action of protease which breaks down the immature virus's long protein chains, transforming it into a mature virus capable of infecting other healthy cells.

Since the introduction of the first antiretroviral therapy (ART), zidovudine (AZT)—a nucleoside reverse transcriptase inhibitor (NRTI), in 1987 [10], HIV treatments have advanced significantly. Highly active antiretroviral therapy (HAART) regimens, also known as combinatorial ART (cART), is now the primary treatment for HIV [11]. Antiretroviral drugs can assist in lowering viral load, fight infections, and enhance the overall quality of life. Anti-HIV medications are classified into six categories: nucleoside or nucleotide reverse transcriptase inhibitors (NRTIs), non-nucleoside reverse transcriptase inhibitors (NNRTIs), protease inhibitors (PIs), integrase inhibitors, fusion inhibitors, and co-receptor inhibitors. We provide a classification of the anti-HIV medications in Table 6.1 [10, 12]. Recent efforts towards development of in HIV-1 reverse transcriptase inhibitors, which are in the clinical phase are summarized in the paper by Shaung et al. [13].

6.3 The Resistance Problem

Despite the progress made in antiretroviral therapy, the management of HIV has been hindered greatly by the emergence of resistance. When the virus is targeted by antiretroviral agents, it may undergo a change in its genetic material, i.e. the naturally occurring or 'wild type' version of the genome. This is referred to as a 'mutation' and may lead to an inability of the drug to block the replication of the virus, thus causing the virus to become 'resistant' to the drug. Due to the advent of drug-resistant viruses, all antiretroviral medications, including those from more recent pharmacological classes, are at danger of becoming partly or completely ineffective [14].

Several aspects associated with the HIV life cycle and replication are important contributors to the organism's fast and widespread establishment of resistance. The HIV-reverse transcriptase (RT) enzyme is known for its 'poor fidelity' (i.e. the enzyme

Table 6.1 Summary of approved drugs for treatment of HIV encompassing the five major classes

	Nucleoside or -tide reverse transcriptase inhibitors (NRTIs)		Non-nucleoside reverse transcriptase inhibitors (NNRTIs)	Protease inhibitors (PIs)	Integrase strand transfer inhibitors (INSTIs)	Entry inhibitors	
						Fusion inhibitors	Co-receptor inhibitors
Mechanism of action	Competitive inhibition of nucleic acid from viruses by causing chain termination during reverse transcription		Alter HIV-RT structure by binding to an allosteric region about 10 Å away from the enzyme's active site, thus preventing reverse transcription	Competitively block the protease enzyme by binding to its catalytic site with a high affinity and inhibiting the enzyme's ability to function, resulting in the production of immature and non-contagious viral particles	Prevent integration of the viral DNA into the host genome	Prevent viral entry into host cell. Fusion inhibitors prevent fusion of the virus with the host cell, whereas CCR5 antagonists prevent infection of the CD4 T-cells by blocking the CCR5 receptor	
FDA approved drugs	Nucleoside	Nucleotide	Nevirapine (NVP)-1996 Delavirdine (DLV)-1997 Efavirenz (EFV)-1998 Etravirine (ETV)-2008 Rilpivirine (RPV)-2011 Elsulfavirine (ESV)–2017 Doravirine (DOR)-2018	Saquinavir-1995 Ritonavir-1996 Indinavir-1996 Nelfinavir-1997 Amprenavir-1999 Atazanavir Fosamprenavir-2003 Tipranavir-2005 Darunavir-2006	Raltegravir-2007 Dolutegravir-2013 Elvitegravir-2014 Cabotegravir-2021	E-Fostemsavir-2020	Maraviroc-2007
	Zidovudine (AZT)-1987 Didanosine (ddI)-1991 Zalcitabine (ddC)-1992 Stavudine (d4T)-1994 Lamivudine (3TC)-1995 Abacavir (ABC)-1998 Emtricitabine (FTC)-2003	Tenofovir disoproxil fumarate (TDF)-2001					

is rather nonselective throughout the copying process) and is prone to introducing errors while transcribing viral RNA into DNA [11, 15]. According to some estimates, HIV-RT introduces one mutation for every viral genome that is transcribed. The high mutation rate of HIV-RT, when combined with the high levels of viral production and turnover, means that the patient may oftentimes have a diverse mixture of viral quasi species within a few weeks of infection. One or more of these viral quasi species may be resistant to medication, and the quasi species which confer the most advantage to the virus (i.e. reducing susceptibility to an antiviral agent) are retained through a process of Darwinian selection. Resistance may also develop independently of ART therapy, this happens when an individual contracts HIV for the first time from a resistant strain, usually transmitted from a HIV-positive person undergoing antiretroviral therapy. This is referred to as primary or acquired resistance. Significant advancements have made been in successfully identifying mutations linked to drug resistance and comprehending the processes through which they confer resistance. Several mechanisms have been discovered, and they vary for drugs both from the same class and from other classes [14, 16]. These have been covered comprehensively in reviews by Cilento et al. [11] and Collier et al. [17]

Clearly, we now face an urgent problem, as the number of persons with HIV resistant strains is rising. To combat medication resistance and reduce the high costs associated with treatment, new drug classes must continuously be investigated and developed.

6.4 Structure-Based Methods

In silico drug design or computer-aided drug design (CADD) has the potential to accelerate the tedious process of designing and developing a drug candidate. With the recent developments in the architecture and algorithms of structure-based drug design, intensive computations can be performed in a time-affordable manner [18]. The structure-based drug design (SBDD) techniques require an experimentally solved structure of the protein or better still the protein–ligand complex, which is obtained using X-ray crystallography, nuclear magnetic resonance (NMR) spectroscopy, and comparative protein modelling or homology modelling.

The SBDD-based methods help not only in identification of hits but also in the ability to optimize them. The docking method, which is central to SBDD, is based on the theory of the lock and key mechanism, where both the protein and ligand are considered as rigid bodies. The introduction of the 'induced-fit' theory, proposed by Koshland gave us an understanding that the active site of the protein continuously undergoes changes as the ligand interacts with the protein [19]. Ideally, both the ligand and protein should be considered flexible bodies. However, keeping in mind the limitations of computational resources, the protein may often be treated as a rigid body. There are multiple algorithms [20] that have been devised to treat the docking problem. To summarize, molecular docking methods are useful in predicting the binding pose of the ligand (drug) to the protein and its relative affinity [21]. Binding

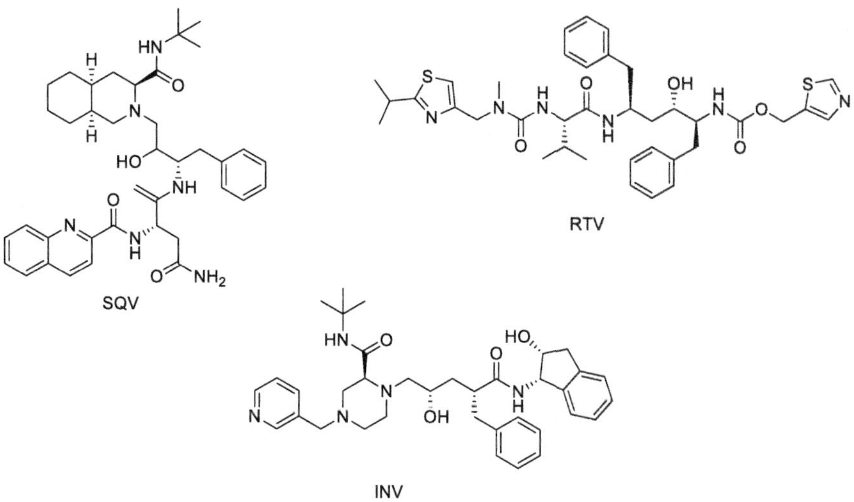

Fig. 6.2 Examples of HIV drugs currently on the market discovered using SBDD methods [saquinavir-SQV, ritonavir-RTV, and indinavir-INV]

studies using molecular dynamics (MD) simulations [22] on the 'best' binding pose can be performed to understand the affinity. MD can help in ligand optimization and provide an insight on the pathways and kinetics of interactions. Virtual screening (VS) is a technique where docking is carried out on a large dataset of molecules to identify potential leads among them. The HIV drugs that have been discovered using the SBDD techniques [23] are given in Fig. 6.2.

The structure-based drug design methodology has been implemented and proved to hasten the drug discovery process in several instances. Recent developments and directions in molecular docking and MD simulations are discussed below.

6.4.1 Molecular Docking

Molecular docking is one of the most widely used in silico techniques for drug design. A set of ligands with optimized geometry is set to position in the binding site of the protein, i.e. to generate a pose of the ligand with a corresponding binding energy [24]. The method is often coupled to a scoring function to calculate a score, which reflects the 'goodness of fit' of the ligand pose in the protein pocket. The best pose is identified as the one making optimal contacts with the amino acids in the active site, as reported experimentally. To predict the binding of ligands with a mutant protein, a molecular modelling software [25] can be used to mutate specific residues (most of the times the ones that are reported experimentally). The major source for 3D protein and protein–ligand structures is the Protein Data Bank (PDB, https://www.rcsb.org). The outline of the molecular docking process is shown in Fig. 6.3.

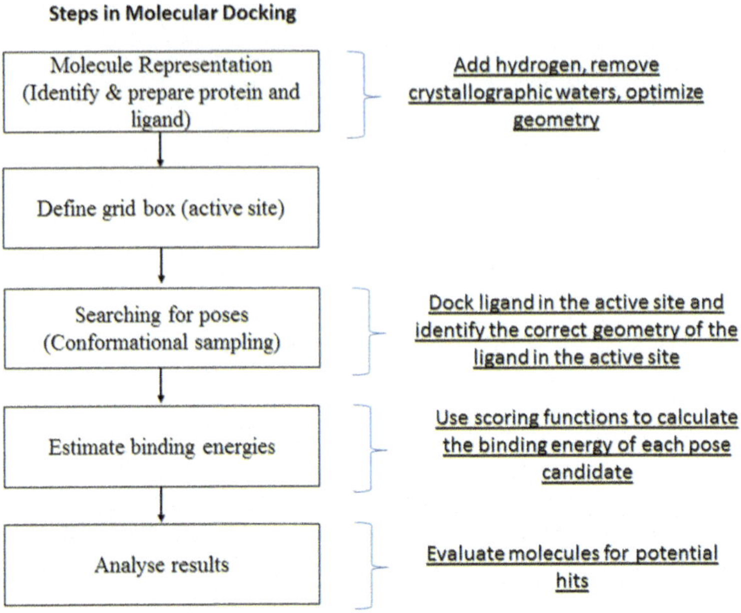

Fig. 6.3 Schematic representation of the molecular docking technique

The molecular docking technique is initially used to understand the interactions between the protein and experimentally confirmed inhibitor. This step acts as a validation of the methodology adopted and is followed by virtual screening, so as to find new inhibitors specific to the target. The technique has also been used to explain the resistance problem [26]. The interactions that are absent in the mutated protein can be identified; this can be exploited to design drugs acting on the resistant strains. The docking programmes most commonly used are AutoDock 4.0 [27], Sybyl-X [28], AutoDock Vina [29], Molegro Virtual Docking [30], AutoDock Racoon [31], and Glide (Schrodinger Inc.) [32]. We now focus on the application of molecular docking as applied to the discovery of HIV-1 inhibitors.

Herbal compounds have paved a way for identification of many drugs, and many researchers have made efforts to identify anti-HIV agents from natural sources. Vora et al. [33] have identified five natural compounds against HIV. These are anolignan B (active against reverse transcriptase), curcumin (an inhibitor of the integrase enzyme), mulberroside C (protease inhibitor), chebulic acid (ribonuclease inhibitor), and neoandrographolide (entry inhibitor). Very few of the natural compounds were found to have an IC_{50} value that is close to the synthetic drugs, so chemical modifications are needed to improve their activity, nevertheless, these could be considered in near future as leads.

A review by Tarasova et al. [34] provides an outline of the molecular docking technique applied to understand structural changes in reverse transcriptase (RT)

associated with HIV-1 resistance. Most of the docking studies reported in literature are performed on the NNRTIs, rather than on the NRTIs. This is attributed to two reasons—first, the NRTIs act via a competitive mechanism and second, the estimation of the binding energy between the NRTIs and the protein is difficult. Also, that there are several possible mechanisms that exist for the resistance, and the role of each mutation in the level of resistance is not known. Perhaps, the discovery of the mechanisms of HIV-1 resistance as revealed by Tarasova et. al. can pave ways for molecular docking applications to develop highly active NRTIs. A compilation of the studies discussed here is presented in Table 6.2.

Singh et al. [35] have explored diarylpyrimidine derivatives as NNRTIs. They docked the compounds at the allosteric site of HIV-RT and identified eight potential ligands having a profile better than the known inhibitor etravirine. Further, molecular dynamics and free energy-based calculations were done to understand the binding affinity and stability of the protein–ligand complexes. Compound **6** in the study showed better stability and inhibition than the reference drug, paving the way for the development of newer second generation NNRTIs. In a study reported by Fraczek et al. [36], a comparison of different molecular docking techniques on a set of potential NNRTIs was carried out. The paper describes a comparison of FlexX, Hyde, Molegro Virtual Docker, Glide, and AutoDock Vina on their ability to predict RT inhibitory activity of 1,2,4-triazoles ($n = 111$) and azoles ($n = 76$) as NNRTIs. They showed that the correlation between the experimentally determined half maximal

Table 6.2 Summary of the docking studies discussed

Target	PDB ID	Ligand	Software	References
Protease, Ribonuclease, IN, RT, gp-120	5KAO, 3QIN, 5EU7, 4G1Q, 1G9M	Natural products	Discovery Studio, Schrodinger, Molegro Virtual Docking	[33]
HIV RT	3MEC	Diarylpyrimidine derivatives	Discovery Studio 3.0	[35]
HIV RT	3DLG	1,2,4-triazole and azole derivatives	Glide, FlexX, Molegro Virtual Docker, AutoDock Vina, Hyde, Sybyl-X	[36]
HIV RT	3MEC	Diarylpyrimidine derivatives	Sybyl-X	[37]
HIV RT	4G1Q	Amino-oxy-diarylquinoline	AutoDock 4.2	[38]
HIV RT	1S1U	Cu (II) ion Schiff base complexes	AutoDock Vina	[39]
HIV RNase	5J1E	Hydroxypyrimidine-2,4-diones	Sybyl-X 2.1	[40]
HIV IN	3OYA	Benzoxazoline, quinazoline, diazocoumarin	AutoDock Vina	[41]

[HIV-RT: HIV-reverse transcriptase, IN: integrase]

effective concentration (EC_{50}) and predicted binding energies were highly dependent on the ligand set. The performance of all the docking programmes was comparable; however, AutoDock Vina, Molegro Virtual Docker, and Hyde indicate that shape matching of ligand and binding sites is the preferred method for identifying inhibitors. However, activity prediction was restricted to only those substances closely akin to a natural ligand. A study on compounds containing the diarylpyrimidine core was reported by Liu et al. [37] as NNRTIs. They performed molecular docking using the Sybyl-X software to generate the 3D binding pose; these structures were then used to calculate various 3D descriptors from which a 3D-QSAR model was built using the comparative molecular field analysis (CoMFA) and comparative molecular similarity indices analysis (CoMSIA) approaches. The phenyl group present in the diarylpyrimidines (Fig. 6.4) was able to engage in π-π stacking interaction with the aromatic residues of the binding site; in contrast, the cycloalkanes are unable to do so. This shows that the phenyl ring at the C4-position of the pyrimidine ring is preferred over cycloalkane motifs for good activity. The best ligands for the 3D-QSAR models were found to be substituents having a 4-isopropyl, 3-hydroxy, 2-fluoro-4-methyl groups. Makarasen et al. [38] designed derivatives containing the amino-oxy-diarylquinoline core as NNRTIs from a pharmacophore model that was constructed on the interaction templates of nevirapine, efavirenz, etravirine, and rilpivirine. Also using molecular docking, they were able to identify important interactions of the ligands with Lys101 and His235 residues via a hydrogen bond and with Tyr318 via π-π stacking. These compounds were synthesized and tested and found to have an inhibition rate of about 39.7% at 1 μM concentration. Shanty et al. [39] identified Cu(II) ion complexes with heterocyclic Schiff bases as inhibitors of HIV-1 RT. These molecules were docked against the protein. The paper summarizes the different types of complexes and their stability towards binding and points out that hydrogen bonding, hydrophobic interactions, and the π-sulphur contacts are crucial interactions. The compounds were synthesized, tested with nevirapine as the reference drug, and were found to be active with an inhibition rate of 86 versus 100% for nevirapine against the target enzyme.

For HIV multiplication, the HIV-1 RT-ribonuclease H (RNase-H) association plays an important role as reported by Gao et al. [40]. In this study, a series of hydroxypyrimidine-2,4-diones ($n = 93$) was curated, and in silico methods like docking followed by MD simulations were performed. The final poses were used to calculate the various physicochemical descriptors, and a 3D-QSAR model was built. For the CoMFA model, the validation metrics are r^2 0.949, q^2 0.908, and F value of 492.826. The steric and electrostatic field contributions are 72.0% and 28.0%, respectively, showing that the steric field contributes more to activity according to the CoMFA model. To sum up, the following substituents may be introduced into appropriate areas to enhance the inhibitory activity of hydroxypyrimidine-2,4-diones: the pyrimidine ring's N1 position is positively charged and can be substituted by small groups; the N3 position is negatively charged and can be substituted by hydrophilic substituents; the linker moiety can be attached with hydrophobic groups; the 2nd or 3rd position of the aromatic moiety can be substituted by bulky, negatively charged, and/or hydrophobic groups; the 2′ or 4′ position of the aromatic moiety can carry bulky, negatively charged, and/or hydrophobic groups; and the 3′ position of the

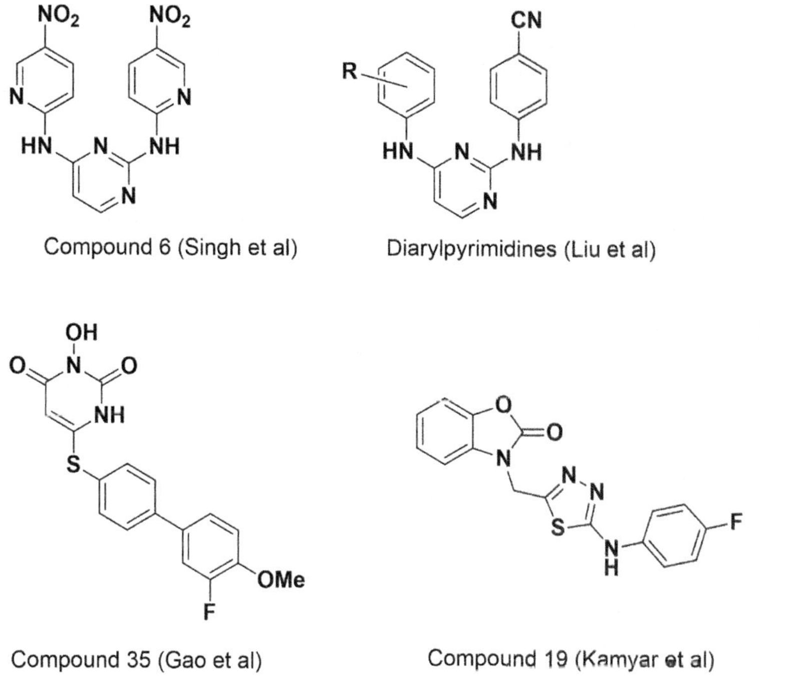

Fig. 6.4 Structure of the ligands—compound 6 [28], 35 [30], 19 [34], and the diarylpyrimidine core [33]

aromatic moiety can accommodate negatively charged groups. The newly designed molecules were proposed as leads for HIV RNase-H inhibitors. An investigation on chalcone derivatives as HIV-1 protease inhibitors was reported by Turkovic et al. [42]. They curated a set of 20 structurally similar chalcones and docked them in the protease enzyme, to decipher the interactions. These molecules were synthesized and tested for anti-HIV-1 activity via a fluorimetric assay. The best molecule exhibited an IC_{50} of 0.001 μM, which is comparable to the commercially available drug Darunavir. Novel 2,3-diaryl-4-quinazolinone derivatives were designed by Hajimahdi et al. [43]; they were docked in the HIV-1 integrase enzyme, and the top ranking molecules were synthesized and assayed for their anti-HIV activity. The study provided novel leads, with the best molecule showing an EC_{50} of 37 μM. Kamyar et al. [41] have explored quinazoline, benzoxazolinone, and diazocoumarin derivatives as anti-HIV agents. The study describes a set of 29 compounds which were docked against the HIV integrase protein using the AutoDock Vina package. Compound 19 in the set binds to the active site of integrase by two major moieties—first is the carbonyl group of the compound which binds to the Mg^{2+} ions and the second is the aryl side chain which fits into the hydrophobic pocket at the protein-DNA interface via a π-stacking interaction. This docking data could provide useful insights for design of new anti-HIV agents.

6.4.2 Molecular Dynamics and Free Energy Calculations

Classical MD by its very nature is able to account for the structural flexibility of the drug-protein system, which is well supported by the induced-fit and the conformational selection theories [44]. This method is a physical model to understand the interactions and motion of the atoms in a molecule as governed by Newton's laws of motion. Generally, a force field is applied to all the atoms present in the system and this is used to estimate the overall energy of the system. When performing an MD simulation, the integration of the laws of motion generates a series of configurations, showcasing a trajectory that provides two crucial pieces of information—positions and velocities of the atoms over time. This is used to calculate the free energies which are correlated with the experimental observations to draw out conclusions about the drug binding process with the target protein (Fig. 6.5).

The core idea of molecular dynamics is the study of the time-dependent behaviour of the system. This is explained by Newton's second law of motion:

$$f_i(\mathrm{t}) = m_i a_i(\mathrm{t}) = -\frac{\partial V[r_i(t)]}{\partial r_i(t)}, \tag{6.1}$$

where m_i is the mass, $a_i(t)$ the acceleration, $f_i(t)$ the total force operating at a certain moment in time t on the ith atom of the system. The vector $r_i(t)$ which depicts the positions of the N interacting atoms in Cartesian space ($r = \{x_1, y_1, z_1; x_2, y_2, z_2; \ldots x_\mathrm{N}, y_N, z_N\}$) represents the configuration of the system at the given instant.

The empirical potential energy equation is given by

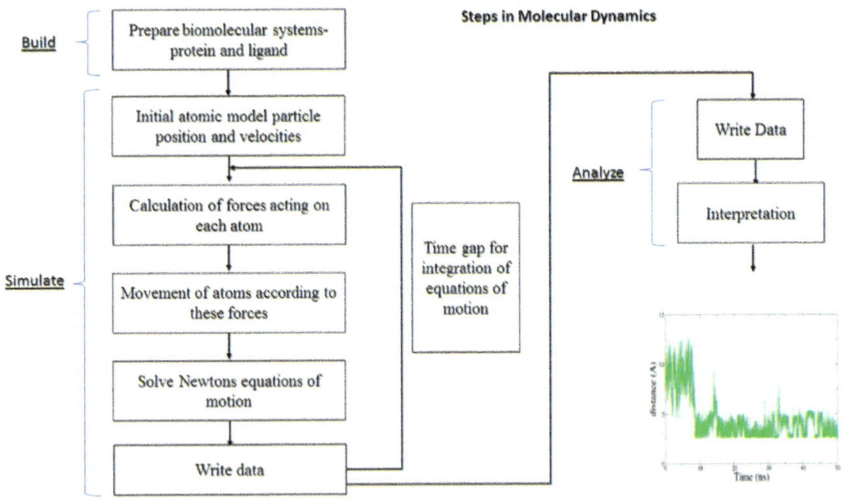

Fig. 6.5 Schematic representation of the MD simulation workflow

$$E_{\text{total}} = \sum_{\text{bonds}} K_r\left(r - r_{\text{eq}}\right)^2 + \sum_{\text{angles}} K_\vartheta\left(\vartheta - \vartheta_{\text{eq}}\right)^2 + \sum_{\text{dihedrals}} \frac{Vo}{2}\left[1 + \cos(n\phi - \gamma)\right] + \sum_{i<j}\left[\frac{A_{ij}}{R_{ij}^2} - \frac{B_{ij}}{R_{ij}^2} + \frac{q_i q_j}{\varepsilon R_{ij}}\right]. \quad (6.2)$$

Equation 6.2 comprises of all forces arising from interactions between the bonded and the non-bonded atoms. The bonded interactions include bonds, angles and dihedrals, while the non-bonded forces are those that arise due to the van der Waals interactions depicted through the Lennard–Jones 6–12 potential and the Coulombic electrostatic forces. These energy terms are parameterized to reproduce the real behaviour of the molecules and are collectively called as the 'force field'. The force fields commonly used in MD simulations are the General Amber Force Field (GAFF) [45], CHARMM [46], and GROMOS [47]. The positions of these atoms are moved according to Newton's laws of motion, using the calculated forces. The time step in the MD simulation is a few (1 or 2) femtoseconds, and the process is repeated several million times which gives the length of the simulation. The most popular software packages for MD simulations are AMBER [48], NAMD [49], CHARMM [50], and GROMACS [51]. Table 6.3 summarizes the different MD methods used on the different targets of HIV discussed here.

The free energy calculations methods can be classified as the end-state free energy methods, also called partitioning-based methods, and the non-partitioning-based methods. The latter group of methods are generally more accurate and computationally exhaustive than the end-state free energy methods. Moreover, end-state free energy methods allow the energy components to be decomposed into electrostatics, van der Waals and bonded energy terms. The philosophy of the non-partitioning-based methods denounces the idea that the free energy can be decomposed into components. The molecular mechanics Poisson-Boltzmann surface area (MM-PB/SA) and molecular mechanics generalized Born surface area (MM-GB/SA) methods are mostly used and belong to the class of end-state free energy methods. Free energy perturbation (FEP) and thermodynamic integration (TI) belong to the class of non-partitioning-based methods. Furthermore, FEP and TI can be used to calculate absolute as well as relative binding free energy, whereas MM-PB/SA and MM-GB/SA methods yield only relative binding free energy [56].

Table 6.3 Summary of the MD simulation-based studies discussed in this section

Target	PDB	Method	Software	References
RNase-H	5J1E	Unbiased all atom MD simulation	GROMACS	[40]
NNRTI	3DLG	Unbiased all atom MD simulation	GROMACS	[52]
HIV protease	1HHP	Unbiased all atom MD simulation	AMBER12	[53]
HIV protease	1HHP	Gaussian accelerated molecular dynamics (GaMD)	AMBER14	[54]
HIV integrase	6C0J	Unbiased all atom MD simulation	GROMACS	[55]

MM-PB/SA and MM-GB/SA methods calculate the free energy of binding by adding a correction for solvation electrostatics to the molecular mechanics gas phase energies. These solvation electrostatics are computed either by Poisson-Boltzmann's method or by the generalized Born model that account for the polar component of the solvation free energy. The non-polar component of the solvation free energy is estimated using the non-polar surface area of the complex, receptor, and ligand. MM-PB/SA and MM-GB/SA energies employ either the single-trajectory or the 3-trajectory approach. In the single-trajectory method, conformational samples are collected from the MD simulations of the complex alone, from which the receptor and ligand components are separated during the energy calculations. In the 3-trajectory approach, separate MD simulations are performed for the complex, the receptor, and the ligand. Irrespective of the approach, the following Eq. 6.3 is used to compute the binding free energy (ΔG_{bind}).

$$\Delta G_{\text{bind}} = \langle \Delta G_{\text{complex}} \rangle - (\langle \Delta G_{\text{protein}} \rangle + \langle \Delta G_{\text{ligand}} \rangle), \tag{6.3}$$

where $\langle \Delta \mathrm{G}_{\text{protein}} \rangle$ and $\langle \Delta \mathrm{G}_{\text{ligand}} \rangle$ are the total energies of the protein and ligand and the total free energy of the protein–ligand complex is given by $\langle \Delta G_{\text{complex}} \rangle$. The angular bracket $\langle \rangle$ indicates the energy is calculated from the structural ensemble derived from MD simulations.

MD simulations, as mentioned earlier, can also be used to understand the binding mechanism of the ligand to the receptor. Using unbiased MD simulations with an explicit solvent model, Huang et al. [53] have made an effort to understand the pathways by which the ligands approach the protein active site and the conformational changes that occur in protein during binding. The ligands xk263 (Fig. 6.6) and ritonavir (Fig. 6.2) in complex with HIV protease were used. The studies reveal that the two ligands bind to the protein by different mechanisms. The xk263 binds fast to the protein with a semi-open flap conformation, indicating the induced-fit mechanism, whereas ritonavir binds slowly to the protein in the open conformation via a conformation selection mechanism. The HIV protease conformational changes and the binding routes of the ligand (xk263) were sampled by Miao et al. [54] using the Gaussian accelerated molecular dynamics (GaMD) method. The HIV protease structure is a homodimer with two loops or 'flaps'. The three main flap conformations—'open', 'semi-open', and 'closed' were found in this study. The apo-protein exhibits the 'semi-open' conformation, whereas the holo-protein predominantly adopts the 'closed' conformation. A number of crucial intermediate states during the ligand binding process were also discovered. The whole pathway of the ligand xk263 which is a fast and tight binder of the HIV protease were successfully reproduced. The GaMD simulation is a useful guide to further probe drug-receptor binding.

The entry of HIV into the T-cell is initiated by the interaction of the viral envelope protein gp120 with the cell surface receptor CD4 as well as the co-receptors CCR5 or CXCR4. R5 virions and X4 virions, respectively, are the names given to viruses that require CCR5 or CXCR4 for entry. According to reports [57], the dendrimer-SPL7013, a microbicide binds to areas on the interface of the gp120-CD4 complex,

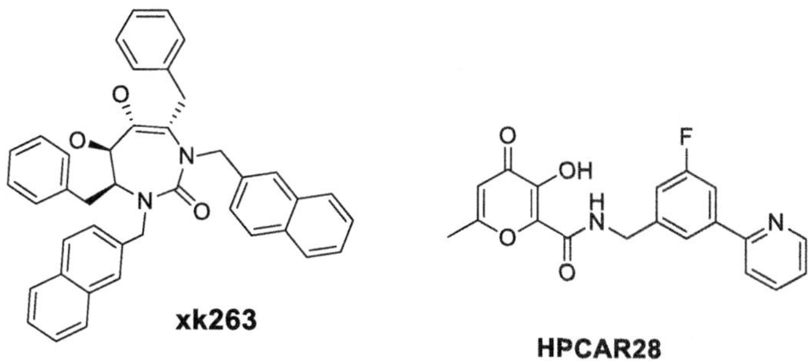

Fig. 6.6 Structures of the ligands xk263 and HPCAR28

blocking viral entry into target cells. Fully atomistic molecular dynamics simulations were employed by Nandy et al. [57] to evaluate the kinetics of dissociation and energetics of the gp120-CD4 complex in the absence and presence of the dendrimer. Molecular docking with steered and fully atomistic MD simulations was able to predict that the dendrimer-SPL7013 does not bind to gp120 alone but binds strongly to the R5 gp120 in the gp120-CD4 complex. This weakens the gp120-CD4 complex and causes its dissociation. As a result, the gp120-CD4 complexes are not formed in adequate number, to form across a virus-cell pair, thereby preventing viral entry. The identification of the contact residues between CD4 and gp120 that were changed by the binding of SPL7013's binding to R5 gp120 were made possible by the atomistic resolution offered by the MD simulations. When the binding energy was decomposed into its component elements, it became clear that the electrostatic component makes the largest contribution to the total binding energy. The study thus provided a mechanism of how SPL7013 prevents R5 HIV-1 from infecting target cells.

A common strategy in drug discovery is to perform an MD simulation of the receptor-ligand complex to understand the binding affinity. Chen et al. [55] designed a novel series of 52 dihydrofuran [3,4-d]pyrimidine (DHPY) analogues as NNRTIs. A systematic in silico study comprising of 3D-QSAR, molecular docking, virtual screening followed by filtering of the top ligands, and lastly MD simulations of the complex were performed. They identified nine lead compounds using this computational strategy. Sirous et al. [58] using the structure-based combinatorial library design method optimized a series of 3-hydroxypyran-4-one derivatives as HIV integrase inhibitors. The method allowed the coupling of the combinatorial library design with the quantum polarized ligand docking (QPLD) and MD simulations. HPCAR28 (Fig. 6.6) was identified as a potential lead in this experiment with an IC_{50} of 0.065 μM. A small library of 93 molecules having the 3-hydroxypyrimidine-2,4-dione core targeting HIV-1 RT associated RNase-H was curated by Gao et al. [40]. To verify the accuracy of the docking methodology, these molecules were docked, and MD simulations were run. The CoMSIA and CoMFA methodologies were used to create 3D-QSAR models. Six new molecules were identified as leads in this study. Wang et al. [52] reported in silico efforts with 38 N1-aryl-benzimidazoles as NNRTIs. The protocol followed in this case was similar to Gao et al.; a 3D-QSAR model was built followed by a pharmacophore model to identify the structural features related to the activity. This study identified positions on the aryl/benzimidazole ring where appropriate substituents will enhance the inhibitory efficacy of the molecules. These are hydrophobic groups at the linker of the C2-position of the benzimidazole moiety; negatively charged and/or hydrogen-bond acceptor groups at the C6-position of the benzimidazole moiety; small, positively charged, and/or hydrogen-bond donor groups at the C2-position of the arylacetamide moiety.

A reliable pharmacophore model was created by Cele et al. [59] using MD simulation ensembles and per-residue energy decomposition. The amino acids that contribute to free energy of binding were the basis for the creation of the pharmacophore library. A pharmacophoric screen for possible reverse transcriptase was also conducted. To verify the system's stability, docking and MD modelling were applied to the complex of GSK952 with the protein. Utilizing the recognized HIV-reverse transcriptase inhibitory activity, the technique was validated. Two hits (ZINC46849657 and ZINC54359621) demonstrated a considerable potential to be further investigated based on the binding free energy.

MD simulations may also be utilized to study the wild type and resistant or different subtypes of the proteins to capture significant ligand interactions that could subsequently be used to design a drug effective against the wild type or resistant species. Halder et al. [60] have reported a study with the FDA approved protease inhibitors—atazanavir, darunavir, and ritonavir. These were analysed for their activity on other HIV protease subtypes like the South African subtype C (C-SA) and B complexes using MD simulations. They have pointed out the specific affinities in the subtypes using PCA, per-residue decomposition analysis, and hydrogen analysis methods. The analysis identified major factors that contribute to increased binding affinity of the compounds against C-SA protease over the B complex. These are stable interactions with catalytic amino acid residues; increased electrostatic interactions, and stable hydrogen-bond formation capacities with amino acid residues in the binding cavity; stability in the flap movements caused by inhibitor binding and decreased entropic cost.

6.5 Quantitative Structure–Activity Relationships (QSARs)

The QSAR formalism attempts to derive a mathematical relationship between the structure of chemicals and their physiological behaviour in biological systems, such as biological activity, disposition, and toxicity. Mathematically, the basis of QSAR is a representation of the biological response of a chemical as a function of its chemical attributes. The linear form of a QSAR equation (model) is given as

$$y = m_0 + m_1x_1 + m_2x_2 + m_3x_3 \ldots m_nx_n, \tag{6.4}$$

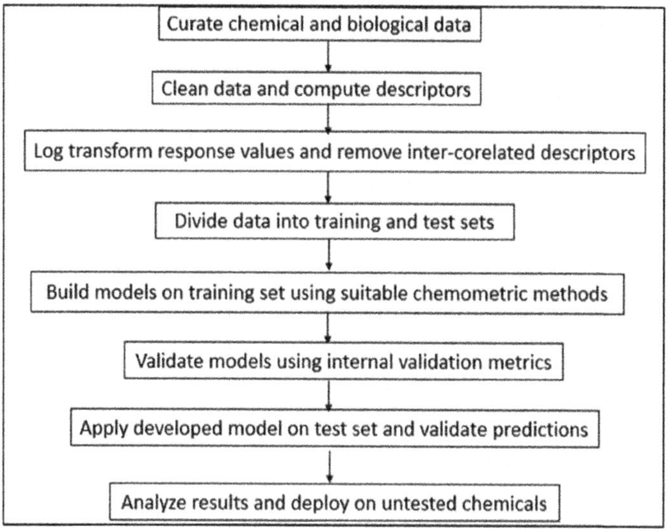

Fig. 6.7 An illustration of the QSAR workflow

where y is the response being modelled and $x_1, x_2, x_3 \ldots x_n$ are numerical representations of the structural features (also known as descriptors), and $m_1, m_2, m_3 \ldots m_n$ are the contributions (weights) of individual descriptors to the response. Once a model of sufficient quality has been developed; it can be tested on hitherto untested or new chemical entities [61]. Oftentimes, QSAR models can also direct lead design and optimization by suggesting structural modifications to obtain the desired pharmacological activity, especially when the model has been developed on a congeneric series of structurally related compounds. The correlation between the response and structural features can be established using various chemometric methods [62]. The original QSAR methodology pioneered by Hansch and Fujita used linear equations to derive correlations between closely related compounds [63]. QSAR models are often tested for their predictive ability on an external set of compounds. If a suitable external set is unavailable, the data is divided into a training set (used for model building) and a test set (for model validation). Prior to deployment on unknown chemicals, any QSAR model must be rigorously validated for reliability, robustness, and predictive ability using suitable validation metrics [61]. We provide an illustration of the QSAR workflow in Fig. 6.7.

6.6 Pharmacophore Modelling

The concept of a pharmacophore was introduced by Paul Ehrlich, who defined it as '*a molecular framework that carries the essential features (phoros) responsible for a drug's biological activity (pharmacon)*' [64]. Since then, the International Union of Pure and Applied Chemistry has redefined a pharmacophore as '*the ensemble of steric and electronic features that is necessary to ensure the optimal supramolecular interactions with a specific biological target and to trigger (or block) its biological response*' [65]. It is noteworthy that a pharmacophore does not refer to particular functional groups, but rather the pattern of features in a molecule such as presence of hydrogen-bond donors or acceptors, hydrophobic (aromatic or aliphatic), cationic, and anionic groups. A pharmacophore model consists of these features (represented as spheres) in a specific three-dimensional pattern and can be used to query test compounds or for screening a library of molecules (pharmacophore-based virtual screening). A molecule that fits the spheres representing the pharmacophore is considered a hit [66]. Depending on the available information regarding the target of interest, a pharmacophore model can either be structure-based or ligand-based. A structure-based pharmacophore model requires prior knowledge of the protein target and comprises of features that best describe the major interactions between the target and its ligands. If the structure of the target is unknown but active molecules are known, a ligand-based pharmacophore model may be developed by mapping the key pharmacophoric features of the active molecule. We outline the key steps in pharmacophore modelling in Fig. 6.8.

There are numerous reports that use QSAR-based methods or a pharmacophore modelling approach to design potent anti-HIV compounds. We summarize some of the recent research in this area in Table 6.4.

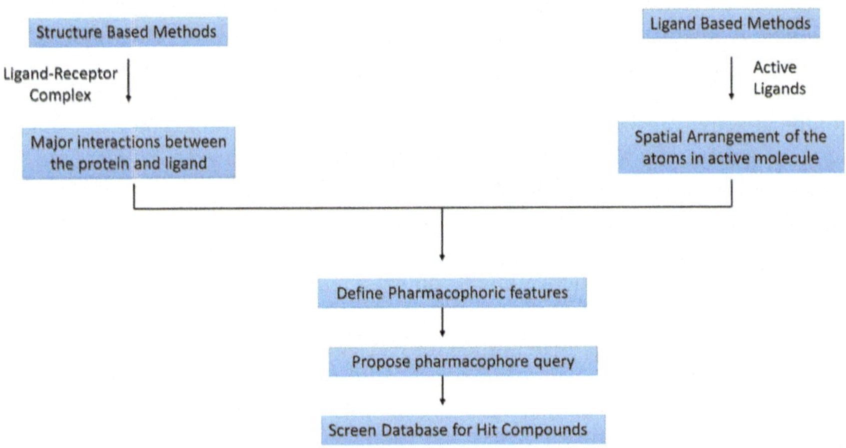

Fig. 6.8 General workflow of pharmacophore modelling

Table 6.4 Some recently reported QSAR and pharmacophore modelling-based studies in drug discovery for HIV

Computational method	Class	Target	Summary	References
2D-QSAR	1238 diverse compounds	HIV protease	Models were built on 1238 protease inhibitors using MLR, SVM, RF, and DNN algorithms, with the DNN model returning r^2_{train} of 0.88 and r^2_{test} of 0.79	[67]
2D-QSAR	346 diverse compounds	HIV protease subtype B mutant	The best model was obtained using XGB consensus prediction, and they postulate that SMILES attributes and graph-based descriptors sufficiently cover the chemical space of the dataset	[68]
HQSAR, CoMFA and CoMSIA	82 tetrahydroimidazo(4,5,1-jk][1,4] benzodiazepinone derivatives	HIV-1 reverse transcriptase	The most robust model was obtained using HQSAR with q^2 of 0.641	[69]
Molecular docking-based-QSAR	73 diarylpyrimidines	HIV-1 reverse transcriptase	The ANN model built with six input layers, four hidden layers, and one output layer has MSE of 0.16 and r^2_{test} of 0.89	[70]

(continued)

Table 6.4 (continued)

Computational method	Class	Target	Summary	References
CoMFA and CoMSIA	38 S-dihydro-alkoxybenzyloxopyrimidines	HIV-1 reverse transcriptase	The CoMFA model returned q^2 of 0.766 and r^2 0.949 for the CoMFA model, while the CoMSIA model returned q^2 of 0.827 and r^2 0.974	[71]
Combined Topomer CoMFA and HQSAR[a]	37 pyrrole derivatives	HIV fusion inhibitors	The final model has a high r^2_{train} of 0.96 and r^2_{test} of 0.96	[72]
Topomer CoMFA, CoMSIA and pharmacophore modelling	58 diarylpyrimidines	Non-nucleoside reverse transcriptase	The CoMSIA model has r^2_{train} of 0.95 and r^2_{test} of 0.73. The pharmacophore hypothesis shows that the left-wing aromatic ring and central pyrimidine on the diarylpyrimidine moiety are essential for activity, and they suggest modifications on the right wing to improve activity	[73]
Pharmacophore modelling and 3D-QSAR	45 1,5-dihydrobenzo[b][1,4] diazepine-2,4-dione analogues	HIV capsid protein	The pharmacophore hypothesis shows presence of a hydrophobic site, two aromatic rings and two acceptor regions in all the active molecules, and the 3D-QSAR model has r^2_{train} of 0.92	[74]

(continued)

Table 6.4 (continued)

Computational method	Class	Target	Summary	References
Pharmacophore modelling, atom-based QSAR, molecular docking, MMGBSA scoring	66 metronidazole derivatives	Non-nucleoside reverse transcriptase	A combined structure and ligand-based strategy has been followed, and the results suggest that metronidazole derivatives may be promising NNRTIs	[75]
Pharmacophore modelling	10 sulfonamide derivatives	HIV-1 glycoprotein 120	The designed molecules were synthesized using a green approach, and the best compound shows tenfold increase in activity compared to the standard	[76]
Combined molecular docking and pharmacophore approach	27 diverse compounds identified as hits	Chemokine receptor 5 and chemokine receptor 4	From the 27 tested compounds, three were identified as inhibitors with IC_{50} values ranging from 10.64 to 64.56 μM	[77]
QSAR, molecular docking and pharmacophore modelling	36 triazolothienopyrimidine derivatives	HIV-1 reverse transcriptase	In addition to the models, several insights on the key structural features of these moieties that contribute to anti-HIV activity have been provided	[78]

[a]HQSAR—Hologram QSAR

6.7 The Emergence of Machine Learning in Drug Discovery for HIV

In the past decade, machine learning (ML) has transformed drug discovery and development, with real-world applications ranging from virtual screening and de novo design to reaction prediction and retrosynthesis [79]. The explosive rise of ML is due to a combination of factors. Recent years have seen vast improvements in computing capacity and chemometric techniques, leading to the emergence of newer and more powerful ML methods. ML-based models have been further aided by the development of general-purpose statistical packages like the R [80] and Python programming languages [81] which enable implementation of a wide range of ML algorithms for classification and regression analyses [82]. Given the high attrition rate in drug development, many pharmaceutical companies have begun to invest their resources in leveraging the power of AI/ML to reduce development costs [83]. There are several powerful AI-based tools for drug discovery and development today such as AlphaFold (protein 3D structure prediction) [84], DeepChem (a python-based tool for predictive modelling in drug discovery—https://github.com/deepchem/deepchem), and DeepTox (toxicity prediction using deep learning—http://www.bioinf.jku.at/research/DeepTox/) [85].

Broadly speaking, ML is the practice of using algorithms to process data, learn, and make a prediction about a desired outcome [83]. ML algorithms may be supervised or unsupervised, based on whether labels are assigned to the training data. It is important to bear in mind that the quality of the model relies largely on the quality of the input data and imposes an upper limit on the accuracy and generalizability of any subsequently developed ML model. Therefore, the raw data must be screened for errors, omissions, missing values, and data type conversions. Once cleaned, the chemical data is represented as input for ML, preferably with open-source implementations such as RDKit (https://www.rdkit.org) or Dscribe (https://singroup.github.io/dscribe), and an appropriate ML algorithm is chosen [86]. There is a plethora of ML algorithms to choose from, ranging from simple linear methods to complex nonlinear neural net architectures. Although their black-box nature oftentimes renders mechanistic interpretation unfeasible, nonlinear ML algorithms have proven to be effective at delineating complicated relationships between biological phenomena and molecular structure [83]. To build a robust ML model, it is essential to avoid both overfitting and underfitting. The model parameters and hyperparameters in the algorithm of choice (for instance, weights, and activation functions in neural nets) must be tuned to their optimum values. Models are built on the training set, and once a set of satisfactory models have been obtained, they are validated on a test set of compounds not part of the training set. Once the models are finalized, it is considered good practice to make the data and code publicly available for ease of reproducibility [86]. We summarize the background of some popular ML algorithms in the following section.

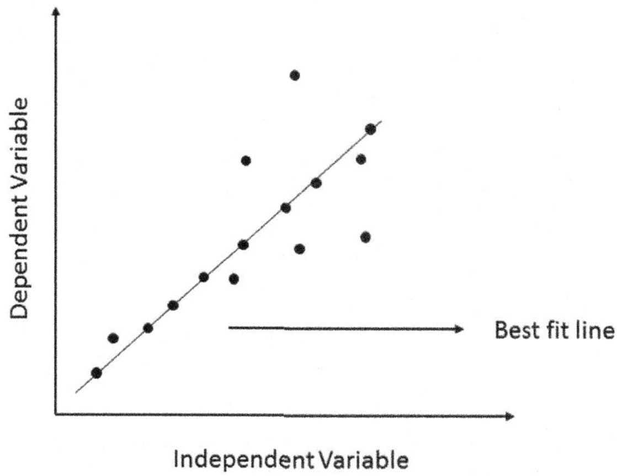

Fig. 6.9 Best fit line obtained using MLR

6.7.1 Multiple Linear Regression

Multiple linear regression (MLR) is one of the most used algorithms owing to its simplicity, reproducibility, and ability to produce models that allow easy interpretation. MLR is an extension of simple linear regression to more than one dimension and attempts to build a linear relationship between the dependent variable (target property) and independent variables (molecular feature space) by fitting the data to a straight line (shown in Fig. 6.9).

The best fit line is calculated using the slope intercept form and is given as follows:

$$y_i = m_1x_1 + m_2x_2 + m_3x_3 \ldots m_nx_n, \tag{6.5}$$

where y_i is the target property, $m_1, m_2, m_3...m_n$ are the regression coefficients and $x_1, x_2, x_3...x_n$ are the descriptors (features or independent variables). The goal of MLR is to find the values of the coefficients in the MLR equation that minimize the mean-squared-error (average of the squared error between the observed target values and the values predicted by the model). In addition to assuming a linear relationship between the dependent and independent variables, MLR also assumes that the variables are not correlated (i.e. they do not show multicollinearity) [61, 62].

6.7.2 Logistic Regression

Logistic regression (LR) is a supervised ML algorithm that, despite its name, is used for classification problems. LR is a transformation of linear regression in that it uses a logistic or sigmoid function to model a binary output variable, which restricts the value of y from 0 to 1, as shown in Eq. 6.6.

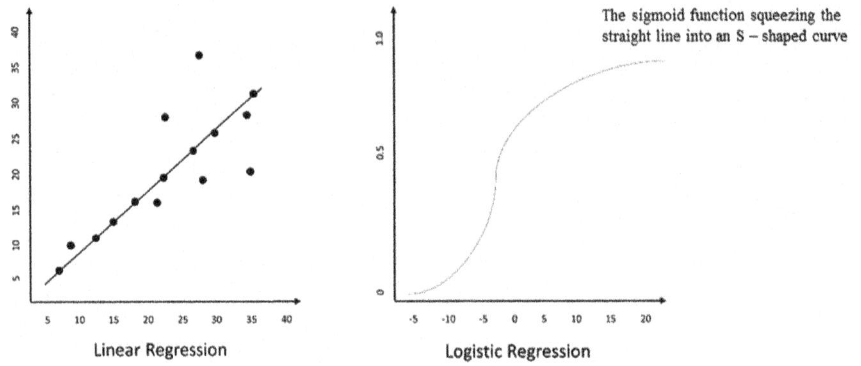

Fig. 6.10 Graphs for linear and logistic regression

$$\text{Sigmoid function}[F(x)] = \frac{1}{1 + e^{-(\beta_0 + \beta_1 x)}}, \tag{6.6}$$

where $\beta_0 + \beta_1 x$ is analogous to $y = mx + c$ for linear regression

In its most basic form, logistic regression is used to predict binary outcomes, but it can also be extended to multiclass labels (multinomial logistic regression). Since LR takes a linear combination of features and applies a nonlinear sigmoidal function, it does not require a linear relationship between the target and predictor variables (shown in Fig. 6.10).

The vertical axis denotes the probability of a given classification, and the horizontal axis contains different values of x. Model predictions are interpreted as the probability of a sample belonging to a particular class. LR is less prone to overfitting, and its interpretability is a major advantage over black-box methods such as neural networks, however, their simplistic nature may be a drawback when working with rich and complex data [87, 88].

6.7.3 *Naïve Bayes*

The Naïve-Bayes is a probabilistic algorithm used primarily for classification tasks and assigns the most likely class for each sample by applying the Bayes' theorem. The Bayes' theorem is a formula for calculating conditional probabilities and is given in Eq. 6.7.

$$P(A/B) = \frac{P(B/A)P(A)}{P(B)}, \tag{6.7}$$

where $P(A/B)$ is posterior probability (probability of hypothesis A given B is an observed outcome), $P(B/A)$ is likelihood probability (probability of event B given A

is an observed outcome), $P(A)$ is prior probability (probability of hypothesis before observing evidence) and $P(B)$ is marginal probability (probability of event B).

The Naïve-Bayes algorithm assumes that all the variables in the feature space are independent of each other (hence the term 'naïve'), an assumption which may not always be true. Despite this, Naïve-Bayes performs well on datasets with non-independent predictors and works well with small and/or noisy datasets. Naïve-Bayes performs especially well when the input data is categorical. It is not the algorithm of choice for high-dimensionality problems or when the feature space comprises of many continuous variables, since the latter case requires mathematical transformations on the input data [90].

6.7.4 *Support Vector Machines*

Support vector machine (SVM) is a supervised ML algorithm that is commonly used for both classification and regression problems. SVM works by plotting each data point in n-dimensional space where n is the number of features in the input data, the numerical value of each feature is a particular coordinate, and the support vectors are the coordinates of each observation [61]. The process of training an SVM model is to identify a 'hyperplane' that maximizes the distance between the support vectors of the two class labels [91] (illustrated in Fig. 6.11).

For data that is not linearly separable, SVM introduces a method known as the kernel trick. The input data is mapped onto a higher dimensional space using a kernel function and separated in that space using a maximum margin hyperplane. The most used kernels include the linear kernel, the polynomial kernel, the radial basis function kernel, and the sigmoid kernel. SVMs are one of the most widely used ML algorithms

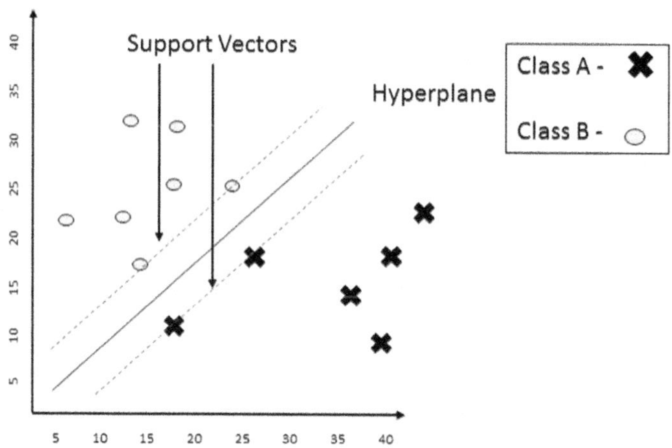

Fig. 6.11 SVM illustration showing the hyperplane that best separates the two classes by maximizing the distance between the support vectors

owing to their effectiveness in high dimensions but are computationally expensive and may not be the method of choice for large datasets [89].

6.7.5 Tree-Based Methods

Decision trees, useful for both classification and regression problems are intuitive ML tools that utilize a tree-like structure for decision making. Analogous to trees in nature, decision trees comprise of three types of nodes: a root node (from which the tree starts), decision nodes (branches) that split the data into subsets, and terminal nodes (the leaves) to assign the data to a target property (shown in Fig. 6.12) [61]. The algorithm starts by searching the feature space and selecting the feature that best separates the classes, and assigns the identified feature to the root node. The subsets created are searched again for identifying features for further separating the data, these features are assigned to the decision nodes [92]. This process is repeated iteratively until all the samples are predicted satisfactorily or if further partitioning does not lead to an improved outcome [61].

Tree-based methods use goodness functions such as Gini scores, gain ratios, and information gain [92]. Various algorithms can be used for constructing decision trees, and among these, the most used is the random forest (RF) algorithm. RF is an ensemble learning method wherein several decision trees (collectively known as a forest) are built using bootstrapped samples of the data and a consensus prediction is made. This solves the problem of overfitting while improving accuracy. RF is commonly used owing to its superior performance and ease of implementation, as there are only two parameters the user needs to define while building the forest: the number of trees and the number of features in each tree [61].

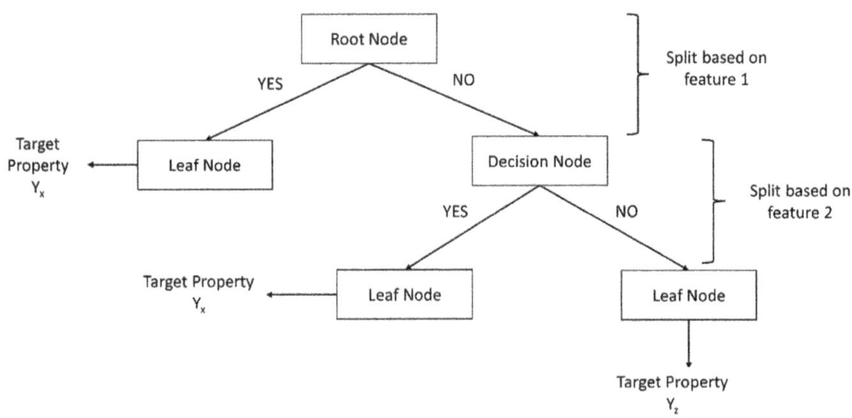

Fig. 6.12 A typical decision tree

6.7.6 Artificial Neural Networks

The idea of artificial neural networks (ANNs) originates from the functioning of the neurons in an animal brain. The architecture of an ANN comprises of three essential layers: the input layer, the hidden layers (may be one or more depending on the complexity of the ANN), and the output layer. The input layer is the first layer in an ANN and receives the training data. The hidden layers perform computations on the input data and recognize patterns and are displayed by the output layer as the results (shown in Fig. 6.13).

Each node in the input layer corresponds to an independent variable and is connected to the hidden layer, and each node in the hidden layer denotes a dependent variable and is connected to the output layer. Each neuron in the neural network has a 'weight' parameter associated with it and receive inputs as signals in accordance with their respective weights. A summation function is used to calculate the combined input signals which are passed through an activation function. The activation function then maps the input to produce an output from the neuron. Examples of activation functions include the hyperbolic tangent function, the softmax function, and the rectified linear unit function. Models learn by adjusting the weights of the neurons, which result in modified outputs for each input. A schematic representation of an artificial neuron is illustrated in Fig. 6.14.

Despite tremendous increase in the use of ANNs in ML tasks, some of its shortcomings remain unsolved. ANNs are referred to as 'black-boxes', because the input and output from the neuron are known, but not what happens inside it. As a result, model interpretation for chemistry problems can oftentimes be challenging [61, 89, 93–95].

There are several reports of ML methods used to tackle the resistance problem in HIV. This chapter will cover some of the more recent work in this area. For a description of previous works, we refer readers to a review by Reimenschneider and Heider [96] (summarized in Table 6.5).

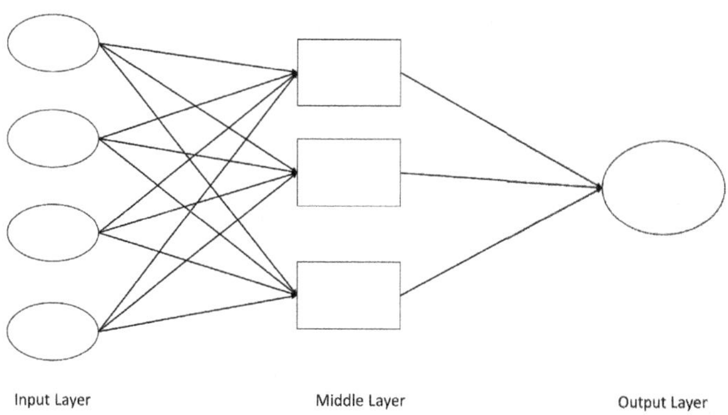

Fig. 6.13 Basic neural network architecture

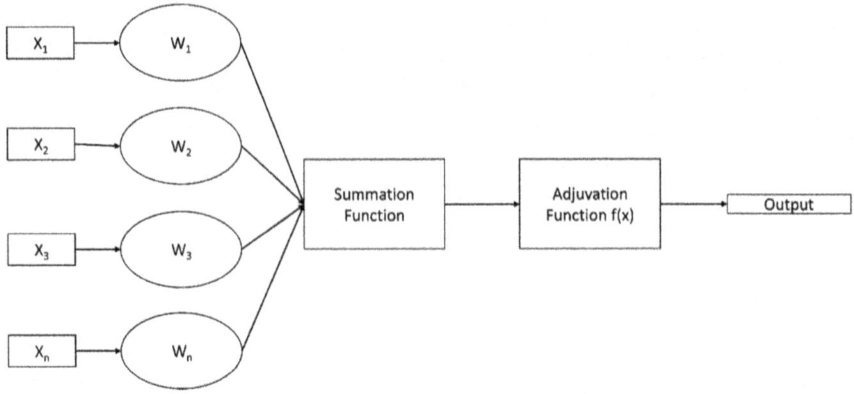

Fig. 6.14 Schematic representation of an artificial neuron

Table 6.5 Summary of reported literature on ML methods

Data	Target/s	ML algorithm	Summary	References
55,000 reverse transcriptase sequences	HIV-1 reverse transcriptase	NB, LR, and RF	An attempt has been made to differentiate between RTI experienced and RTI naïve population, along with the discovery of six new mutations associated with drug resistance	[97]
Genotype–phenotype data for 21 drugs obtained from StanfordDB	HIV-1 reverse transcriptase, HIV protease, HIV integrase	RF and SVM	Drug fold values were modelled as a function of protein sequence represented as physicochemical properties using a weighted ML approach. The model was able to build satisfactorily predictive models for 13 out 21 HIV approved drugs	[98]

(continued)

Table 6.5 (continued)

Data	Target/s	ML algorithm	Summary	References
NIAID ChemDB HIV, Opportunistic infection and tuberculosis Therapeutics database	HIV-1 wild-type cell-based and reverse transcriptase DNA polymerase inhibition	NB, DT, RF, SVM, kNN[a], DNN[a], consensus	Models for HIV inhibition were developed using publicly available data, and a comparison of different ML methods demonstrated that SVM, deep learning, and consensus models showed the most promising results	[99]
Stanford HIV drug resistance database	HIV-1 reverse transcriptase, HIV protease, HIV integrase	RF and SVM	Sequences from HIV isolates and their susceptibility to antiretroviral drugs were modelled using weighted categorical kernel functions, which resulted in superior models for predicting drug resistance, especially in the case of HIV-1 RT	[100]
Stanford HIV Drug Resistance Database	HIV-1 reverse transcriptase and HIV protease	RF	Amino acid sequences were represented as short fragments and used as descriptors to model resistance of RT and protease to marketed drugs. They conclude that model performance is more sensitive to certain drugs than descriptor type	[101]
Stanford HIV Drug Resistance Database	HIV-1 reverse transcriptase and HIV protease	MLP, BRNN[a] and CNN	CNNs were reported as the best performing DL method. The black-box problem was addressed by feature importance analysis for identification of known and novel mutations as biologically relevant features, giving an interpretable DL model	[102]

(continued)

Table 6.5 (continued)

Data	Target/s	ML algorithm	Summary	References
HIV-1 RT mutant susceptibility data from ChEMBL	HIV-1 reverse transcriptase	NB	Loss of HIV-RT activity on mutations in three major residues: Y181, K103, and L100 was studied. The models had an average ROC AUC of 0.920	[103]
Darunavir-bound HIV-1 protease variants	HIV protease	LR	Data collected from atomistic simulations of HIV protease variants in complex with darunavir were used as input variables in ML, and mechanistic insignts were provided on how alterations in the darunavir-protease complex can affect drug binding	[104]
Darunavir-bound HIV-1 protease variants	HIV protease	Ordinary least squares	ML was coupled with parallel MD simulations, and the linear regression model correlating non-covalent drug-receptor interactions of darunavir for HIV protease and its mutants was accurate and performed well on a test set of protease variants	[105]
Resistance data for six antiretroviral drug classes comprising of 23 drugs	HIV-1 reverse transcriptase, HIV protease, HIV integrase	RF	A web server named SHIVA was developed for resistance prediction of some commonly used antiretroviral drugs. SHIVA was found to be superior to other popular server-based prediction tools such as geno2pheno, HIVdb, and WebPSSM	[106]

(continued)

Table 6.5 (continued)

Data	Target/s	ML algorithm	Summary	References
Protein sequences with data for ten drugs were collected from the HIV drug resistance database	HIV-1 reverse transcriptase, HIV protease	Binary relevance classifiers, classifier chains, and ensembles of classifier chains	The cross-resistance problem in HIV was addressed by building multi-label classification models. Multi-label learning was able to improve classification accuracy as compared to binary classifiers	[107]
Stanford HIV drug resistance database	HIV protease	PLS[a], RF, LGBM[a], and SVR	A model for resistance prediction was developed based on a homology modelling and molecular field mapping approach. The model based on the CoMFA methodology was robust, with the LGBM algorithm performing the best	[108]
Stanford HIV Drug Resistance database	HIV-1 reverse transcriptase, HIV protease	ANN	A subtype-specific approach for the prediction of fold resistance was followed, with the ANN model comparable to reported models in the literature	[109]

[a]*kNN* k-nearest neighbour, *DNN* deep neural network, *BRNN* bidirectional recurrent neural networks, *PLS* partial least squares, *LGBM* light gradient boosting machine

6.8 Conclusion

The surge in computational power along with the abundance of available biological data has paved the way for in silico methods to play an instrumental role in anti-HIV drug discovery and resistance mapping. Remarkable results have been achieved in our understanding of HIV biology, the impact of mutations, and subsequent development of resistance and in the design of novel molecules that are active against these resistant strains of the virus. The identification and elucidation of new drug targets has greatly benefitted structure-based methods such as molecular docking and MD simulations, whereas QSAR modelling and other ML-based methods have been aided by the development of publicly available databases such as the StanfordDB. However, several challenges remain. Data quality is an ongoing concern, and any theoretical model must be validated rigorously and evaluated for their usefulness in a real-life scenario. In this regard, it is essential that a collaborative framework is established that enables rapid experimental validation of hypotheses generated in silico. Hopefully, computational approaches will continue to play an important role and will enable improved treatment of HIV and eventually, its eradication.

References

1. Charneau P, Borman AM, Quillent C, Guétard D, Chamaret S, Cohen J, Rémy G, Montagnier L, Clavel F (1994) Isolation and envelope sequence of a highly divergent HIV-1 isolate: definition of a new HIV-1 group. Virology 205(1):247–253. https://doi.org/10.1006/viro.1994.1640
2. HIV/AIDS. https://www.who.int/news-room/fact-sheets/detail/hiv-aids
3. Seitz R (2016) Human immunodeficiency virus (HIV). Transfus Med Hemotherapy 43(3):203–222. https://doi.org/10.1159/000445852
4. Waymack J, Sundareshan V (2021) Acquired immune deficiency syndrome; StatPearls Publishing
5. How Is HIV Transmitted? https://www.hiv.gov/hiv-basics/overview/about-hiv-and-aids/how-is-hiv-transmitted
6. Rossi E, Meuser ME, Cunanan CJ, Cocklin S (2021) Structure, function, and interactions of the Hiv-1 capsid protein. Life 11(2):1–25. https://doi.org/10.3390/life11020100
7. Kirchhoff F (2016) Encyclopedia of AIDS. Encycl AIDS 2016 (January). https://doi.org/10.1007/978-1-4614-9610-6
8. Ugolini S, Mondor I, Sattentau QJ (1999) HIV-1 attachment : another look 99:144–149. https://doi.org/10.1016/S0966-842X(99)01474-2
9. HIV/AIDS Glossary. https://clinicalinfo.hiv.gov/en/glossary/life-cycle
10. De Clercq E (2009) Anti-HIV drugs: 25 compounds approved within 25 years after the discovery of HIV. Int J Antimicrob Agents 33(4):307–320. https://doi.org/10.1016/j.ijantimicag.2008.10.010
11. Cilento ME, Kirby KA, Sarafianos SG (2021) Avoiding drug resistance in HIV reverse transcriptase. Chem Rev 121(6):3271–3296. https://doi.org/10.1021/acs.chemrev.0c00967
12. Portegies P (2002) Antiretroviral therapeutics. J Neurovirol 8(SUPPL. 2):148–150. https://doi.org/10.1080/13550280290167966
13. Gu SX, Zhu YY, Wang C, Wang HF, Liu GY, Cao S, Huang L (2020) Recent discoveries in HIV-1 reverse transcriptase inhibitors. Curr Opin Pharmacol 54:166–172. https://doi.org/10.1016/j.coph.2020.09.017

14. Maldarelli F (2006) HIV drug resistance. Handb Pediatr HIV Care, 2nd ed, pp 397–414. https://doi.org/10.1017/CBO9780511544781.016
15. Preston BD, Poiesz BJ, Loeb LA (1998) Fidelity of HIV-1 reverse transcriptase. Science (80-.) 242(4882):1168–1171. https://doi.org/10.1126/science.2460924
16. Vandamme AM, Van Laethem K, De Clercq E (1999) Managing resistance to anti-HIV drugs: an important consideration for effective disease management. Drugs 57(3):337–361. https://doi.org/10.2165/00003495-199957030-00006
17. Collier DA, Monit C, Gupta RK (2019) The impact of HIV-1 drug escape on the global treatment landscape. Cell Host Microbe 26(1):48–60. https://doi.org/10.1016/j.chom.2019.06.010
18. Anderson A (2003) The process of structure- based drug design. Chem Biol 10:787–797. https://doi.org/10.1016/j.chembiol.2003.09.002
19. Koshland DE (1995) The key-lock theory and the induced fit theory. Angew Chemie Int Ed English 33(23–24):2375–2378. https://doi.org/10.1002/anie.199423751
20. Saikia S, Bordoloi M (2019) Molecular docking: challenges, advances and its use in drug discovery perspective. Curr Drug Targets 20(5):501–521. https://doi.org/10.2174/1389450119666181022153016
21. Śledź P, Caflisch A (2018) Protein structure-based drug design: from docking to molecular dynamics. Curr Opin Struct Biol 48:93–102. https://doi.org/10.1016/j.sbi.2017.10.010
22. Karplus M, McCammon JA (2010) Molecular dynamics simulations of biomolecules. Mol Simul 36(13):1035–1044. https://doi.org/10.1080/08927022.2010.501797
23. Talele T, Khedkar S, Rigby A (2010) Successful applications of computer aided drug discovery: moving drugs from concept to the clinic. Curr Top Med Chem 10(1):127–141. https://doi.org/10.2174/156802610790232251
24. Fan J, Fu A, Zhang L (2019) Progress in molecular docking. Quant Biol 7(2):83–89. https://doi.org/10.1007/s40484-019-0172 y
25. Pettersen EF, Goddard TD, Huang CC, Couch GS, Greenblatt DM, Meng EC, Ferrin TE (2004) UCSF chimera—a visualization system for exploratory research and analysis. J Comput Chem 25(13):1605–1612. https://doi.org/10.1002/jcc.20084
26. Almerico AM, Tutone M, Lauria A (2008) Docking and multivariate methods to explore HIV-1 drug-resistance: a comparative analysis. J Comput Aided Mol Des 22(5):287–297. https://doi.org/10.1007/s10822-008-9186-7
27. Morris GM, Huey R, Lindstrom W, Sanner MF, Belew R, Goodsell D, Olson A (2009) AutoDock4 and AutoDockTools4: automated docking with selective receptor flexibility. J Comput Chem. https://doi.org/10.1002/jcc.21256
28. Sybyl X Molecular Modelling Software Packages. TRIPOS Associates, Inc.
29. Trott O, Olson AJ (2009) Software news and update AutoDock vina: improving the speed and accuracy of docking with a new scoring function, efficient optimization, and multithreading. J Comput Chem 31(2). https://doi.org/10.1002/jcc.21334
30. Bitencourt-Ferreira G, Filgueira de Azevedo Jr W (2019) How docking programs work. In: Docking screens for drug discovery, Springer, pp 35–50
31. Forli SR (2010) AutoDock VS: an automated tool for preparing autodock virtual screenings
32. Friesner RA, Banks JL, Murphy RB, Halgren TA, Klicic JJ, Mainz DT, Repasky MP, Knoll EH, Shelley M, Perry JK, Shaw DE, Francis P, Shenkin PS (2004) Glide: a new approach for rapid, accurate docking and scoring. 1. method and assessment of docking accuracy. J Med Chem 47(7):1739–1749. https://doi.org/10.1021/jm0306430
33. Vora J, Patel S, Sinha S, Sharma S, Srivastava A, Chhabria M, Shrivastava N (2019) Molecular docking, QSAR and ADMET based mining of natural compounds against prime targets of HIV. J Biomol Struct Dyn 37(1):131–146. https://doi.org/10.1080/07391102.2017.1420489
34. Tarasova O, Poroikov V, Veselovsky A (2018) Molecular docking studies of HIV-1 resistance to reverse transcriptase inhibitors: mini-review. Molecules 23(5):11–13. https://doi.org/10.3390/molecules23051233
35. Singh VK, Srivastava R, Gupta PSS, Naaz F, Chaurasia H, Mishra R, Rana MK, Singh RK (2021) Anti-HIV potential of diarylpyrimidine derivatives as non-nucleoside reverse transcriptase inhibitors: design, synthesis, docking, TOPKAT analysis and molecular dynamics

simulations. J Biomol Struct Dyn 39(7):2430–2446. https://doi.org/10.1080/07391102.2020.1748111
36. Fraĭczek T, Siwek A, Paneth P (2013) Assessing molecular docking tools for relative biological activity prediction: a case study of triazole HIV-1 NNRTIs. J Chem Inf Model 53(12):3326–3342. https://doi.org/10.1021/ci400427a
37. Liu G, Wang W, Wan Y, Ju X, Gu S (2018) Application of 3D-QSAR, pharmacophore, and molecular docking in the molecular design of diarylpyrimidine derivatives as HIV-1 nonnucleoside reverse transcriptase inhibitors. Int J Mol Sci 19(5). https://doi.org/10.3390/ijms19051436
38. Makarasen A, Kuno M, Patnin S, Reukngam N, Khlaychan P, Deeyohe S, Intachote P, Saimanee B, Sengsai S, Boonsri P, Chaivisuthangkura A, Sirithana W, Techasakul S (2019) Molecular docking studies and synthesis of amino-oxy-diarylquinoline derivatives as potent non-nucleoside HIV-1 reverse transcriptase inhibitors. Drug Res (Stuttg) 69(12):671–682. https://doi.org/10.1055/a-0968-1150
39. Shanty AA, Raghu KG, Mohanan PV (2019) Synthesis, characterization: spectral and theoretical, molecular docking and in vitro studies of copper complexes with HIV RT enzyme. J Mol Struct 1197:154–163. https://doi.org/10.1016/j.molstruc.2019.06.097
40. Gao Y, Chen Y, Tian Y, Zhao Y, Wu F, Luo X, Ju X, Liu G (2019) In Silico study of 3-hydroxypyrimidine-2,4-diones as inhibitors of HIV RT-associated RNase H using molecular docking, molecular dynamics, 3D-QSAR, and pharmacophore models. New J Chem 43(43):17004–17017. https://doi.org/10.1039/c9nj03353j
41. Faghihi K, Safakish M, Zebardast T, Hajimahdi Z, Zarghi A (2019) Molecular docking and QSAR study of 2-benzoxazolinone, quinazoline and diazocoumarin derivatives as anti-HIV-1 agents. Iran J Pharm Res 18(3):1253–1263. https://doi.org/10.22037/ijpr.2019.1100746
42. Turkovic N, Ivkovic B, Kotur-Stevuljevic J, Tasic M, Marković B, Vujic Z (2020) Molecular docking, synthesis and anti-HIV-1 protease activity of novel chalcones. Curr Pharm Des 26(8):802–814. https://doi.org/10.2174/1381612826666200203125557
43. Hajimahdi Z, Zabihollahi R, Aghasadehi MR, Zarghi A (2019) Design, synthesis, docking studies and biological activities novel 2,3-Diaryl-4-quinazolinone derivatives as anti-HIV-1 agents. Curr HIV Res 17(3)
44. McCammon JA, Gelin BR, Karplus M (1977) Dynamics of folded proteins. Nature 267. https://doi.org/10.1038/267585a0
45. Wang J, Wolf RM, Caldwell JW, Kollman PA, Case DA (2004) Development and testing of a general amber force field. J Comput Chem 56531(9):1157–1174. https://doi.org/10.1002/jcc.20035
46. Vanommeslaeghe K, Hatcher E, Acharya C, Kundu S, Zhong S, Shim J, Darian E, Guvench O, Lopes P, Vorobyov I, Mackerell AD Jr (2010) CHARMM general force field: a force field for drug-like molecules compatible with the CHARMM all-atom additive biological force fields. J Comput Chem 31(4):671–690. https://doi.org/10.1002/jcc.21367
47. Schmid N, Eichenberger AP, Choutko A, Riniker S, Winger M, Mark AE, Van Gunsteren WF (2011) Definition and testing of the GROMOS force-field versions 54A7 and 54B7. Eur Biophys J 40(7):843–856. https://doi.org/10.1007/s00249-011-0700-9
48. Case DA, Cheatham TE, Darden T, Gohlke H, Luo R, Merz KM, Onufriev A, Simmerling C, Wang B, Woods RJ (2005) The amber biomolecular simulation programs. J Comput Chem 26(16):1668–1688. https://doi.org/10.1002/jcc.20290
49. Phillips JC, Braun R, Wang W, Gumbart J, Tajkhorshid E, Villa E, Chipot C, Skeel RD, Kalé L, Schulten K (2005) Scalable molecular dynamics with NAMD. J Comput Chem 26(16):1781–1802. https://doi.org/10.1002/jcc.20289
50. Brooks BR, Bruccoleri RE, Olafson BD, States DJ, Swaminathan S, Karplus M (1983) CHARMM: a program for macromolecular energy, minimization, and dynamics calculations. J Comput Chem 4(2):187–217. https://doi.org/10.1002/jcc.540040211
51. Christen M, Hünenberger PH, Bakowies D, Baron R, Bürgi R, Geerke DP, Heinz TN, Kastenholz MA, Kräutler V, Oostenbrink C, Peter C, Trzesniak D, Van Gunsteren WF (2005) The GROMOS software for biomolecular simulation: GROMOS05. J Comput Chem 26(16):1719–1751. https://doi.org/10.1002/jcc.20303

52. Wang W, Tian Y, Wan Y, Gu S, Ju X, Luo X, Liu G (2019) Insights into the key structural features of N1-ary-benzimidazols as HIV-1 NNRTIs using molecular docking, molecular dynamics, 3D-QSAR, and pharmacophore modeling. Struct Chem 30(1):385–397. https://doi.org/10.1007/s11224-018-1204-3
53. Huang YMM, Raymundo MAV, Chen W, Chang CEA (2017) Mechanism of the association pathways for a pair of fast and slow binding ligands of HIV-1 protease. Biochemistry 56(9):1311–1323. https://doi.org/10.1021/acs.biochem.6b01112
54. Miao Y, Huang YMM, Walker RC, McCammon JA, Chang CEA (2018) Ligand binding pathways and conformational transitions of the HIV protease. Biochemistry 57(9):1533–1541. https://doi.org/10.1021/acs.biochem.7b01248
55. Chen Y, Tian Y, Gao Y, Wu F, Luo X, Ju X, Liu G (2020) In silico design of novel HIV-1 NNRTIs based on combined modeling studies of Dihydrofuro[3,4-d]Pyrimidines. Front Chem 8(March):1–17. https://doi.org/10.3389/fchem.2020.00164
56. Martis EAF, Coutinho EC (2019) Free energy-based methods to understand drug resistance mutations, 1–24. https://doi.org/10.1007/978-3-030-05282-9_1
57. Nandy B, Saurabh S, Sahoo AK, Dixit NM, Maiti PK (2015) The SPL7013 dendrimer destabilizes the HIV-1 Gp120-CD4 complex. Nanoscale 7(44):18628–18641. https://doi.org/10.1039/c5nr04632g
58. Sirous H, Chemi G, Gemma S, Butini S, Debyser Z, Christ F, Saghaie L, Brogi S, Fassihi A, Campiani G, Brindisi M (2019) Identification of novel 3-hydroxy-pyran-4-one derivatives as potent HIV-1 integrase inhibitors using in silico structure-based combinatorial library design approach. Front Chem 7(August):1–20. https://doi.org/10.3389/fchem.2019.00574
59. Cele FN, Ramesh M, Soliman MES (2016) Per-residue energy decomposition pharmacophore model to enhance virtual screening in drug discovery: a study for identification of reverse transcriptase inhibitors as potential anti-hiv agents. Drug Des Devel Ther 10:1365–1377. https://doi.org/10.2147/DDDT.S95533
60. Halder AK, Honarparvar B (2019) Molecular alteration in drug susceptibility against subtype B and C-SA HIV-1 proteases: MD study. Struct Chem 30(5):1715–1727. https://doi.org/10.1007/s11224-019-01305-0
61. Roy K, Kar S, Das RN (2015) Understanding the basics of QSAR for applications in pharmaceutical sciences and risk assessment https://doi.org/10.1016/C2014-0-00286-9
62. Verma J, Khedkar V, Coutinho E (2010) 3D-QSAR in drug design—a review. Curr Top Med Chem 10(1):95–115. https://doi.org/10.2174/156802610790232260
63. Hansch C, Fujita T (1964) ρ-σ-π analysis. a method for the correlation of biological activity and chemical structure. J Am Chem Soc, 86(8):1616–1626. https://doi.org/10.1021/ja01062a035
64. Langer T, Hoffmann RD Pharmacophores and Pharmacophore Searches
65. Wermuth CG, Ganellin CR, Lindberg P, Mitscher LA (1998) Glossary of terms used in medicinal chemistry. Pure Appl Chem 70(5):1129–1143
66. Qing X, Lee XY, De Raeymaeker J, Tame JR, Zhang KY, De Maeyer M, Voet AR (2014) Pharmacophore modeling: advances, limitations, and current utility in drug discovery. J Receptor Ligand Channel Res 7:81–92. https://doi.org/10.2147/JRLCR.S46843
67. Tian Y, Zhang S, Yin H, Yan A (2020) Quantitative structure-activity relationship (QSAR) models and their applicability domain analysis on HIV-1 protease inhibitors by machine learning methods. Chemom Intell Lab Syst 196:103888. https://doi.org/10.1016/j.chemolab.2019.103888
68. Halder AK (2018) Finding the structural requirements of diverse HIV-1 protease inhibitors using multiple QSAR modelling for lead identification. SAR QSAR Environ Res 29(11):911–933. https://doi.org/10.1080/1062936X.2018.1529702
69. Tong J, Lei S, Qin S, Wang Y (2018) QSAR studies of TIBO derivatives as HIV-1 reverse transcriptase inhibitors using HQSAR. CoMFA and CoMSIA J Mol Struct 1168:56–64. https://doi.org/10.1016/j.molstruc.2018.05.005
70. Beglari M, Goudarzi N, Shahsavani D, Arab Chamjangali M, Dousti R (2020) QSAR modeling of anti-HIV activity for DAPY-like derivatives using the mixture of ligand-receptor binding

information and functional group features as a new class of descriptors. Netw Model Anal Heal Inf Bioinforma 9(1). https://doi.org/10.1007/s13721-020-00261-8
71. Wang Y, Chang J, Wang J, Zhong P, Zhang Y, Lai CC, He Y (2018) 3D-QSAR studies of S-DABO derivatives as non-nucleoside HIV-1 reverse transcriptase inhibitors. Lett Drug Des Discov 16(8):868–881. https://doi.org/10.2174/1570180815666180810112321
72. Han D, Tan J, Zhou Z, Li C, Zhang X, Wang C (2018) Combined Topomer CoMFA and hologram QSAR studies of a series of pyrrole derivatives as potential HIV fusion inhibitors. Med Chem Res 27(7):1770–1781. https://doi.org/10.1007/s00044-018-2190-0
73. Liu G, Wan Y, Wang W, Fang S, Gu S, Ju X (2019) Docking-based 3D-QSAR and pharmacophore studies on diarylpyrimidines as non-nucleoside inhibitors of HIV-1 reverse transcriptase. Mol Divers 23(1):107–121. https://doi.org/10.1007/s11030-018-9860-1
74. Bhole RP, Bonde CG, Bonde SC, Chikhale RV, Wavhale RD (2021) Pharmacophore model and atom-based 3D quantitative structure activity relationship (QSAR) of human immunodeficiency virus-1 (HIV-1) capsid assembly inhibitors. J Biomol Struct Dyn 39(2):718–727. https://doi.org/10.1080/07391102.2020.1715258
75. Cutinho PF, Roy J, Anand A, Cheluvaraj R, Murahari M, Chimatapu HSV (2020) Design of metronidazole derivatives and flavonoids as potential non-nucleoside reverse transcriptase inhibitors using combined ligand- and structure-based approaches. J Biomol Struct Dyn 38(6):1626–1648. https://doi.org/10.1080/07391102.2019.1614094
76. Vangala R, Sivan SK, Peddi SR, Manga V (2020) Computational design, synthesis and evaluation of new sulphonamide derivatives targeting HIV-1 Gp120. J Comput Aided Mol Des 34(1):39–54. https://doi.org/10.1007/s10822-019-00258-0
77. Mirza MU, Saadabadi A, Vanmeert M, Salo-Ahen OMH, Abdullah I, Claes S, De Jonghe S, Schols D, Ahmad S, Froeyen M (2020) Discovery of HIV entry inhibitors via a hybrid CXCR4 and CCR5 receptor pharmacophore-based virtual screening approach. Eur J Pharm Sci 155(July):105537. https://doi.org/10.1016/j.ejps.2020.105537
78. Ravichandran V, Rohini K, Harish R, Parasuraman S, Sureshkumar K (2019) Insights into the key structural features of triazolothienopyrimidines as anti-HIV agents using QSAR, molecular docking, and pharmacophore modeling. Struct Chem 30(4):1471–1484. https://doi.org/10.1007/s11224-019-01304-1
79. Deng J, Yang Z, Ojima I, Samaras D, Wang F (2022) Artificial intelligence in drug discovery: applications and techniques. Brief Bioinform 23(1):1–65. https://doi.org/10.1093/bib/bbab430
80. R: A language and environment for statistical computing (2013)
81. Sanner MF (1999) Python : a programming language for software integration and development. J Mol Graph Model 17(1):57–61
82. Dixon SL, Duan J, Smith E, Von Bargen CD, Sherman W, Repasky MP (2016) AutoQSAR: an automated machine learning tool for best-practice quantitative structure-activity relationship modeling. Future Med Chem 8(15):1825–1839. https://doi.org/10.4155/fmc-2016-0093
83. Vamathevan J, Clark D, Czodrowski P, Dunham I, Ferran E, Lee G, Li B, Madabhushi A, Shah P, Spitzer M, Zhao S (2019) Applications of machine learning in drug discovery and development. Nat Rev Drug Discov 18(6):463–477. https://doi.org/10.1038/s41573-019-0024-5
84. Jumper J, Evans R, Pritzel A, Green T, Figurnov M, Ronneberger O, Tunyasuvunakool K, Bates R, Žídek A, Potapenko A, Bridgland A, Meyer C, Kohl SAA, Ballard AJ, Cowie A, Romera-Paredes B, Nikolov S, Jain R, Adler J, Back T, Petersen S, Reiman D, Clancy E, Zielinski M, Steinegger M, Pacholska M, Berghammer T, Bodenstein S, Silver D, Vinyals O, Senior AW, Kavukcuoglu K, Kohli P, Hassabis D (2021) Highly accurate protein structure prediction with AlphaFold. Nature 596(7873):583–589. https://doi.org/10.1038/s41586-021-03819-2
85. Mayr A, Klambauer G, Unterthiner T, Hochreiter S (2016) DeepTox: toxicity prediction using deep learning. Front Environ Sci 3(FEB). https://doi.org/10.3389/fenvs.2015.00080
86. Artrith N, Butler KT, Coudert FX, Han S, Isayev O, Jain A, Walsh A (2021) Best practices in machine learning for chemistry. Nat Chem 13(6):505–508. https://doi.org/10.1038/s41557-021-00716-z

87. Belyadi H, Haghighat A (2021) Machine learning guide for oil and gas using python; Gulf Professional Publishing. https://doi.org/10.1016/c2019-0-03617-5
88. Subasi A (2020) Practical machine learning for data analysis using python. https://doi.org/10.1016/B978-0-12-821379-7.00008-4
89. Carracedo-Reboredo P, Liñares-Blanco J, Rodríguez-Fernández N, Cedrón F, Novoa FJ, Carballal A, Maojo V, Pazos A, Fernandez-Lozano C (2021) A review on machine learning approaches and trends in drug discovery. Comput Struct Biotechnol J 19:4538–4558. https://doi.org/10.1016/j.csbj.2021.08.011
90. Rhys HI (2020) Machine learning with R, the Tidyverse and MLR; Manning Publications
91. Pisner DA, Schnyer DM (2019) Support vector machine; Elsevier Inc., 2019. https://doi.org/10.1016/B978-0-12-815739-8.00006-7
92. Djuris J, Ibric S, Djuric Z (2013) Neural computing in pharmaceutical products and process development; Woodhead Publishing Limited https://doi.org/10.1533/9781908818324.91
93. Yacim JA, Boshoff DGB (2018) Impact of artificial neural networks training algorithms on accurate prediction of property values. J Real Estate Res 40(3):375–418. https://doi.org/10.1080/10835547.2018.12091505
94. Puri M, Pathak Y, Sutariya VK, Tipparaju S, Moreno W (2015) Artificial neural network for drug design, delivery and disposition. https://doi.org/10.1016/C2014-0-00253-5
95. Krogh A (2008) What are artificial neural networks? Nat Biotechnol 26(2):195–197. https://doi.org/10.1038/nbt1386
96. Riemenschneider M, Heider D (2016) Current approaches in computational drug resistance prediction in HIV. Curr HIV Res 14(4):307–315
97. Blassel L, Tostevin A, Villabona-Arenas CJ, Peeters M, Hué S, Gascuel O (2021) Using machine learning and big data to explore the drug resistance landscape in HIV. PLoS Comput Biol 17(8):1–21. https://doi.org/10.1371/journal.pcbi.1008873
98. Cai Q, Yuan R, He J, Li M, Guo Y (2021) Predicting HIV drug resistance using weighted machine learning method at target protein sequence-level. Mol Divers 25(3):1541–1551. https://doi.org/10.1007/s11030-021-10262-y
99. Zorn KM, Lane TR, Russo DP, Clark AM, Makarov V, Ekins S (2019) Multiple machine learning comparisons of HIV cell-based and reverse transcriptase data sets. Mol Pharm 16(4):1620–1632. https://doi.org/10.1021/acs.molpharmaceut.8b01297
100. Ramon E, Belanche-Muñoz L, Pérez-Enciso M (2019) HIV drug resistance prediction with categorical kernel functions. BMC Bioinformatics 410(20):233–244. https://doi.org/10.1007/978-3-030-17935-9_22
101. Tarasova O, Biziukova N, Filimonov D, Poroikov V (2018) A computational approach for the prediction of HIV resistance based on amino acid and nucleotide descriptors. Molecules 23(11). https://doi.org/10.3390/molecules23112751
102. Steiner MC, Gibson KM (2020) Techniques on HIV-1 sequence data, pp 1–24
103. Kaiser TM, Burger PB, Butch CJ, Pelly SC, Liotta DC (2018) A machine learning approach for predicting HIV reverse transcriptase mutation susceptibility of biologically active compounds. J Chem Inf Model 58(8):1544–1552. https://doi.org/10.1021/acs.jcim.7b00475
104. Whitfield TW, Ragland DA, Zeldovich KB, Schiffer CA (2020) Characterizing protein-ligand binding using atomistic simulation and machine learning: application to drug resistance in HIV-1 protease. J Chem Theory Comput 16(2):1284–1299. https://doi.org/10.1021/acs.jctc.9b00781
105. Leidner F, Kurt Yilmaz N, Schiffer CA (2021) Deciphering complex mechanisms of resistance and loss of potency through coupled molecular dynamics and machine learning. J Chem Theory Comput 17(4):2054–2064. https://doi.org/10.1021/acs.jctc.0c01244
106. Riemenschneider M, Hummel T, Heider D (2016) SHIVA—A web application for drug resistance and tropism testing in HIV. BMC Bioinfo 17(1):1–6. https://doi.org/10.1186/s12859-016-1179-2
107. Riemenschneider M, Senge R, Neumann U, Hüllermeier E, Heider D (2016) Exploiting HIV-1 protease and reverse transcriptase cross-resistance information for improved drug resistance prediction by means of multi-label classification. BioData Min. 9(1):1–6. https://doi.org/10.1186/s13040-016-0089-1

108. Ota R, So K, Tsuda M, Higuchi Y, Yamashita F (2021) Prediction of HIV drug resistance based on the 3D protein structure: proposal of molecular field mapping. PLoS One 16(8 August):1–15. https://doi.org/10.1371/journal.pone.0255693
109. Sheik Amamuddy O, Bishop NT, Tastan Bishop Ö (2017) Improving fold resistance prediction of HIV-1 against protease and reverse transcriptase inhibitors using artificial neural networks. BMC Bioinf 18(1):1–7. https://doi.org/10.1186/s12859-017-1782-x

Chapter 7
Recent Insight of the Emerging Severe Fever with Thrombocytopenia Syndrome Virus: Drug Discovery, Therapeutic Options, and Limitations

Shilpa Chatterjee, Arindam Maity, and Debanjan Sen

Abstract Severe fever with thrombocytopenia syndrome virus (SFTSV) also known as Dabie bandavirus of the family Phenuiviridae is a negative-strand RNA virus and a tick-borne virus. Replication of SFTSV into systemic circulation and occurrence of viremia cause cytokine storm and T-cell overstimulation. The event of viremia-induced thrombocytopenia causes reduced platelet count and splenic macrophages, followed by endothelial damages and compromised immune system that cause multi-organ damages. Limited options for specific anti-SFTSV drugs pose significant challenges associated with clinical management of SFTSV infection. This book chapter chiefly emphasizes upon the genetic diversity, geographical distribution, pathogenesis associated with various clinical aspects like symptoms, diagnosis, and available clinical management options. In addition, current research linked with anti-SFTSV drug development is comprehensively portrayed in this review.

Keywords Severe Fever with Thrombocytopenia Syndrome virus (SFTSV) · Dabie bandavirus · Huaiyangshan Banyangvirus · Phenuiviridae · Clinical symptoms of SFTSV infection · Clinical diagnosis of SFTSV · SFTS L protein · SFTSV drug target · Molecular Docking · Molecular Dynamics

---

S. Chatterjee
Department of Biomedical Science, College of Medicine, Chosun University, Gwangju, Republic of Korea
e-mail: shilpachatterjee@chosun.kr

A. Maity
Department of Pharmaceutical Technology, JIS University, Kolkata, India

D. Sen (✉)
Department of Pharmaceutical Technology, BCDA College of Pharmacy & Technology, 78 Jessore Road, Hridaypur, Kolkata 700127, India
e-mail: debanjansen48@gmail.com

S. Kar and J. Leszczynski (eds.), *Current Trends in Computational Modeling for Drug Discovery*, Challenges and Advances in Computational Chemistry and Physics 35,
https://doi.org/10.1007/978-3-031-33871-7_7

7.1 Introduction

Dabie bandavirus, also called severe fever with thrombocytopenia syndrome virus (SFTSV), is a tick-borne virus of the genus *Bandavirus*, belonging to the family *Phenuiviridae*, order *Bunyavirales* [1]. Synonymously SFTSV is also known as *Huaiyangshan Banyangvirus*. The clinical condition caused by SFTSV is known as severe fever with thrombocytopenia syndrome (SFTS) [1]. Although the new nomenclature was accepted by the International Committee on Taxonomy of Viruses (ICTV), SFTSV is the most widely used term in the scientific community. The SFTSV can be classified as follows: Realm (Riboviria), Kingdom (Orthornavirae), Phylum (Negarnaviricota), Class (Ellioviricetes), Order (Bunyavirales), Family (Phenuiviridae), Genus (Bandavirus), Species (Dabie bandavirus).

Figure 7.1 describes the schematic diagram of SFTSV. SFTSV is a negative-strand RNA that has been divided into large (L), medium (M), and small (S) segments [2]. The RNA-dependent RNA polymerase (RdRp), which acts as a viral transcriptase/replicase, is encoded by the L segment. The M segment codes for a membrane protein precursor that matures into the envelope's two glycoproteins, Gn and Gc. The S segment is a two-protein ambisense RNA; the antisense RNA encodes Np and the sense RNA encodes NSs. Np is involved in the encapsidation of viral RNA and the creation of the RNP complex, while NSs interfere with the synthesis of host interferon [3].

Aside from the fact that SFTSV-related health problems are becoming more prevalent in people, the pathogenesis of the SFTS virus in humans is still unknown, and no cure for the virus exists. Avoiding tick bites is a simple approach to protect ourselves from infection. As a result, this disease has developed to cause major health problems in humans in a variety of places around the world. In this chapter, we have discussed about SFTS disease and its causative agent, epidemiology, pathogenesis, diagnosis, and recent development in the treatment.

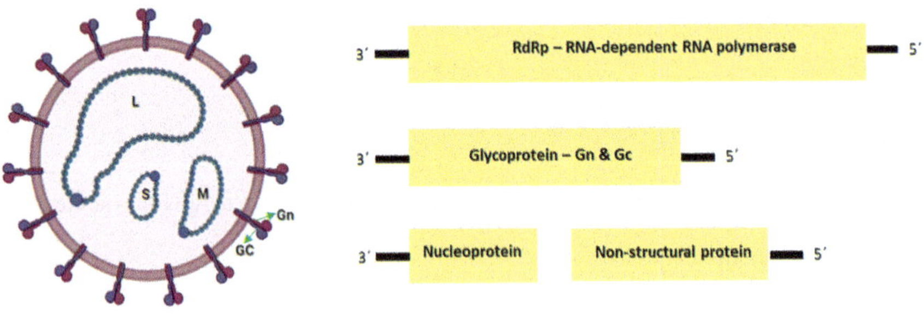

Fig. 7.1 Schematic diagram of SFTSV

7.2 Geographical Distribution and Its Genetic Diversity

The first case of SFTS was recorded in Henan and Hubei provinces of China in 2009, and it quickly spread to neighboring provinces in the country's central, eastern, and northeastern regions [2]. In 2012, SFTS cases were also recorded in Japan and Korea, as well as Vietnam and Taiwan [4–7]. The mechanisms behind the spread of SFTS are unknown; however, the spread of emerging viruses is commonly attributed to two main mechanisms: increased contact between wildlife and human populations and geographical spread of the hematophagous arthropod vector or their vertebrate host outside of the endemic area. Ticks carrying the parasite *H. longicornis* are a widespread parasite of migratory birds that breed and travel between endemic sites in China, Korea, and Japan [2]. Furthermore, the Asia–Pacific range of *H. longicornis* corresponds to the migration route of birds in the East Asian-Australasian flyway. This suggests that migrating birds are involved in the spread of *H. longicornis* [8].

7.3 Mechanism and Pathogenesis of SFTSV

1—SFTSV human transmission via tick bite. 2—SFTSV carrying tick target nearest lymph nodes generating impaired immune response via B-cell differentiation. 3—further replication of SFTSV into systemic circulation and occurrence of viremia causing cytokine storm and T-cell overstimulation. 4—viremia induced thrombocytopenia; causing reduced platelet count and splenic macrophages. 5—endothelial damages and compromised immune system cause multi-organ damages (Fig. 7.2).

The virus propagates inside the cytoplasm of host cells after infection by SFTSV, and the release of RNPs by SFTSV begins transcription catalyzed by viral RdRp. Complementary RNAs (cRNAs) and viral RNAs are produced because of viral RNA replication aided by protein synthesis (vRNAs). The duplication of all three segments (S, M, and L) varies. Due to the interaction between viral protein N and RdRp, the produced cRNAs and vRNAs are then packed in RNP. RNPs that have just been produced are employed to make viral mRNA and protein [9].

The etiology of SFTSV is unknown, however according to bunyavirus pathogenicity, SFTSV suppresses the host's immune response, resulting in intense virus proliferation and organ failure. After investigating SFTSV patients, it was shown that CD3-positive and CD4-positive T lymphocytes, which play a role in immune function, are in lower numbers than normal, and the number of natural killer cells is higher, especially in the acute and severe stages of the infection [10]. By stimulating the patient's health situation, immune function suppression promotes the spread of secondary infection. Natural killer cells conduct immunoregulatory tasks by generating cytokines such as interferon, tumor necrosis factor (TNF), interleukin 10, and granulocyte colony-stimulating factors (G-CSF). These cytokines are proportional to the severity of the condition. Inflammatory cytokines also play a role in the pathophysiology of viral infections, and some pro-inflammatory cytokines are

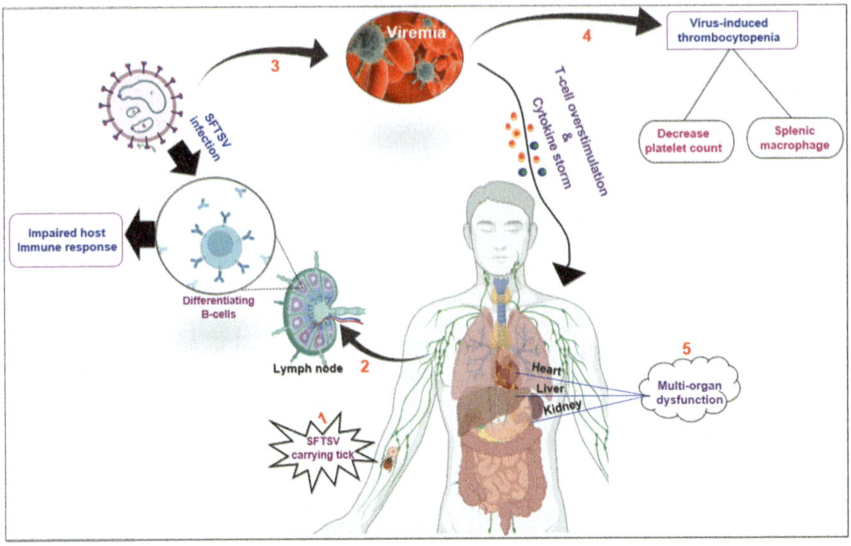

Fig. 7.2 Mechanism of action of SFTSV pathogenesis

overexpressed in the cytokine pool, indicating a severe form of SFTS [11]. Interferon is produced by the innate immune system to guard against viruses, but people with SFTS lack interferon. In SFTSV-infected monocytes, all interferon-related transcription factors are moderately upregulated, whereas upstreaming molecules like TNF-receptor-associated factors 3, 6 and antiviral signaling protein of mitochondria are either unaffected or decreased, further inhibiting interferon induction [12]. Unbalanced cytokines such as interleukins 6 and 10, interleukin-1 receptor antagonist, G-CSF, interferon-inducible protein, and monocyte chemotactic protein 1 have exhibited three distinct patterns in fatal instances of SFTS compared to nonfatal cases. Platelet-derived growth factor (PDGF) and regulated on activation and generally expressed by T-cells (RANTES) expression, on the other hand, are low. All of these cytokines recover to normal levels throughout the convalescent phase. Only in fatal cases as well as in the convalescent phase of survivors does the expression of interleukins 1 and 8 and inflammatory proteins 1 and 1 of macrophages rise [13]. Fever symptoms are linked to elevated TNF in SFTS patients, which works on the endothelium, boosts vasodilating chemicals, nitric oxide synthase, and increases vascular permeability [14]. SFTS virus attaches to platelets, which are recognized and destroyed by circulating macrophages in the spleen, resulting in thrombocytopenia [15]. SFTSV can replicate mostly in reticular cells, although it can also multiply in other cells [16]. Several organ failures arose from SFTSV's targeting of multiple organs. Therefore, starting with a feverish illness in the acute phase, the patient develops multiple organ failure (severe form) and eventually dies (fatal form). In fatal human SFTS cases, SFTSV infects B-cells, lymphocytes, and several lymphoid or nonlymphoid organs, including the blood, spleen, liver, adrenal glands, gut, heart, lungs, and kidneys [17].

7.4 Clinical Symptoms

The incubation period for SFTS can extend anywhere from 5 to 14 days, depending on viral levels and the point of infection [18]. The regular tick bite skin markings do not have eschar, which is typical of scrub typhus patients [19]. Fever, gastrointestinal symptoms (e.g., nausea, vomiting, stomach pain, and diarrhea), and neurological symptoms (e.g., altered mental status) characterize the majority of patients [2, 18]. Thrombocytopenia (100,000/mm^3) and leukopenia (4000/mm^3) were found in the majority of SFTS patients, along with raised alanine aminotransferase (ALT), aspartate aminotransferase (AST), and alkaline phosphatase (ALP) levels and acute renal injury. Lactate dehydrogenase (LDH) and ferritin levels rise as well, as does the activated partial thromboplastin time (aPTT), as well as proteinuria with or without hematuria [2, 18, 20]. Cardiomegaly with or without pericardial effusion and patchy consolidation with ground-glass opacity (GGO) are the most common findings on chest radiographs in patients with SFTS, which aids in the early differentiation of SFTS from scrub typhus, which is characterized by interstitial pneumonia on chest radiographs [21]. During the second week of illness, most patients with severe SFTS die from MOF, which includes acute renal injury, myocarditis, arrhythmia, and meningoencephalitis [22, 23]. The average time from the commencement of illness to death is 9 days [24]. The fatality rate for SFTS varies between 6 and 21% [25–27]. Advanced age, altered mental condition, higher serum LDH and AST levels, prolonged aPTT, and high viral RNA loads in the serum are all poor prognostic markers [26–30]. Similar to cytokine, LDH, AST, and blood urea nitrogen (BUN) levels, viral RNA load has been found to provide useful information for treatment strategies or the prognosis of patients with SFTS [31]. These findings are in line with what has been observed in humans.

7.5 Diagnosis

SFTS is a disorder that is difficult to detect if medical personnel are unaware of it. Fever, low platelet counts, and low white blood cell counts are common symptoms in patients. If the patients have a history of tick bites in endemic locations such as central and eastern China, rural South Korea, or southern Japan, SFTS should be considered. The importance of early diagnosis of SFTSV infection for patient survival cannot be overstated. Because the clinical indications of SFTS are nonspecific, laboratory confirmation is required; also, other tick-borne diseases such as scrub typhus and anaplasmosis generate comparable symptoms [19, 22]. For laboratory diagnosis of SFTS, reverse transcriptase (RT) real-time PCR for the detection of viral RNA in the serum during the first week of illness is a very sensitive and specific diagnostic technique [32]. In the acute phase and for up to 20 days after the onset of symptoms, viral RNA can be found in the blood; however, analyzing serum samples within 2 weeks of the onset of sickness is recommended [32]. RT-PCR approaches based on the nucleotide sequence of SFTSV strains reported in China may be less susceptible to diagnoses of the SFTS lineage identified in other countries due to significant genetic differences among SFTSV inhabitants. To overcome the aforementioned obstacle, Yoshikawa et al. devised a sensitive and specific conventional one-step RT-PCR method as well as a quantitative one-step RT-PCR that can detect both strains [33]. In addition, a number of PCR approaches are being developed to diagnose SFTSV more readily and swiftly. Huang et al. devised a reverse transcription-loop-mediated isothermal amplification (RT-LAMP) approach that has 99% sensitivity and 100% specificity for detecting novel bunyaviruses [34]. Baek et al. also demonstrated that RT-LAMP may provide quick diagnosis in 30–60 min with a sensitivity 10 times higher than traditional RT-PCR [35]. IFA or an enzyme-linked immunosorbent assay (ELISA) are effective diagnostic methods for detecting viral-specific IgM and IgG in the serum 7 days after the onset of the disease; SFTS is diagnosed when IgM antibodies are detected, IgG antibody seroconversion is observed, or the antibody titer increases by at least fourfold [18]. However, IFA sensitivities for IgM and IgG detection after 2 weeks following onset of symptoms are 32–62% and 63–76%, respectively, while ELISA sensitivities are 53–62% and 58–86% [36]. As a result, IFA or ELISA may be insufficient for SFTS diagnosis in the early stages. Hemorrhagic fever with renal syndrome (HFRS), severe dengue fever, thrombocytopenic purpura (TTP), leptospirosis, human granulocytic anaplasmosis (HGA), and Lyme disease are all viral infections with hemorrhagic fever. Patients with these disorders have clinical symptoms and test results that are comparable to those who have SFTS. As a result, in locations where illnesses coexist with SFTS, differential diagnosis is critical (e.g., South Korea, China, and Japan). Scrub typhus and SFTS, in example, cause identical clinical symptoms and test findings in endemic locations. When a score of 2 was obtained after the evaluation of four variables (i.e., altered mental status, leukopenia, prolonged aPTT, and normal C-reactive protein levels), all of which weighed one point, Kim et al. proposed a scoring system that showed 100% sensitivity and 97% specificity [19]. Li et al. proposed a multiplex real-time RT-PCR assay to undertake successful screening for early SFTS diagnosis and to differentiate it from other diseases (such as those caused by the Hantaan, Seoul, and dengue viruses) in the acute phase to more easily and rapidly identify the infections [37]. Virus isolation for laboratory diagnosis, on the other hand, is currently challenging to implement in the clinic because it requires a BSL3 laboratory and takes 2–5 days.

7.6 SFTS Therapeutic Options

There have been no prospective randomized studies on treatment options that have shown to be beneficial in the treatment of SFTS. Symptoms such as fever, diarrhea, dehydration, bleeding propensity, and shock are treated with conservative measures such as hydration, transfusion, and the use of antipyretics, inotropic drugs, and G-CSF. However, in the acute stage, rapidly advancing cases of the disease are difficult to treat appropriately; many patients with severe SFTS are thought to have developed sepsis or septic shock due to verified MOF before being recognized. As a result, early detection of SFTS is critical. Due to the difficulties of therapy and the high mortality rate of SFTS, various treatments have been tried. In the following paragraphs, we will discuss therapy approaches that have been presented.

Antiviral Drugs

1. **Favipiravir**

Toyama Chemical Co., Ltd. developed and produced favipiravir (T-705), which has broad antiviral activity against RNA viruses including as influenza, arenaviruses, bunyaviruses, West Nile virus, yellow fever virus, and foot-and-mouth disease virus [38]. Host enzymes convert favipiravir to its active form, ribofuranosyl-5-triphosphate, which inhibits viral RNA polymerase in the host cells. In vitro favipiravir resistance has only been reported in a few cases [39, 40]. Furthermore, in Vero cells [41], the IC90 of favipiravir (22 μM) was lower than that of ribavirin (263 μM) [42]. Animal models have been used to test the efficacy of favipiravir in vivo. Favipiravir, given intraperitoneally (i.p.) at doses of 60 or 300 mg/kg/day for 5 days, totally protected mice from death caused by SFTSV infection, with only a minor weight loss [41]. When favipiravir treatment began on or before 3 days after infection, all mice survived, whereas animals treated at 4 and 5 days after infection had 83% and 50% survival, respectively [41]. These findings suggested that favipiravir could be used as a preventative as well as a treatment for SFTSV infections. In most cases, favipiravir is taken orally by humans. In a mouse model, favipiravir given orally (p.o.) had equal efficacy to favipiravir given intravenously (i.p). [43]. Furthermore, in a STAT2 deletion golden Syrian hamster model, therapy with favipiravir (300 or 150 mg/kg/day) offered complete protection against a deadly SFTSV challenge [44].

2 **Ribavirin**

Ribavirin is a nucleotide analog having broad-spectrum antiviral function against various viruses. It can be given intravenously, orally, or by a nebulizer [45]. Ribavirin has both direct and indirect modes of action against viruses, including the suppression of inosine monophosphate dehydrogenase and immunomodulatory effects [46]. It has also been investigated whether ribavirin can be used to treat SFTS sufferers. A study on the effects of ribavirin on SFTSV was published in 2017 by Lee et al. which reported that ribavirin decreased cytopathic effects and replication of SFTSV at an

IC_{50} ranging from 3.69 to 8.72 g/mL [47]. So far, several studies have performed to identify the effects of ribavirin on SFTS but most of these are combination therapies along with ribavirin used for SFTS treatment. Additionally, anemia and hyperamylasemia are two adverse effects of ribavirin that have been documented [48]. Therefore, ribavirin administration is not a proven viable therapy option [49, 50].

3 Hexachlorophene

Yuan et al. (2019) screened an FDA-approved drug library containing 1528 drug compounds and found five that inhibited SFTSV replication at 10 μM concentrations, including two antibacterial and antifungal disinfectants (hexachlorophene and triclosan), a multi-kinase inhibitor for the treatment of advanced solid organ tumors (regorafenib), and a small molecule agonist of the C-mannosylation of thrombo (broxyquinoline) [51]. Hexachlorophene was the most potent of them all, with an IC50 of 1.3 ± 0.3 μM (RNA load) and 2.6 ± 0.14 μM (plaque reduction) and the highest selectivity index (50% cytotoxic concentration [CC50]/IC50, 18.7), which was lower than the other four antiviral medicines. Furthermore, the findings revealed that hexachlorophene treatment inhibited SFTSV entrance while having no effect on virus-host cell adhesion or virus infectivity [51]. Hexachlorophene was anticipated to attach to the deep hydrophobic pocket between domains I and III of the SFTSV Gc glycoprotein, causing cell membrane fusion to be disrupted. Hexachlorophene is an antibacterial chemical that is commonly found in soaps and scrubs, as well as an experimental cholinesterase inhibitor [52]. In vitro, hexachlorophene suppressed the viral replication of a coronavirus linked to severe acute respiratory syndrome by blocking 3C-like protease, which is required for the virus's lifecycle [52].

4 2′-Fluoro-2′-deoxycytidine

The nucleoside inhibitor 2′-fluoro-2′-deoxycytidine (2′-FdC) is employed in anticancer medications. Borna virus [53], Lassa virus [54], Crimean-Congo hemorrhagic fever virus [55], influenza virus [56], and herpesviruses are among the RNA and DNA viruses that it suppresses in vitro [57] 2′-FdC has been reported to have antiviral action against a variety of bunyaviruses, including La Crosse virus, Maporal virus, Punta Toro virus, Rift Valley fever virus, San Angelo virus, Heartland virus, and SFTSV, according to [58]. In an in vitro test, the IC90 of 2′-FdC against SFTSV was 3.7 μM. A 100 mg/kg/day therapy with 2′-FdC was 100% protective against death caused by SFTSV in an in vivo research utilizing IFNAR/mice. However, after SFTSV inoculation, all mice treated with 2′-FdC lost a significant amount of weight, whereas mice treated with favipiravir lost very little weight, suggesting that favipiravir was more efficient than 2′-FdC in limiting morbidity during infection [58].

5 Calcium Channel Blockers

Calcium channel blockers (CCBs) lower intracellular Ca^{2+} levels and are commonly used to treat hypertension, angina, and supraventricular arrhythmias, among other cardiovascular conditions. Antiviral activity of CCBs has recently been reported against ebolavirus, marburgvirus, Junn virus, West Nile virus, and Japanese encephalitis virus [59–63]. The CCBs benidipine hydrochloride and nifedipine were discovered as inhibitors of SFTSV replication in vitro by limiting viral internalization and lowering genome replication during the post-entry phase, according to a screening of 700 FDA-approved medicines [64]. The viral binding, fusion, and budding were not affected by this mechanism. Treatment with benidipine hydrochloride or nifedipine decreased SFTSV replication by lowering virus-induced Ca^{2+} influx, according to the findings of an in vitro investigation. In C57BL/6 mice and humanized mouse models, the anti-SFTSV effect of these two CCBs was further investigated, demonstrating treatment results of reduced viral load, improved platelet count, and lower fatality rate in the humanized mouse model. Notably, nifedipine is one of the most commonly prescribed medications in China for the treatment of hypertension and atherosclerosis. As a result, Li et al. (2019) conducted a retrospective clinical investigation on a large cohort of 2087 SFTS patients, including 83 nifedipine-treated patients who received nifedipine before admission and during hospitalization, 48 non-nifedipine-treated patients who received nifedipine before admission but not during hospitalization, and 249 general SFTS patients who did not receive nifedipine at all [64]. The case fatality rate in the nifedipine-treated group (3.6%) was less than half that of the overall SFTS group (19.7%) or the non-nifedipine-treated group (20.8%) [64]. In contrast to ribavirin, nifedipine-treated patients with a high viral load (> 106 copies/mL) had a significantly lower-case fatality rate (2.4%) as compared to general SFTS patients (29%) and non-nifedipine-treated patients (34.5%). Hematemesis was shown to be less common in the nifedipine-treated group, which is one of the hemorrhagic symptoms that is closely linked to death. The authors demonstrated the inhibitory effect of benidipine hydrochloride or nifedipine in cultured cells in an animal model in this article. Most importantly, it was discovered that nifedipine treatment boosted viral clearance and clinical recovery.

6 Caffeic Acid

Caffeic acid (CA) is a polyphenol chemical component connected to coffee that can be found in a variety of plants, including coffee beans. Chlorogenic acid, the ester of caffeic acid, is found in 70–350 mg per cup of coffee [65]. It has a number of biological effects, including cancer cell suppression and antiviral activities [66–71]. In an in vitro test employing Huh 7.5.1–8 cells, a highly tolerant derivation of human hepatoma Huh7 cells, found that CA suppressed SFTSV replication dose-dependently [72]. CA had an IC50 of 48 μM and a CC50 of 7.6 mM against SFTSV. Surprisingly, pretreatment of SFTSV with CA before inoculation lowered the virus copy number in the supernatant of infected cells at 72 h after infection, and the inhibitory impact was greatly diminished when the cells were treated with CA after SFTSV inoculation. As a result, the scientists hypothesized that CA worked mostly on viral particles or influenced the early stages of SFTSV infection, while it might also limit viral genome replication in host cells.

7 **Amodiaquine**

Amodiaquine, a new antimalarial medication, has been shown to have antiviral effects against ebolavirus, dengue virus, and zika virus [72–76]. The mechanism of amodiaquine's inhibitory effect against malaria and those viruses is unknown. Amodiaquine and other halogen compounds (fluorine, bromine, and iodine) were tested against SFTSV replication in vitro by [77]. The IC50 for fluorine, bromine, and iodine, respectively, was 36.6, 31.1, and 15.6 μM for fluorine, bromine, and iodine compound. Amodiaquine was found to be a selective inhibitor of SFTSV replication among the drugs examined. Amodiaquine had a CC50 of >100 and an IC50 of 19.1 μM, respectively. Amodiaquine IC50 was lower than ribavirin's (40.1 μM) and favipiravir's (25.0 μM).

8 **IFN-γ**

Type II IFNs only have one member, IFN-γ. By modulating antigen processing and presentation pathways, it encourages macrophages and dendritic cells to provide direct antimicrobial activity. Activated T-cells and activated natural killer cells were assumed the only important sources of IFN-γ, but under certain conditions, macrophages and dendritic cells can also be driven to create IFN-γ in vitro [78]. IFN-γ plays a crucial function in viral infection because it can directly increase the development of several putative antiviral IFN-stimulating proteins via STAT1 signaling.

9. **Monoclonal antibodies (Mab)**

Monoclonal antibodies are regarded as new therapeutic agents for SFTS across a variety of treatment options. According to Guo et al., monoclonal antibodies worked to neutralize SFTSV infection in Vero cells by attaching to a linear epitope in the glycoprotein Gn's ectodomain [79]. This neutralizing activity results from the suppression of interactions between glycoprotein Gn and cellular receptors, which prevents viral cell attachment. Additionally, Kim et al. revealed that their chosen antibody was reactive to the SFTSV's envelope glycoprotein Gn and protected 80% of mice and host cells [80], indicating that monoclonal antibodies could be able to defend against SFTSV. These findings imply that monoclonal antibodies may provide SFTS patients a promising therapy alternative.

7.7 Structure-Based Drug Design Approach Guided Identification of Potential Binders

The viral L protein can be considered as one of the emerging targets for developing therapeutic agents. The L protein synthesizes three different RNA species during the viral replication [81] and pose significant importance. The cap binding

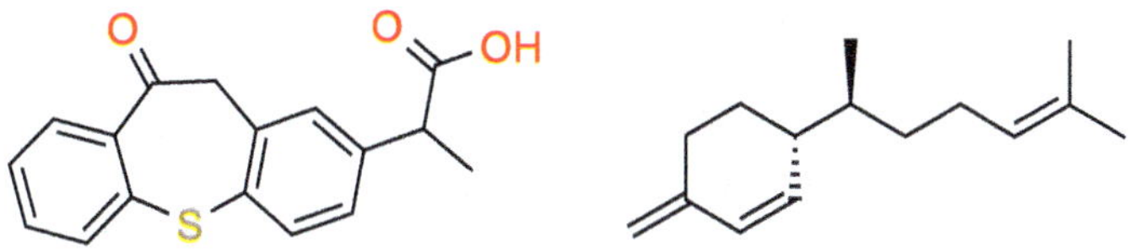

Fig. 7.3 Potential SFTS-L protein binder identified by computational methods

domain of the SFTS-L protein extensively utilized to identify potential binders. The drug re-purposing approach identified Zaltoprofen as a potential SFTS-L protein binder Fig. 7.3 [82]. The Phe1703, Tyr1719; Gln1707, Asp1771, Leu1772; Pro1706, Ile1738, and Ile1774; Phe1703 and Tyr1719; Phe1703, Gln1707, Asp1771, and Trp1725; Pro1706, Ile1738, Ile1774, Leu1768, Leu1772 residues of SFTS-L protein (PDB ID:6XYA) are the crucial amino acids present in the binding site. The molecular docking followed by molecular dynamics investigation identified β-sesquiphellan-drene as a binder of membrane glycoprotein polyprotein of SFTS virus [83].

Few more chemical compounds like Bromfenac, Cinchophen, and Elliptinium depict stable binding with SFTS-L protein with acceptable docking score. The molecular dynamics study of these compounds bound systems was conducted.

The researchers calculated various parameters like RMSD, RMSF, Radius of Gyration (Rg), Solvent accessible surface area, etc., from the molecular dynamic's trajectory can be found in Fig. 7.4. Each holo-protein (ligand bound protein) depicts lower RMSD, RMSF, and Rg profile in compared to apo-protein (ligand free protein). The RMSD values were found to be less then 3.0 Å. RMSD profile for a globular protein less then 3.0 Å indicated a stable system [84]. On the basis of RMSD profile associated with other parameters, it can be stated these chemical compounds identified from Drugbank Database exhibited stable binding with SFTS-L protein.

Virtual screening of Indian natural products presents in Indian medicinal plants identified few potential hits against SFTS-L protein. The name of the hits are Gamma-glutamylaspartic acid, 2′-Deoxymugineic acid, Traumatic acid, Betalamic acid, Epoxyoleic acid, respectively [85]. Studies depict presence of divalent Mn^{2+} ions in the SFTS-L protein binding site required for the activity [85].

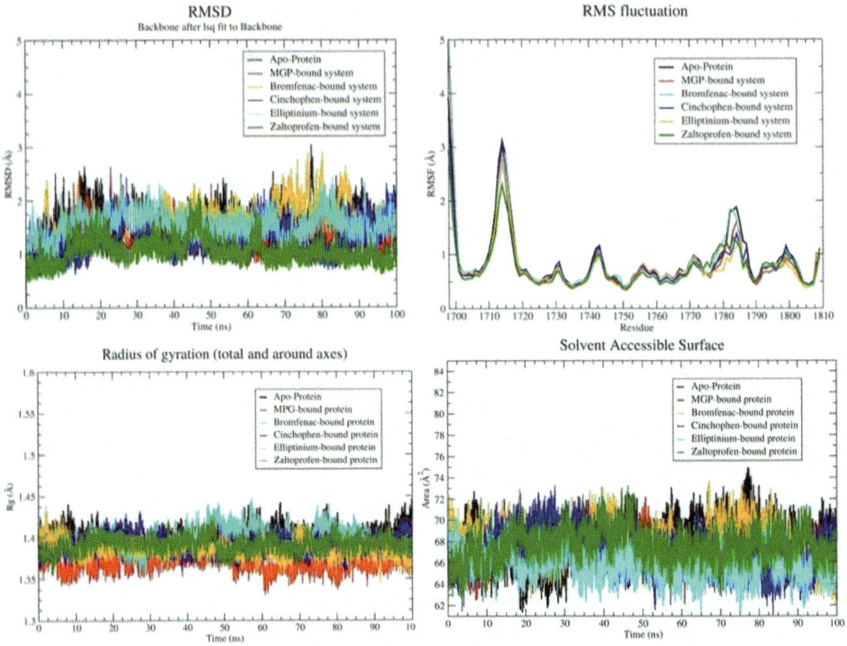

Fig. 7.4 Various parameters calculated from molecular dynamics trajectory. The figure is reproduced from Ref. [82] open access under a CC BY 4.0 license, https://creativecommons.org/licenses/by/4.0/

7.8 Conclusion

The SFTSV can be considered as one of the most life-threatening infections. Till date lack of appropriate medications pose significant emergence to develop new therapeutic agents. Unavailability of large number of chemical entities exhibiting distinct anti-SFTS properties, structure-based drug design approach was largely adopted over ligand-based drug design approach, to identify potential hits. Recent molecular docking followed by molecular dynamics guided approach identified sets of hits against SFTS-1 protein. However, there is a huge scope to identify more relevant hits against this virus in order to complete the journey of a small molecules from bench to bed side.

References

1. ICTV. ICTV Taxonomy History: SFTS virus. https://talk.ictvonline.org/taxonomy/p/taxonomyhistory?taxnode_id=20141803&src=NCBI&ictv_id=20141803 (ICTV, 2020)
2. Yu X-J et al (2011) Fever with thrombocytopenia associated with a novel bunyavirus in China. N Engl J Med 364:1523–1532

3. Wiwanitkit S, Wiwanitkit V (2015) Acute viral hemorrhage disease: a summary on new viruses. J Acute Dis 4:277–279
4. Takahashi T et al (2014) The first identification and retrospective study of severe fever with thrombocytopenia syndrome in Japan. J Infect Dis 209:816–827
5. Kim K-H et al (2013) Severe fever with thrombocytopenia syndrome, South Korea, 2012. Emerg Infect Dis 19:1892
6. Tran XC et al (2019) Endemic severe fever with thrombocytopenia syndrome. Vietnam Emerg Infect Dis 25:1029
7. Lin T-L et al (2020) The first discovery of severe fever with thrombocytopenia syndrome virus in Taiwan. Emerg Microbes Infect 9:148–151
8. Yun Y et al (2015) Phylogenetic analysis of severe fever with thrombocytopenia syndrome virus in South Korea and migratory bird routes between China, South Korea, and Japan. Am J Tropical Med Hyg. 93:468–474
9. Lei X-P, Liu M, Yu X (2015) Severe fever with thrombocytopenia syndrome and its pathogen SFTSV. Microbes Infect 17:149–154
10. Sun L, Hu Y, Niyonsaba A et al (2013) Detection and evaluation of immunofunction of patients with severe fever with thrombocytopenia syndrome. Clin Exp Med. https://doi.org/10.1007/s10238-013-0259-0
11. Deng B, Zhang S, Geng Y et al (2012) Cytokine and chemokine levels in patients with severe fever with thrombocytopenia syndrome virus. PLoS ONE 7:41365
12. Qu B, Qi X, Wu X et al (2012) Suppression of the interferon and NF-κB responses by severe fever with thrombocytopenia syndrome virus. J Virol 86:8388–8401
13. Sun Y, Jin C, Zhan F et al (2012) Host cytokine storm is associated with disease severity of severe fever with thrombocytopenia syndrome. J Infect Dis 206:1085–1094
14. Seynhaeve ALB, Vermeulen CE, Eggermont AMM, Hagen TLMT (2006) Cytokines and vascular permeability: an in vitro study on human endothelial cells in relation to tumor necrosis factor-alpha-primed peripheral blood mononuclear cells. Cell Biochem Biophys 44:157–169
15. Jin C, Liang M, Ning J et al (2012) Pathogenesis of emerging severe fever with thrombocytopenia syndrome virus in C57/BL6 mouse model. Proc Natl Acad Sci USA 109:10053–10058
16. Liu Q, Biao H, Si-Yang H, Feng W, Xing-Quan Z et al (2014) Severe fever with thrombocytopenia syndrome, an emerging tick-borne zoonosis. Lancet Infect Dis 14:763–772
17. Suzuki T, Sato Y, Sano K, Arashiro T, Katano H, Nakajima N, Morikawa S et al (2020) Severe fever with thrombocytopenia syndrome virus targets B cells in lethal human infections. J Clin Investig 130(2):799–812
18. Liu Q, He B, Huang SY, Wei F, Zhu XQ (2014) Severe fever with thrombocytopenia syndrome, an emerging tick-borne zoonosis. Lancet Infect Dis 14:763–772
19. Kim MC, Chong YP, Lee SO, Choi SH, Kim YS, Woo JH, Kim SH (2018) Differentiation of severe fever with thrombocytopenia syndrome from scrub typhus. Clin Infect
20. Kim UJ, Oh TH, Kim B, Kim SE, Kang SJ, Park KH, Jung SI, Jang HC (2017) Hyperferritinemia as a diagnostic marker for severe fever with thrombocytopenia syndrome. Dis Markers 2017:6727184
21. Yun JH, Hwang HJ, Jung J, Kim MJ, Chong YP, Lee SO, Choi SH, Kim YS, Woo JH, Kim MY et al (2019) Comparison of chest radiographic findings between severe fever with thrombocytopenia syndrome and scrub typhus: single center observational cross-sectional study in South Korea. Medicine 98:e17701
22. Miyamoto S, Ito T, Terada S, Eguchi T, Furubeppu H, Kawamura H, Yasuda T, Kakihana Y (2019) Fulminant myocarditis associated with severe fever with thrombocytopenia syndrome: a case report. BMC Infect Dis 19:266
23. Park SY, Kwon JS, Kim JY, Kim SM, Jang YR, Kim MC, Cho OH, Kim T, Chong YP, Lee SO et al (2018) Severe fever with thrombocytopenia syndrome-associated encephalopathy/encephalitis. Clin Microbiol Infect 24:432.e1–432.e4
24. Ding F, Zhang W, Wang L, Hu W, Soares Magalhaes RJ, Sun H, Zhou H, Sha S, Li S, Liu Q et al (2013) Epidemiologic features of severe fever with thrombocytopenia syndrome in China, 2011–2012. Clin Infect Dis 56:1682–1683

25. Sun J, Lu L, Wu H, Yang J, Ren J, Liu Q (2017) The changing epidemiological characteristics of severe fever with thrombocytopenia syndrome in China, 2011–2016. Sci Rep 7:9236
26. Choi SJ, Park SW, Bae IG, Kim SH, Ryu SY, Kim HA, Jang HC, Hur J, Jun JB, Jung Y et al (2016) Severe fever with thrombocytopenia syndrome in South Korea, 2013–2015. PLoS Negl Trop Dis 10:e0005264
27. Kato H, Yamagishi T, Shimada T, Matsui T, Shimojima M, Saijo M, Oishi K (2016) Epidemiological and clinical features of severe fever with thrombocytopenia syndrome in Japan, 2013–2014. PLoS ONE 11:e0165207
28. Li H, Lu QB, Xing B, Zhang SF, Liu K, Du J, Li XK, Cui N, Yang ZD, Wang LY et al (2018) Epidemiological and clinical features of laboratory-diagnosed severe fever with thrombocytopenia syndrome in China, 2011–2017: a prospective observational study. Lancet Infect Dis 18:1127–1137
29. Wang L, Wan G, Shen Y, Zhao Z, Lin L, Zhang W, Song R, Tian D, Wen J, Zhao Y et al (2019) A nomogram to predict mortality in patients with severe fever with thrombocytopenia syndrome at the early stage-A multicenter study in China. PLoS Negl Trop Dis 13:e0007829
30. Zhang YZ, He YW, Dai YA, Xiong Y, Zheng H, Zhou DJ, Li J, Sun Q, Luo XL, Cheng YL et al (2012) Hemorrhagic fever caused by a novel Bunyavirus in China: pathogenesis and correlates of fatal outcome. Clin Infect Dis 54:527–533
31. Hwang J, Kang JG, Oh SS, Chae JB, Cho YK, Cho YS, Lee H, Chae JS (2017) Molecular detection of severe fever with thrombocytopenia syndrome virus (SFTSV) in feral cats from Seoul Korea. Ticks Tick Borne Dis 8:9–12
32. Sun Y, Liang M, Qu J, Jin C, Zhang Q, Li J, Jiang X, Wang Q, Lu J, Gu W et al (2012) Early diagnosis of novel SFTS bunyavirus infection by quantitative real-time RT-PCR assay. J Clin Virol 53:48–53
33. Yoshikawa T, Fukushi S, Tani H, Fukuma A, Taniguchi S, Toda S, Shimazu Y, Yano K, Morimitsu T, Ando K et al (2014) Sensitive and specific PCR systems for detection of both Chinese and Japanese severe fever with thrombocytopenia syndrome virus strains and prediction of patient survival based on viral load. J Clin Microbiol 52:3325–3333
34. Huang XY, Hu XN, Ma H, Du YH, Ma HX, Kang K, You AG, Wang HF, Zhang L, Chen HM et al (2014) Detection of new bunyavirus RNA by reverse transcription-loop-mediated isothermal amplification. J Clin Microbiol 52:531–535
35. Baek YH, Cheon HS, Park SJ, Lloren KKS, Ahn SJ, Jeong JH, Choi WS, Yu MA, Kwon HI, Kwon JJ et al (2018) Simple, rapid and sensitive portable molecular diagnosis of SFTS virus using reverse transcriptional loop-mediated isothermal amplification (RT-LAMP). J Microbiol Biotechnol 28:1928–1936
36. Ra SH, Kim MJ, Kim MC, Park SY, Park SY, Chong YP, Lee SO, Choi SH, Kim YS, Lee KH et al (2020) Kinetics of serological response in patients with severe fever with thrombocytopenia syndrome. Viruses 13:6
37. Li Z, Qi X, Zhou M, Bao C, Hu J, Wu B, Wang S, Tan Z, Fu J, Shan J et al (2013) A two-tube multiplex real-time RT-PCR assay for the detection of four hemorrhagic fever viruses: Severe fever with thrombocytopenia syndrome virus, Hantaan virus, Seoul virus, and dengue virus. Arch Virol 158:1857–1863
38. Furuta Y, Takahashi K, Shiraki K, Sakamoto K, Smee DF, Barnard DL et al (2009) T-705 (favipiravir) and related compounds: novel broad-spectrum inhibitors of RNA viral infections. Antiviral Res 82:95–102. https://doi.org/10.1016/j.antiviral.2009.02.198
39. Delang L, Guerrero NS, Tas A, Quérat G, Pastorino B, Froeyen M et al (2014) Mutations in the chikungunya virus non-structural proteins cause resistance to favipiravir (T-705), a broad-spectrum antiviral. J Antimicrob Chemother 69:2770–2784. https://doi.org/10.1093/jac/dku209
40. Goldhill DH, Te Velthuis AJW, Fletcher RA, Langat P, Zambon M, Lackenby A et al (2018) The mechanism of resistance to favipiravir in influenza. Proc Natl Acad Sci USA 115:11613–11618. https://doi.org/10.1073/pnas.1811345115
41. Tani H, Fukuma A, Fukushi S, Taniguchi S, Yoshikawa T, Iwata-yoshikawa N et al (2016) Efficacy of T-705 (Favipiravir) in the treatment of infections with lethal severe fever with

thrombocytopenia syndrome virus. mSphere 1:e00061–e00015. https://doi.org/10.1128/mSphere.00061-15
42. Shimojima M, Fukushi S, Tani H, Yoshikawa T, Fukuma A, Taniguchi S et al (2014) Effects of ribavirin on severe fever with thrombocytopenia syndrome virus in vitro. Jpn J Infect Dis 67:423–427. https://doi.org/10.7883/yoken.67.423
43. Tani H, Komeno T, Fukuma A, Fukushi S, Taniguchi S, Shimojima M et al (2018) Therapeutic effects of favipiravir against severe fever with thrombocytopenia syndrome virus infection in a lethal mouse model: dose-efficacy studies upon oral administration. PLoS ONE 13:e0206416. https://doi.org/10.1371/journal.pone.0206416
44. Gowen BB, Westover JB, Miao J, Van Wettere AJ, Rigas JD, Hickerson BT et al (2017) Modeling severe fever with thrombocytopenia syndrome virus infection in golden syrian hamsters: importance of STAT2 in preventing disease and effective treatment with favipiravir. J Virol 91:e01942-e11916. https://doi.org/10.1128/JVI.01942-16
45. Snell NJ (2001) Ribavirin—current status of a broad-spectrum antiviral agent. Expert Opin Pharmacother 2:1317–1324
46. Graci JD, Cameron CE (2006) Mechanisms of action of ribavirin against distinct viruses. Rev Med Virol 16:37–48. https://doi.org/10.1002/rmv.483
47. Lee MJ, Kim KH, Yi J, Choi SJ, Choe PG, Park WB, Kim NJ, Oh MD (2017) In vitro antiviral activity of ribavirin against severe fever with thrombocytopenia syndrome virus. Korean J Intern Med 32:731–737
48. Lu QB, Zhang SY, Cui N, Hu JG, Fan YD, Guo CT, Qin SL, Yang ZD, Wang LY, Wang HY et al (2015) Common adverse events associated with ribavirin therapy for severe fever with thrombocytopenia syndrome. Antivir Res 119:19–22
49. Oh WS, Heo ST, Kim SH, Choi WJ, Han MG, Kim JY (2014) Plasma exchange and ribavirin for rapidly progressive severe fever with thrombocytopenia syndrome. Int J Infect Dis 18:84–86
50. Park I, Kim HI, Kwon KT (2017) Two treatment cases of severe fever and thrombocytopenia syndrome with oral ribavirin and plasma exchange. Infect Chemother 49:72–77
51. Yuan S, Chan JFW, Ye ZW, Wen L, Tsang TGW, Cao J et al (2019) Screening of an FDA-approved drug library with a two-tier system identifies an entry inhibitor of severe fever with thrombocytopenia syndrome virus. Viruses 11:E385
52. Hsu JTA, Kuo CJ, Hsieh HP, Wang YC, Huang KK, Lin CPC et al (2004) Evaluation of metal-conjugated compounds as inhibitors of 3CL protease of SARS-CoV. FEBS Lett 574:116–120
53. Bajramovic JJ, Volmer R, Syan S, Pochet S, Gonzalez-Dunia D (2004) 2′-fluoro-2′-deoxycytidine inhibits Borna disease virus replication and spread. Antimicrob Agents Chemother 48:1422–1425
54. Welch SR, Guerrero LW, Chakrabarti AK, McMullan LK, Flint M, Bluemling GR et al (2016) Lassa and Ebola virus inhibitors identified using minigenome and recombinant virus reporter systems. Antiviral Res 136:9–18
55. Welch SR, Scholte FEM, Flint M, Chatterjee P, Nichol ST, Bergeron É et al (2017) Identification of 2′-deoxy-2′-fluorocytidine as a potent inhibitor of Crimean-Congo hemorrhagic fever virus replication using a recombinant fluorescent reporter virus. Antiviral Res 147:91–99
56. Kumaki Y, Day CW, Smee DF, Morrey JD, Barnard DL (2011) In vitro and in vivo efficacy of fluorodeoxycytidine analogs against highly pathogenic avian influenza H5N1, seasonal, and pandemic H1N1 virus infections. Antiviral Res 92:329–340
57. Wohlrab F, Jamieson AT, Hay J, Mengel R, Guschlbauer W (1985) The effect of 2′-fluoro-2′-deoxycytidine on herpes virus growth. Biochim Biophys Acta 824:233–242
58. Smee DF, Jung KH, Westover J, Gowen BB (2018) 2′-Fluoro-2′-deoxycytidine is a broad-spectrum inhibitor of bunyaviruses in vitro and in phleboviral disease mouse models. Antiviral Res 160:48–54. https://doi.org/10.1016/j.antiviral.2018.10.013
59. Sakurai Y, Kolokoltsov AA, Chen CC, Tidwell MW, Bauta WE, Klugbauer N et al (2015) Two-pore channels control Ebola virus host cell entry and are drug targets for disease treatment. Science 347:995–998. https://doi.org/10.1126/science.1258758
60. Dewald LE, Dyall J, Sword JM, Torzewski L, Zhou H, Postnikova E et al (2018) The calcium channel blocker bepridil demonstrates efficacy in the murine model of marburg virus disease. J Infect Dis 22:S588–S591

61. Lavanya M, Cuevas CD, Thomas M, Cherry S, Ross SR (2013) siRNA screen for genes that affect Junín virus entry uncovers voltage-gated calcium channels as a therapeutic target. Sci Transl Med 5:204ra131
62. Scherbik SV, Brinton MA (2010) Virus-induced Ca^{2+} influx extends survival of west nile virus-infected cells. J Virol 84:8721–8731. https://doi.org/10.1128/JVI.00144-10
63. Wang S, Liu Y, Guo J, Wang P, Zhang L, Xiao G et al (2017) Screening of FDA-approved drugs for inhibitors of Japanese encephalitis virus infection. J Virol 91:e01055-e11017
64. Li H, Zhang LK, Li SF, Zhang SF, Wan WW, Zhang YL et al (2019) Calcium channel blockers reduce severe fever with thrombocytopenia syndrome virus (SFTSV) related fatality. Cell Res 29:739–753
65. Clifford MN (1999) Chlorogenic acids and other cinnamates–nature, occurrence and dietary burden. J Sci Food Agric 79:362–372
66. Tang H, Yao X, Yao C, Zhao X, Zuo H, Li Z (2017) Anti-colon cancer effect of caffeic acid p-nitro-phenethyl ester in vitro and in vivo and detection of its metabolites. Sci Rep 7:7599
67. Bułdak RJ, Hejmo T, Osowski M, Bułdak Ł, Kukla M, Polaniak R et al (2018) The impact of coffee and its selected bioactive compounds on the development and progression of colorectal cancer in vivo and in vitro. Molecules 23:E3309
68. Wang GF, Shi LP, Ren YD, Liu QF, Liu HF, Zhang RJ et al (2009) Anti-hepatitis B virus activity of chlorogenic acid, quinic acid and caffeic acid in vivo and in vitro. Antiviral Res 83:186–190
69. Utsunomiya H, Ichinose M, Ikeda K, Uozaki M, Morishita J, Kuwahara T et al (2014) Inhibition by caffeic acid of the influenza a virus multiplication in vitro. Int J Mol Med 34:1020–1024
70. Ding Y, Cao Z, Cao L, Ding G, Wang Z, Xiao W (2017) Antiviral activity of chlorogenic acid against influenza A (H1N1/H3N2) virus and its inhibition of neuraminidase. Sci Rep 7:1–11
71. Langland J, Jacobs B, Wagner CE, Ruiz G, Cahill TM (2018) Antiviral activity of metal chelates of caffeic acid and similar compounds towards herpes simplex, VSV-Ebola pseudotyped and vaccinia viruses. Antiviral Res 160:143–150. https://doi.org/10.1016/j.antiviral.2018.10.021
72. Ogawa M, Shirasago Y, Ando S, Shimojima M, Saijo M, Fukasawa M (2018) Caffeic acid, a coffee-related organic acid, inhibits infection by severe fever with thrombocytopenia syndrome virus in vitro. J Infect Chemother 24:597–601
73. Gignoux E, Azman AS, De Smet M, Azuma P, Massaquoi M, Job D et al (2016) Effect of artesunate.amodiaquine on mortality related to Ebola virus disease. N Engl J Med 374:23–32
74. Sakurai Y, Sakakibara N, Toyama M, Baba M, Davey RA (2018) Novel amodiaquine derivatives potently inhibit Ebola virus infection. Antiviral Res 160:175–182
75. Boonyasuppayakorn S, Reichert ED, Manzano M, Nagarajan K, Padmanabhan R (2014) Amodiaquine, an antimalarial drug, inhibits dengue virus type 2 replication and infectivity. Antiviral Res 106:125–134
76. Balasubramanian A, Teramoto T, Kulkarni AA, Bhattacharjee AK, Padmanabhan R (2017) Antiviral activities of selected antimalarials against dengue virus type 2 and Zika virus. Antiviral Res 137:141–150
77. Baba M, Toyama M, Sakakibara N, Okamoto M, Arima N, Saijo M (2017) Establishment of an antiviral assay system and identification of severe fever with thrombocytopenia syndrome virus inhibitors. Antivir Chem Chemother 25:83–89
78. Thäle C, Kiderlen AF (2005) Sources of interferon-gamma (IFN-γ) in early immune response to Listeria monocytogenes. Immunobiology 210:673–683. https://doi.org/10.1016/j.imbio.2005.07.003
79. Guo X, Zhang L, Zhang W, Chi Y, Zeng X, Li X, Qi X, Jin Q, Zhang X, Huang M et al (2013) Human antibody neutralizes severe fever with thrombocytopenia syndrome virus, an emerging hemorrhagic Fever virus. Clin Vaccine Immunol 20:1426–1432
80. Kim KH, Kim J, Ko M, Chun JY, Kim H, Kim S, Min JY, Park WB, Oh MD, Chung J (2019) An anti-Gn glycoprotein antibody from a convalescent patient potently inhibits the infection of severe fever with thrombocytopenia syndrome virus. PLoS Pathog 15:e1007375

81. Vogel D, Thorkelsson SR, Quemin ERJ, Meier K, Kouba T, Gogrefe N, Busch C, Reindl S, Günther S, Cusack S, Grünewald K, Rosenthal M (2020) Structural and functional characterization of the severe fever with thrombocytopenia syndrome virus L protein. Nucleic Acids Res 48:5749–5765. https://doi.org/10.1093/NAR/GKAA253
82. Chatterjee S, Kim CM, Kim DM (2021) Potential efficacy of existing drug molecules against severe fever with thrombocytopenia syndrome virus: an in silico study. Sci Rep 11(1):1–8. https://doi.org/10.1038/s41598-021-00294-7
83. Joshi A, Sunil Krishnan G, Kaushik V (2020) Molecular docking and simulation investigation: effect of beta-sesquiphellandrene with ionic integration on SARS-CoV2 and SFTS viruses. J Genetic Eng Biotech 18. https://doi.org/10.1186/S43141-020-00095-X
84. Chatterjee S, Maity A, Chowdhury S, Islam A, Muttinini RK, Sen D (nd) In silico analysis and identification of promising hits against 2019 novel coronavirus 3C-like main protease enzyme. https://doi.org/10.1080/07391102.2020.1787228
85. Vivek-Ananth RP, Sahoo AK, Srivastava A, Samal A (2022) Virtual screening of phytochemicals from Indian medicinal plants against the endonuclease domain of SFTS virus L polymerase. RSC Adv 12:6234–6247. https://doi.org/10.1039/D1RA06702H

Chapter 8
Computational Toxicological Aspects in Drug Design and Discovery, Screening Adverse Effects

Emilio Benfenati, Gianluca Selvestrel, Anna Lombardo, and Davide Luciani

Abstract Toxicological aspects represent a fundamental step in the process of drug design and discovery. There are multiple platforms available, and recently freely available tools provided results comparable with those obtained from the commercial ones. We will present examples of models for the different endpoints which can be used. In addition, the future perspectives are to take into account in an earlier stage the adverse effects, in order to simplify the long process of drug design and discovery, and to optimize the selection of preferable features present in a new pharmaceutical. In this new vision, a more holistic approach can apply multiple methodologies and not only the screening of the adverse effects.

Keywords Drug · In silico · Read-across · Toxicology · VEGAHUB

8.1 Introduction

The use of in silico models is a fundamental component of all areas of science for decades. The use of computers offers unique opportunities and opens new avenues in research, in the development of new substances and products, and it may help our society in many ways. Here, we address the sector of the applications related to the evaluation of toxicity and how the computation tools may help. We will address these specific topics. (i) Which models can be used to address toxicological endpoints. Here, we will speak about in silico models, as opposed to in vivo and in vitro models. In silico model, however, is a broad term which is also used in other areas related to pharmaceuticals, such as clinical studies. In our case, we focus on models for toxicity. (ii) We will address read-across, particularly the approaches where computer models are relevant—read-across can be also done manually, and indeed historically, this was the way to proceed. (iii) We will discuss how these two non-testing methods, in

E. Benfenati (✉) · G. Selvestrel · A. Lombardo · D. Luciani
Istituto di Ricerche Farmacologiche Mario Negri IRCCS, Milano, Italy
e-mail: emilio.benfenati@marionegri.it

S. Kar and J. Leszczynski (eds.), *Current Trends in Computational Modeling for Drug Discovery*, Challenges and Advances in Computational Chemistry and Physics 35,
https://doi.org/10.1007/978-3-031-33871-7_8

silico models and read-across, should be combined, and in general how to address a weight-of-evidence approach, for the purpose to get a robust evaluation of the toxicity value for a substance, taking advantage of multiple, heterogeneous data. (iv) We will extend our discussion to the case when the evaluation of the toxicity goes beyond one single endpoint and tries to get a comprehensive view. (v) As a further exploration of the potential that a substance may be risky, we will see how to integrate in silico tools for hazard and exposure. (vi) The tools and approaches related to safety by design will be presented with some examples. (vii) We will see that beyond pre-built models, there are software packages offering the possibility to develop specific models on purpose. (viii) Finally, we will derive conclusions and mention the new perspectives which are expected in future. Within this discussion, we will provide reference to existing in silico tools or future perspectives and will often refer to the architecture of software packages as in VEGAHUB (www.vegahu b.eu), to show examples.

8.2 Tools for Individual Endpoints

The computational models to cope with toxicological aspects are more and more numerous and sophisticated. There are multiple tools which can predict a large set of endpoints, as listed here:

- Mutagenicity (Ames test)
- In vitro micronucleus
- In vivo micronucleus
- Chromosomal aberration
- Carcinogenicity
- Developmental toxicity
- Reproductive toxicity
- Repeated-dose toxicity
- Acute systemic toxicity
- Skin sensitization
- Skin irritation
- Eye irritation
- Liver toxicity
- Cardiotoxicity
- Neurotoxicity
- Nephrotoxicity
- Endocrine disruption.

This list is not exhaustive and limited to human health effects; some points include several endpoints. We will introduce an example where in silico models are well established, and then we will touch on some other cases to mention only the key aspects. The reader interested in a more systematic discussion on the specific endpoints may refer to a recent book [1]. Let us consider the case of the models for mutagenicity, determined with the Ames test. There are tens of in silico models commercially available and other tens of models free. In the case of their use for pharmaceutical impurities, the International Conference on Harmonization (ICH) M7 guideline [2] includes recommendations on the use of in silico models, and in particular, methods based on two approaches should be used together: One approach refers to the rules defined by experts codifying structural alerts (SAs) associated to mutagenic effect, and the second approach should be based on statistical methods [3]. In this way, the authorities recommend using two orthogonal methods. The reason is that none of them is considered perfect, while the two methods may cover different aspects, and thus, it is safer to have multiple tools to identify possible reasons of concern. Although we agree with this perspective, we notice further aspects to be discussed. One point refers to the regulatory context. In another area, of industrial substances, within the REACH regulation (the European regulation on Registration, Evaluation, Authorization and Restriction of Chemicals) [4], this strategy is also welcome, but it has been also mentioned that, ideally, it should be preferable to have ten separate values, one for each of the ten conditions to be used with different strains with and without metabolic activation. We will not address this point here.

The use of the models based on SAs is quite convincing because it "explains" the reason for the adverse effect. The user should be aware that the simple presence of a SA in the molecule is not sufficient to label a substance. Each SA is present in a certain number of mutagenic substances, but there are also non-mutagenic substances which contain the SA. In some cases, most of the substances are not mutagenic. Indeed, depending on the SA, there may be false positives, i.e. substances which are predicted mutagenic while they are not. Another critical aspect is that some SAs are based on very few substances. Furthermore, the different in silico models based on the SAs contain different numbers of SAs, and they overlap only partially; thus, there is no agreement in the community of experts. Finally, there is not a complete list of the SAs, and thus, a substance may be mutagenic even if it does not contain a SA (so far identified at least).

For all these reasons, the authorities rather recommend an additional, empirical approach, so that in silico model may identify hazardous, substances which may escape the identification of the adverse effect using SAs. We will come back on the ways to integrate multiple values from different models later.

Of course, models are different according to the method employed. The models may also differ with respect to other perspectives, and indeed we mentioned tens of models for mutagenicity. In the case of the expert-based models, the differences are quite subjective, since the rules have been codified by human experts, based on their personal opinion, the paper they reviewed, the assumptions done. For instance, the model derived by Benigni-Bossa lists tens of SAs for mutagenicity, and all of them are also used to label carcinogenicity. This assumption may have not been adopted by other models in the case of SAs for carcinogenicity. In the case of the statistical models, the differences generally derive from three sources:

1. The chemicals at the basis of the model. There may be differences in their number, their heterogenicity, and the nature of the chemicals. For instance, in the

case of models for Ames mutagenicity, some models are based on about 20,000 substances [5], while others are based on much smaller collections [6]. This last reference, for instance, relates to azo compounds, and this offers an example of the difference related to the nature of the substances. It is obvious that the training set is smaller for azo compounds. However, in some cases, it is useful to have focused models for specific chemical classes. For instance, the Benigni-Bossa model identifies the presence of the aromatic azo moiety and immediately assigns as mutagenic the substance. Actually, as shown in Gadaleta et al. [6], from an experimental point of view, half of the aromatic azo dyes are non-mutagenic. This model was the basis of the effort to develop more accurate models, able to provide better predictions for the specific chemical category. Thus, in the case of azo compounds, the model, even if based on hundreds of substances, and not tens of thousands, provides better results, because quite focused. The heterogenicity of the training set is another obvious aspect. It is easier to obtain good models if the set is homogeneous, but the applicability domain of the model is more limited.

2. The chemical information. The way to address the chemical information may provide very different results. This refers both to the format used as input and to the descriptors used. In some cases, the ways to describe the molecule are quite compact and simple. For some models, in particular old models, simple physico-chemical information was used. Nowadays, there are thousands of chemical descriptors and of fingerprints which are used to capture relevant features. The format to represent the molecule can be simple, as the SMILES string, or bi- or tri-dimensional representations. We have to mention that typically the highest level of uncertainty and variability in toxicity models is associated to the experimental values. Thus, in several cases, we found no advantages in the use of tri-dimensional representation, versus bi-dimensional ones. Even SMILES are sometimes used as direct input of the in silico models for toxicity, and this is a convenient approach since there is no need of calculating molecular descriptors. This is the case of the models based on CORAL [7, 8].
3. The algorithm used to build up the model is the third main component in the model, which surely influences the model. Today there are more and more models using advanced algorithms. Machine learning is largely applied, and deep learning is used too. These recent algorithms surely provide novel solutions and better possibilities to cope with large collections of data, and nonlinear phenomena. Multitask modelling is also a recent interesting approach. The use of these sophisticated algorithms is useful provided that there are enough data. For smaller sets traditional methods are equivalent. These recent approaches have been described recently [9, 10]. The multitude of tools also made clear that there is not one single approach. Conversely, there are multiple ways to get quite similar, equivalent results. This aspect implies a shift from the assumption and the effort to identify the "perfect", ideal model, because based on the proper structural components and the correct equations at the basis of the biochemical

process. Modelling toxicology is moving towards a probabilistic perspective, as well as risk assessment in general [11].

There are other differences quite important when considering the in silico models for toxicity predictions. Some models are regression models, i.e. provide quantitative values as outcome, while others are classifiers. Most typically, for human toxicity endpoints, the classifiers are binary ones, such as toxic or not. In some cases, models have been developed to address the potency of the adverse level, such as high, medium, or low. Quite often the models are indeed classifiers for most of the endpoints listed at the beginning of this section. However, there are quantitative models also for mutagenicity and carcinogenicity, taking into account the potency [12, 13]. On the other hand, even if the lethal dose which kills 50% of the animal (LD50) is a quantitative value, there are models which refer to threshold values to classify substances as toxic or not, or with different levels of toxicity. In this case, attention should be paid that the threshold values may differ depending on the regulation and the country.

We mentioned above that there are commercial and free models, publicly available on the internet. Of course, a clear difference is economic. Exercises which have been done on different tools and endpoints did not identify differences in performance between commercial and free models [14–17]. A main difference, beyond the price, is the access level and the transparency and documentation. Free models have wide access without restrictions. Quite typically the documentation is good, such as the information on the substances in the training set and the algorithm. Some of these tools, such as VEGA (www.vegahub.eu), are also open source. Conversely, typically this kind of information is not available for commercial software. The algorithm is proprietary, and the availability of the structures and toxicological data of the substances within the training set may be difficult to obtain and report. Thus, the commercial software is quite opaque. The documentation is requested within certain regulations (like REACH [4]), and it is on the basis of the confidence in the results obtained. The different models for the different endpoints certainly have a different level of reliability depending on the endpoint. We discussed above the case of Ames mutagenicity. This is an endpoint where the results are usually good. The property is relatively simple, and there are in a few cases models based on about 20,000 substances, as we said. The general performance of the different models has been compared and may vary depending on the kind of substances [16, 18, 19].

Other endpoints have different performance, and some large exercises have been done [15–17]. The endpoints which refer to chronic toxicity or involve complex toxicological processes are more difficult; this is the case of developmental toxicity and reproductive toxicity. In this case, the data available are for a few hundreds of compounds, and surely, the toxicological process is quite complex, involving in some cases more than one generation. Great caution should be used to evaluate the results of the models for these endpoints; multiple models should be used, and it is recommended to carefully verify if there are similar substances with experimental data supporting the final evaluation, according to a weight-of-evidence approach which will be discussed below.

Quite often recent in silico models do contain tools to evaluate if the prediction is reliable. This is done by referring to the information present in the substances in the training set and considering the so-called applicability domain. For instance, in the case of the models in VEGA, the applicability domain is measured in a quantitative way, and the software provides the applicability domain index (ADI) ranging from 0 to 1. This index is calculated by the software based on the chemical and toxicological information, and on the algorithm [20]. Thus, in practice, the ADI looks at the most similar compounds present in the set of substances at the basis of the model. This is a first contribution of the ADI. This piece of information, which is purely chemical, is addressed considering the similarity values of the most similar substances (see also the following section regarding similarity), and the presence of unusual chemical moieties. Another component of the ADI is specific for the property of the model, such as toxicity. In this case, the software compares the predictions obtained on the most similar compounds present in the set used to build the model. Of course, the predictions are specific for the specific endpoint and are closely related to the local situation represented by the most similar compounds. This value is called the accuracy of predictions and is integrated within the ADI. Furthermore, the software compares the predicted value of the target compound, with the experimental values of the most similar compounds. This value is also very closely related to the property, and thus it may change for different endpoints, even looking at the same substances. Finally, for the ADI calculation, the software takes into account some factors related to the algorithm, but these components usually have a lower impact on the final ADI.

Not all the in silico models have this complex approach as VEGA. In some cases, the applicability domain is addressed only in a qualitative way, and the substance is assigned only as inside or outside the applicability domain, as in the case of the T.E.S.T. (https://www.epa.gov/chemical-research/toxicity-estimation-software-tool-test) and Danish QSAR Database (https://qsar.food.dtu.dk/). Furthermore, the applicability domain is typically calculated by comparing the target substance with those in the training set, and this is done using tools for the chemical similarity but not so often considering other factors related to the endpoint and the algorithm, as done in VEGA. We will discuss more in detail this point later.

8.3 Tools for Read-Across

In silico models and read-across belong to the so-called non-testing methods. Read-across has been used for decades by experts to evaluate substances. Even the process of the identification of the SAs discussed above is somehow related to the concept that some similar compounds present a common toxicological effect, simply because they share some similar molecular moieties. The expert systems derived from this strategy originate from a concept implicit in the read-across process. However, exploring more cases, it has been found that there are substances which apparently are similar but present different toxicological profiles. On the one hand, several rules of exceptions to the SAs have been introduced; on the other hand, other features, not only based on the chemical similarities, have been introduced, to supplement the approach with further inputs. Beyond the chemical similarity, toxicological aspects (e.g. mode of action), physico-chemical properties, and toxicokinetic properties have been proposed. Experimental values and in silico predictions can be used and assessed manually [21]. Here, we are more interested in the approaches where computers have a higher role. In some cases, the read-across tools refer to collections of data or programs which are heterogeneous, such as in vitro data or information on metabolism, integrated with information based on the structure [22, 23]. In other cases, all the information necessary for the read-across derives from the chemical structure, and the additional information related to the toxicological aspects, for instance, is also derived from the chemicals structure, in the form of SA or information on the mode of action [24, 25]. In this perspective, the advantage is that the approach can aim to obtain the complete matrix of the values to be used for read-across, solving a main critical aspect of traditional read-across: to strongly rely on the data availability. Indeed, read-across is opportunistic and, traditionally, depends on the experimental data. However, if we imagine using predictions, whenever experimental data are missing, we can potentiate the approach and have a more reproducible strategy. Programs such as ToxRead and ToxDelta, both available within the VEGAHUB platform (www.vegahub.eu), aim to address this aspect [24, 25].

Let us consider ToxRead as an example. It applies two processes of similarity search: The first one is based on the structural similarity and the second one relates to the specific property of interest, which can be a toxicological endpoint or another endpoint. The structural similarity is calculated using the software developed for the VEGA in silico models [20]. Particularly, this tool is based on a combination of ways to represent and compare structures, related to the presence of certain components, with the relative weights. These components and weights have been optimized on millions of substances, and in general, the approach is quite robust. However, it is important to comment a fundamental aspect of similarity. It is not an objective property of a chemical, or in general of a certain item. Similarity always implies the presence of at least two items, and it is related to the purpose of similarity. In practice, in our case of substances and toxicity properties, for instance, the definition of similarity is necessarily related to the endpoint. For instance, two substances may have a similar fish bioconcentration factor value, but very different genotoxicity. Indeed, in the case of the genotoxicity, there are peculiar SAs—i.e. peculiar fragments, which may represent the occurrence of a toxicological process or not—but these fragments may be neutral or not so relevant for bioconcentration. Let us consider the presence of an epoxide versus an ether group. From the "point of view" of the bioconcentration factor, the epoxy group is a kind of ether group, while regarding genotoxicity the epoxy may imply a potential toxic effect, which is not observed in the case of the ether group.

Based on this concept, ToxRead includes a second series of metrics for similarity which are specific for a defined endpoint. Thus, in practice, ToxRead has tens of modules, one for each endpoint. The tool for the chemical similarity is applied for all modules, and then there are collections for rules, each collection specific for a certain endpoint.

For instance, in the case of Ames mutagenicity, ToxRead contains a collection of more than 800 rules. These rules include the SAs of Benigni-Bossa, plus other collections, calculated with SARpy [26] or other algorithms, or also extracted manually. An interesting observation is that the Benigni-Bossa SAs only represent fragments associated with toxicity, while the other collections contain also fragments associated with the lack of toxicity. In the case of properties with continuous values, ToxRead has rules which are associated with threshold values.

Thus, ToxRead combines the two similarity measures, the structural one and another related to the property. For the target compound, the user can visualize all the SAs and rules associated with the effect. Thus, the software provides a general view of the factors related to the presence and absence of effect. As we said, ToxRead contains both kinds of fragments, pointing towards effect or lack of effect. This fact is an improvement compared to the expert-based approach, which is quite subjective, and based on personal experience. Conversely, ToxRead is moving towards a more systematic approach, which is more objective and reproducible, compared to the "manual" read-across.

It is also important to notice that if A is similar to B and B is similar to C, we cannot conclude that A is similar to C. In practical terms, this means that the similarity application is quite local, and it loses its utility moving away from the very similar substances. There are multiple algorithms to measure similarity, and in many cases, the similarity is normalized between 1 and 0. The different metrics do not overlap; thus, the comparison should be done internally, within each software. For instance, in the case of the structural similarity within the VEGA tools (VEGA in silico models and ToxRead), 1 means identity, and substances should be considered with good similarity if the similarity is above 0.9 or 0.85. If the similarity is lower than 0.75, the two substances contain important dissimilar parts, but these values are not "official", unique thresholds; conversely, there is a tendency, and these values are only indicative and vary by the endpoint and the substance. Similarity value is a key factor for read-across, but the number of similar substances is also very important. If the similar substance is only one, this implies uncertainty. If there are more similar substances with quite close property values, this is much better. Read-across is very sensitive to noise, indeed, and if it is based on a single substance, the quality of the value of the source substance should be high (in read-across we call target substance the substance to be evaluated, and source substances the substances with the experimental values used for read-across). This is a limitation associated with read-across compared to the in silico models. In silico models are using multiple substances, as we have seen, even tens of thousands, and thus in this case if there is noise, and substances with data of lower reliability, this is not a critical aspect. In the case of read-across, we must be quite sure about the data quality of the source substances, because for read-across, very few substances are typically used and in some cases even one.

If we use more than one substance, which is preferable, we should use interpolation preferably, avoiding extrapolation: thus, if we have a set of substances with a carbon chain of different lengths, we should have substances with the lengths longer and shorter than the chain of the target substance. Compared to in silico models, read-across may also have advantages. If we have very similar compounds, the overall assessment may be more robust if based on read-across compared to the in silico models. The in silico models relate to the global population; thus, the read-across may be more robust in a local situation. We will discuss this point later in more details.

Another important point related to read-across is that the results are dependent on the similarity metrics, associated thresholds, and number of similar substances that we use. For instance, ToxRead allows selecting the number of substances for read-across. If all the similar substances have the same label (in the case of classifiers) or close property values (in the case of continuous values), we can derive our conclusion quite easily. Conversely, if the similar compounds in the cluster for read-across are not homogeneous, regarding the property value, we are in a critical situation. This will be discussed later addressing weight-of-evidence.

8.4 Weight-of-Evidence

In the case of non-testing methods, and in general considering experimental values, it is common to use multiple values; in several cases, the values may derive from hetero-geneous sources. This represents an issue, regarding the process of comparing and integrating multiple values. The European Food Safety Authority (EFSA) addressed this within a specific guidance document [27]. This guidance indicates that the user should proceed sequentially (1) gathering all data, (2) evaluating the data separately, and then (3) integrating the results of the process. The process of the evaluation is detailed in the guidance, and basically, for the integration the user should evaluate these three aspects: (a) the relevance, (b) the reliability, and (c) the consistency of the multiple data.

The relevance should be evaluated regarding the specific purpose, thus with reference to the problem formulation. This aspect is quite important for instance in the case of read-across. Let us imagine we have a source compound, which is mutagenic and contains a SA related to mutagenicity. In this case, we should check if the SA is also present in the target compound. If the SA is not present in the target compound, this similar compound is not relevant in our case. Indeed, the mutagenicity is due, very probably, to the SA, but if this SA is not present in the target substance, this information is irrelevant. Conversely, if the SA is present in both the source and in the target compounds, this source is surely relevant. At this point, we can investigate the reliability of the information regarding the SA. To do this, we can check if the similar substances with the SA are mutagenic or not (we already commented that there is a certain number of substances with the SA which are not active, depending on the specific SA, and there may be rules of exception for a certain SA). Thus, if we observe that similar substances are not active, we can conclude that for the specific case the reliability of the SA is low.

Similarly, in the case of in silico models, it is often possible to evaluate the reliability of the result since several models do provide measurements regarding this. We have already discussed above that models quite often apply tools to evaluate the applicability domain, and we said that in some cases this measurement is a continuous value, while in other cases, this is addressed as inside or outside. Of course, in the case of a quantitative value, as for the VEGA models, we have a refined appreciation of the reliability domain. If the value is quantitative, it is also possible to assign a weight to the individual in silico models, and then integrate the results by applying different weights for each model. Thus, in the case of an individual in silico models, considering the reliability, one approach is to use both the value of the prediction and its reliability by applying a weight, and this is used for instance in the case of the VEGA consensus model for mutagenicity where four individual models are combined according to the scheme described in the literature [28]. Other platforms of in silico models integrate the results of the individual models in a different way, because they cope with the applicability domain in a categorical way, as inside or outside the applicability domain. This is the case of the T.E.S.T. software of the US EPA (https://www.epa.gov/chemical-research/toxicity-estimation-software-tool-test), and the Danish QSAR Database (https://qsar.food.dtu.dk/). These systems accept or not the models based on threshold values for the applicability domain, and then all the accepted models are considered equivalent regarding their reliability. There are different possibilities to integrate the values from in silico models and read-across, and we discussed them [29].

Once the relevance and the reliability of each line of evidence have been characterized, the last point, as we discussed above, is to integrate the separate lines of evidence. Referring to the EFSA guidance above mentioned [27], and to the identification of separate lines of evidence to be integrated, we notice that the in silico models (at least some of them) provide three lines of evidence:

1. The prediction. This is the value given by the model, which is supported by the descriptors, algorithms, etc.
2. Similar compounds. They are shown by some models, such as VEGA and T.E.S.T., for instance. This line of evidence should be used as read-across, as discussed above.
3. The potential mechanism involved in the process. This of course is a piece of information which may not be present. This depends on the model, the substance, and the endpoint. Indeed, some models do not contain the indication about the mechanism, because they are not built using this piece of information. For instance, a model based on the kNN algorithm (in which the prediction is based on the k most similar compounds of the training set, combined usually by a mean or a median) ignores the mechanism.

Not all the in silico models are so descriptive and detailed. The VEGA format, as an example, is quite rich regarding these pieces of information. Thus, the user should analyse these three lines of evidence separately and then compare them. In case of conflicting lines of evidence, it is useful to refer to the process we discussed above regarding the relevance and the reliability of each line of evidence. For instance, we had the example above of a similar substance which was not relevant, because it contained a SA that is not present in the target compound. In this case, the read-across based on this substance should be disregarded. Conversely, it may be that the read-across indicates the presence of a very similar compound which is toxic, while the prediction is for non-toxicity. In this case, the high similarity implies high relevance, and thus, this line of evidence prevails unless we have a clear explanation why the similar compound is toxic, for a certain reason, which does not apply to the target substance.

8.5 Tools for Integrating Multiple Endpoints

In [30], the European Commission indicated a general strategy to reach a toxic-free environment, minimizing and substituting the substances of concern and promoting the development of chemicals sustainable by design. This means that assessors have to identify the riskiest substances. Typically, this is done by analysing the properties that lead considering a chemical as a substance of very high concern (SVHC). How identify an SVHC is defined by laws or regulations. Depending on the regulation of reference (even among the European regulations), the thresholds may differ [31]. In this paragraph, we will refer to the REACH [4] definition that considers SVHC the chemicals that are persistent, bioaccumulative and toxic (PBT), very persistent and very bioaccumulative (vPvB), carcinogenic, mutagenic, or reprotoxic (CMR), or endocrine disruptors (ED).

From a computational point of view, the identification of the SVHC means the integration of various evidence for several endpoints. It is a further step in the integration of the available information (see the previous paragraph). The assessor has to integrate the information available for each endpoint (e.g. for the endpoint persistence in water, he/she has to integrate half-life in water, ready biodegradability information, hydrolysis, etc.), several endpoints to evaluate a property (e.g. persistence in water, sediment and soil to evaluate the persistence), and then combine all the properties (e.g. persistence, bioaccumulation and toxicity for the PBT/vPvB assessment).

Tools like VEGA can process several chemicals and several endpoints at the same time, but the results are not integrated; the user has to do it manually or using other tools. Moreover, different users may obtain different integrated results, depending on the integration strategy used. For this reason, in the last years, several methods and tools to assess the PBT/vPvB, the CMR, and/or the ED properties were developed. We can divide them into two categories, the screening, and the prioritization (or ranking) tools. The screening tools divide the list of chemicals into two (e.g. toxic and non-toxic) or a few classes (e.g. toxic, moderately toxic, non-toxic). The prioritization tools assign a score to each chemical, which allows to order them from the most to the less concerning. Both can be useful, depending on the purpose. Industries may want to evaluate several possible substances in an early design phase, before synthesizing them, to decide which ones can proceed in the development process. In this case, a screening tool may be sufficient. If a regulatory body wants to decide on the most concerning substances to plan management strategies and has to focus its attention on a subgroup of the concerning chemicals (e.g. the most concerning ones), it needs a prioritization tool.

Here we present, as an example, the JANUS tool (https://www.vegahub.eu/), developed to prioritize chemicals based on seven properties—i.e. persistence (P), bioaccumulation (B), toxicity (T), carcinogenicity (C), mutagenicity (M), reprotoxicity (R), and ED—and the REACH [4] thresholds. It responds to the specific requirements of the German Umweltbundesamt (UBA) to rank the registered chemicals from the most to the less hazardous. This means that a screening approach, like the one proposed in [32], is not sufficient. With JANUS, the user has a tool that runs QSAR models for more than 20 endpoints, integrates these predictions in a sort of automatic weight-of-evidence with the experimental values, if available, to assess the seven properties, and integrates them into three scores for the prioritization always considering the reliability of the values. In this way, two chemicals with the same assessment value but with different reliabilities will have different prioritization scores. Moreover, it offers the possibility to evaluate the microbial metabolites in the same way as the parental.

More in detail, JANUS considers firstly the seven properties separately. The properties are evaluated considering the presence of experimental values; they can be inserted by the user or retrieved by the VEGA models implemented in JANUS. Indeed, JANUS runs 48 VEGA models that can be used as key predictions or as supporting information (e.g. to modulate the reliability depending on their concordance with the key prediction). In some cases, like P and B, there are screening classes. They are classes of substances (e.g. perfluoroalkylic compounds) recognized as hazardous (e.g. P) but not well predicted by the models. For these chemicals, an arbitrary assessment and reliability are assigned. In the case of multiple values with the same reliability, their agreement is also considered (i.e. the disagreement reduces the reliability of the property).

The output of the first part is one assessment value and its reliability that are combined into property scores. The property scores are then integrated into three different prioritization scores. The first one is based only on the P and B properties (with equal weight), the second is based on the human health-related properties (C, M, R and ED, combined on a worst-case approach), and the last is based on all the properties (combining the human health properties and T on a worst-case approach). The scores range from 0 (non-hazardous) to 1 (hazardous). They will be close to 1 for hazardous chemicals with good reliability and close to 0 for non-hazardous chemicals with good reliability. The chemicals with scores around 0.5 may be moderately hazardous with good reliability or chemicals (hazardous or not) with low reliability.

As mentioned above, an advantage of the JANUS tool is the possibility to run the metabolism module. It is based on the public EAWAG Biocatalysis/Biodegradation Database (https://envipath.org) and generates the metabolites of the first step of microbial biodegradation. The metabolites are then processed as the parental compound to allow the user to analyse the possible concerns derived from the metabolites.

Literature reports several other tools or methods for screening and prioritization. Some of them are summarized below.

- Strempel et al. [33] report a screening method for PBT and vPvB properties based on predicted and experimental values. The output is four classes—PBT, nonPBT0 (none of the properties is of concern), nonPBT1 (one of the properties is of concern), and nonPBT2 (two properties are of concern). For each property, the value is divided by the assigned threshold to obtain a property score. The PBT and the vPvB score are the average of the property scores.
- Böhnhardt [32] reports a strategy based on predicted log Kow and biodegradation. The authors, applying a screening approach, identified 132 chemicals to be deeply analysed (starting from a list of 4445 chemicals).
- Another approach (https://repository.qsartoolbox.org/Tools/List/Profilers) was developed by the European Chemicals Agency (ECHA); it is based on three screening profilers (one for each of the three properties, P, B and T) that use predicted and experimental values applying a workflow through the OECD QSAR toolbox (https://qsartoolbox.org/). They allow classifying chemicals as persistent or not, bioaccumulative or not, or toxic or not (considering ecotoxicity only), respectively.
- In [34], the authors developed a tool to identify the SVHC based on chemical similarity. This screening method was tested on an external set in [35] and became a freely available tool [36].
- Carlsen and Walker [37] proposed a ranking based on partial order theory that uses as input predicted values for P, B, and T.
- Shin et al. [38] developed a scoring system based on exposure and hazard indicators, collected from databases, to rank chemicals for the occupational environment, therefore, only for human health risk.
- Papa and Gramatica [39] proposed the PBT index, a screening tool for the PBT assessment based on a multiple-linear regression equation that uses four descriptors. It considers the cumulative PBT behaviour. The revised version is freely available (http://www.qsar.it/). The output is an index with a threshold to classify chemicals as PBT or non-PBT. The authors defined it as a precautionary approach [40].

8.6 Tools for Integrating Hazard and Exposure

Benefit-risk assessment is the core task for marketing applications for new drugs and to decision making throughout the life cycle of any medicinal product [41]. The ever-increasing inclination to place safe products on the market has led to an evolution of the methodologies adopted to assess drugs but also other consumer products. The change was sealed by the next generation risk assessment (NGRA), an approach characterized by decision making without the use of animal testing [42], in line with the paradigm shift induced by the new European Cosmetic Regulation [43, 44].

Within the European LIFE VERMEER project (LIFE16 ENV/IT/00016) (https://www.life-vermeer.eu/), an innovative strategy to integrate hazard and exposure assessment for human and environmental risks was designed, with the ambitious goal to harmonize and facilitate the risk assessment process in Europe, increasing human and environmental health prevention. A battery of new software was developed and made freely available to the scientific community worldwide. These tools were developed ad hoc for specific case studies, such as cosmetics, food contact materials (FCM), solvents, biocides, dispersants, and oil fractions. However, it is important to highlight that the architecture of these tools is reproducible and adaptable to drugs, or for instance to medical devices, which could be target categories for the after-life plan foreseen within the project. Indeed, despite the project ended in April 2022, it was thought in a future-oriented way, with flexible capabilities for future improvements in mind.

For the chemical assessment, a great number of models and tools exist, some of them freely available, others commercial. Individual and separate models for hazard and exposure assessment must be run in order to perform a risk assessment. Moreover, the application and the interpretation of these models could be intricate, making the integration of various information really challenging [44]. Within the LIFE VERMEER project, new comprehensive and holistic systems were built, integrating models for hazard and exposure within the same platform, offering an innovative and forward-looking solution that can substantially increase the perspective in the field of risk assessment. These novel systems represent an inducement to the use of new approach methodologies, often belittled, or at least not fully included in the daily routine of risk assessors.

As previously indicated, the software we have developed is focused on specific commercial categories. Thus, they were designed following the regulatory framework, including information retrieved from European legislations and guidance drafted by authorities, such as the European Chemical Agency (ECHA), the European Food Safety Authority (EFSA), or the Scientific Committee on Consumer Safety (SCCS) and containing specific thresholds or conditions of use. This represents another key point, which distinguishes these new tools from those already available on the market. Moreover, these new software were developed trying to get inside the user's mind, replicating with an "in silico" design, the expert system approach. For specific sectors, such as cosmetics, where animal testing was banned, in silico methodologies represent the new frontier for risk assessment and VERMEER tools fit perfectly in this context, providing a ground-breaking solution to assess products. VERMEER tools are freely downloadable from the VERMEER website (https://www.life-vermeer.eu/) and VEGAHUB (https://www.vegahub.eu/).

For the cosmetics case study, we have developed VERMEER Cosmolife, an innovative tool for the risk assessment of cosmetic products, which represents, for this sector, the first prototype ever able to integrate within the same platform the two pillars of risk assessment, hazard, and exposure [44]. VERMEER Cosmolife replicates the risk assessment procedure followed by regulators, covering the four main steps of risk assessment (hazard identification, exposure assessment, dose–response assessment, and risk characterization) [45]. Therefore, the expert system approach is partially embedded in the framework of the tool; moreover, several statistical-based tools were incorporated into the structure. The tool, indeed, has some QSAR models to predict mutagenicity, genotoxicity, skin sensitization, and no observed adverse effect level (NOAEL), as well as a tool for the threshold of toxicological concern (TTC) [44].

The software allows evaluating the toxicological profile of cosmetics ingredients, providing at the same time a well-defined indication of exposure scenarios related to specific product types in order to characterize risk for consumers. For the exposure, the calculation provided by the SCCS Notes of Guidance 11th revision [45] was adopted, with a refinement based on new models for skin permeation. VERMEER Cosmolife was built considering the regulatory framework for cosmetics; in particular, it complies with the requirements of the European Regulation 1223/2009 [43]. An important aspect of this system is that it manages simultaneously multiple ingredients enabling a comprehensive evaluation of a typical cosmetic formulation and projecting the attention to real and practical applications. Moreover, the tool is extremely user-friendly, helping, even more, the end user. Finally, the tool has been designed with flexible capabilities for future extension. Additional features will be added in future, taking into account new approaches and different lines of evidence, by exploiting other in-house models and tools, already described in the previous paragraphs. Some examples of how the tool can be used are present in the work of Selvestrel et al. [44].

Whereas VERMEER Cosmolife is a Java stand-alone application, the other VERMEER tools developed within the LIFE VERMEER project were implemented within MERLIN-Expo, a platform for simulating the fate of a chemical in the environment and the human body [46] (https://merlin-expo.eu/). Some VEGA models have been included into MERLIN-Expo in order to create an integrated tool for risk assessment. One of these "MERLIN-Expo-based" tools is VERMEER FCM which provides information with respect to exposure (i.e. migration) and hazard endpoints for chemicals (e.g. additives, etc.) intended to be used in plastics food contact materials. VERMEER FCM was developed taking into consideration the European Regulation 10/2011 [47] and the EFSA Notes for Guidance [48].

With this software, it is possible to predict the concentration of chemical migrants in food in contact with FCM. The predicted concentrations depend on several parameters such as the contact time between food and FCM, the contact temperature, the material type (e.g. type of plastic polymer) as well as important physico-chemical properties of the chemical migrants (e.g. lipophilicity). According to the EFSA Notes for Guidance, toxicological requirements depend on the migration of the chemical into the food. The tool allows predicting various toxicological endpoints required for regulatory purposes. Among them: in vitro mutagenicity, in vitro micronucleus formation, sub-chronic oral toxicity, carcinogenicity, and developmental toxicity. The tool allows running both deterministic and probabilistic simulations, the last of them based on the Monte Carlo algorithm. The tool is freely available on the VERMEER (https://www.life-vermeer.eu/) and VEGAHUB (https://www.vegahub.eu/) websites.

Shifting the attention to the environmental sphere, let us start to consider the VERMEER Rodenticides tool. The VERMEER Rodenticides tool is used to provide exposure and hazard assessments regarding the release of rodenticides in surface waters. This tool is able to predict the concentration of rodenticides in aquatic organisms considering at the same time ecotoxicological endpoints. Also in this case, the regulatory framework plays a fundamental role. Rodenticides are considered as biocidal products and are then submitted to the related European regulation (Regulation (EU) No 528/2012) [49]. In order to facilitate the evaluation of environmental risks associated with rodenticides, ECHA [50] has defined a set of generic scenarios, i.e. a set of conditions about sources, pathways and use patterns of active compounds. One of these scenarios concerns the application of rodenticides on bank slopes of watercourses like rivers, drainage channels, lakes, ponds, lagoons, etc. However, rodenticides can be flushed away due to high rainfall directly into surface waters. The structure of VERMEER Rodenticides is very similar to the FCM one with the difference that other kinds of parameters and physico-chemical properties are needed for the simulation. As in the case of VERMEER FCM, VERMEER Rodenticides allows running both deterministic and probabilistic simulations.

Another sector considered within the VERMEER project is that of solvents. In the synthesis of active pharmaceutical ingredients, solvents play a crucial role and, in the context of the green chemistry, the choice of sustainable and "greener" solvents is essential to preserve the environmental impact [51]. According to this assumption, a new tool for the environmental risk assessment of solvents was developed, and it is freely available. The identification of green solvents is demanding because a lot of parameters (health effect, environmental impact, physico-chemical properties, etc.) have to be taken into account [52]. The VERMEER Solvents tool represents a first attempt to build an innovative system for the assessment of solvents, useful also for pharmaceutical industries, focused on this first prototype, on the environmental health. New features including other critical aspects, previously mentioned, will be implemented in future versions of the tool. The structure reflects that of the VERMEER Rodenticides tool.

Finally, a new tool called VERMEER Dispersants was developed, which allows simulating the distribution of oil components within an aquatic ecosystem under different environmental conditions. It allows focusing on the comparison of the distribution of components with or without the addition of chemical surfactants that support the dispersion of the oil. Oils spills can severely impact the marine environment; therefore, measures to reduce potential damages shall be taken. The application of dispersants, which boost the transformation of floating oil into small droplets, is a valid and effective option to at least reduce this problem [53]. With VERMEER Dispersants, it is possible to predict the concentration of oil components in different environmental compartments of the marine ecosystem over time. Even in this case, both deterministic and probabilistic simulations can be run.

These tools represent an impressive achievement in the field of risk assessment because they move forward on the real case application and because they have a forward-looking fingerprint. Some of these new tools are already used by industries, and they are representing the starting point for new projects, such as SILIFOOD (https://www.sciensano.be/en/projects/applicability-silico-tools-support-risk-assessment-non-evaluated-substances-migrating-food-contact) funded by Belgian Authorities and FANGHI, funded by the Lombardy Region in Italy (https://www.openinnovation.regione.lombardia.it/it/b/38399/forme-avanzate-di-gestione-dei-fanghi-di-depurazione-in-un-hub-innovat). Moreover, from these outcomes, new tools will take shape for other case studies.

8.7 Innovation and Caution in Safe-by-Design Drug Production

At the heart of Safe-by-Design is the idea that when innovating a production process, all risks related to a target product should be as much as possible anticipated. In such an attempt, the precautionary principle plays a central role. According to The Rio Declaration on Environment and Development (1992) [54], "lack of full scientific certainty shall not be used as a reason for postponing cost-effective measures to prevent environmental degradation". The practical implementation of a Safe-by-Design approach, however, does also call for innovative methods, so that the innovation principle is on an equal footing with the precautionary principle [55]. But what should be the appropriate balance between the two principles when pharmaceutical production is considered? In this case, it is possible to tackle the issue from a double perspective. The first concerns the safety of the patient, who is supposed to take the drug for a specific medical reason. The second covers the environmental and societal risks involved in the manufacturing process through which the drug is synthesized. From the medical perspective, innovation could play a positive role. Looking at the most promising information technology advances, artificial intelligence techniques are expected to lead to a deeper critical assessment of protein structure prediction (CASP) in the early steps of drug design and discovery. Campos et al. [56] proposed a concept of molecular editing capable of insertion, removal, or modification of atoms in extremely functionalized chemicals at will and in a precise fashion with computational tools' involvement. The same authors have shown how analogues of a complex lead scaffold might be edited via heteroaromatic reduction, site-selective C–H functionalization, ring contraction, or ring expansion, evading a hypothetically lengthy synthesis of analogues followed by synthetic hurdles. Integrated chemical databases with these Web servers like the use of comparative toxicogenomic database (CTD) (http://ctdbase.org/) for human health may lead to anticipate potential toxicological problems. Notwithstanding, the cautionary principles underlying medical ethic generally leads to recommend the use of old, rather than new and innovative drugs. In general, from the medical perspective, we observe how innovation is often treated with a legitimate suspicion. Particularly, knowledge concerning drug toxicological profiles is regarded well consolidated only after years of monitoring its post-market adverse effects, so a new drug is taken into account only after more conventional remedies failed to solve the clinical problem. These criteria can also operate in non-pharmaceutical contexts. Likewise medical protocols are often applied to discourage the use of novel drugs whenever it is possible; the approach that we implemented in ToxEraser to select a cosmetic targeted to a specific functional use, does also encourage the substitution of an ingredient on the grounds of the evidence consolidated through the systematic assessments of authoritative and regulatory institutions (https://www.life-vermeer.eu/). On the other hand, when the attention is turned to the manufacturing processes, innovation looks rather universally welcome. These processes are more and more blamed for the environmental risk to which society is exposed, as they emerge to be some of the less safe and sustainable of all industrial processes. The medical and regulatory requirements of pharmaceutical purity are the main reasons leading to more waste per kilogram product as compared to making less sophisticated compounds of less stringent purity [55]. To some extent, this problem may be related to the emphasis given to the patient's safety, that is, just the first of the two addressed perspectives. The idea of green chemistry and its principles has been known since 1990, but the real implementation of rules in drug designing and synthesis is still limited. In our experience, medicinal chemists are generally interested in contemplating environmental parameters in pharmaceutical production, but they did not understand the parameters that play a role in the "greenness" of a molecule, the kind of assays that exist or may be developed that measure environmentally relevant parameters, or how such parameters could be used in compound selection or lead optimization. Among the parameters of interest to control environmental risks, we may find the drug persistence in the environment, either due to lack of biodegradation processes or to the mobility of the drug. Further endpoints of eco-toxicological can also be informative, particularly those depending on membrane permeability. Other useful screening can be focused on P450, kinase assays, and reactive metabolites, and the bioaccumulation potential may represent worthwhile targets of assessment. In turn, by impacting on mobility via solubility, permeability, absorption via modification of chemical binding forces, bioaccumulation, even a phiscochemical property as simple as the lipophilicity may inform on multiple endpoints of environmental interest. Though in the beginning, the improvement of methods of synthesis and purification of a targeted molecule is under scrutiny for an expanding class of pharmaceutical substances. By looking at a molecule of interest as the final output of a multi-step process, the attention is first focused on the manufacturing steps which involve the highest environmental and societal risks. Several indicators can be used for this purpose. Some of them are based on the ratio of the total mass of waste and the mass of the final product (E factor), which considers all wastes (reagents, solvent losses, aids, and fuel) included in the process, excluding the final product and water, which are generally not considered under waste. Furthermore, the conversion efficiency of a chemical process in terms of all atoms involved and the desired products (Atom Economy factor) was also advocated to anticipate the potential impact on the environment during drug production [51]. All analyses and investigations point at the use of solvents as the major detrimental environmental factor and hazard in the workplace. Solvents are volatile organic compounds employed in large volumes and leading to high waste, pollution, and health hazards. The concern did lead to several classification systems that can be adopted for practical reference, like the ICH Q3 Class, the Concentration Limit in Pharmaceuticals, and Ranking of Major Organic Solvents in the Form of Hazardous Impact [52]. The use of as few solvents as possible is certainly a guiding principle. Water may be more often employed for the same purposes; other dangerous solvents are deemed necessary at a first superficial glance. However, when the use of water cannot be an option, we find that the criteria underlying the "greenness" metric of the classification of solvents cover many parameters, like occupational health, quality (risk and amount of impurities), utilization of complete reagent, recyclability issues, risk of residual solvents in the pharmaceutical product and final cost. To support the choice of solvents within such a complex multidimensional framework, several tools are made available for designing the green synthesis process under the US EPA and Green Chemistry Expert System (GCES), which offers the GC process guidelines and support information as well as the green solvents/reaction conditions module [57]. More generally, the selection of appropriate solvents, reaction conditions, substrates, and types of reactions can be checked for millions of compounds by permutation and combination, utilizing in silico approaches even before practical synthesis within a period of an hour. This is likely to offer much greater and broader options to synthetic chemists [51].

8.8 Tools for Building New Models

Above we presented many tools which can be used for different purposes (hazard identification, risk assessment, prioritization, etc.), related to different endpoints. However, there are many endpoints which are covered by any model, not with standing the user would have an interesting collection data about a set of substances to deveolp a new, specific model. There are several tools which can be easily applied to develop new models, without a high level of experience. The tools listed within the VEGAHUB site represent just an example (in the section download, https://www.vegahub.eu/download/), but there are more tools available. A great advantage of many of these tools is that they can be downloaded and thus used internally, by the industry. Frequently, industries have collections of data which are proprietary, of restricted use, to be exploited within the research and development phase. Conversely, other tools, such as those discussed above, have dual access, since ideally the same system for the toxicity assessment should be used by industry and regulators.

8.8.1 aiQSAR

This tool, which contains a set of pre-built models, can be also used to develop new models, both as classifier and regression models. The system builds up a battery of local models, obtained with a few tens for substances which are selected based on the similarity of the substance to be predicted. Thus, in a certain way, the system is something between the read-across and the in silico models.

8.8.2 DTC LAB Tools

Developed by the team led by Prof Roy at the Jadavpur University in India, they offer a series of chemometric tools to build in silico models. This system includes models for regression, classification, read-across, as well as specific tools for mixtures and nanomaterials.

8.8.3 SARpy

Within VEGA some of the implemented models have been developed using SARpy, but this tool can be used to develop classifiers for the desired endpoint. The software uses the SMILES format. It starts fragmenting the molecule into fragments which are smaller and smaller. Each fragment is associated with an effect label, based on its prevalence in active or inactive substances. It involves a sequential process, since the substances already predicted by the model are eliminated, and then new fragments are searched.

8.8.4 QSARpy

QSARpy allows developing models for continuous endpoints using the chemicals of the training set and a list of fragments, named modulators, to which a quantitative value (positive, negative or null) is assigned. Modulators are extracted by fragmenting the structures and calculating the difference between the couples of chemicals of the training set that differs only by the fragment. If subtracting and/or adding one or more modulators to a chemical of the training set is possible to obtain the target, the prediction is done by subtracting and/or adding the value assigned to the modulators to the chemical of the training set.

8.8.5 CORAL

CORAL uses the SMILES of the molecule to build up models. It identifies if a certain combination of characters, present in the SMILES, are associated with the effect. These combinations of characters contain few symbols, thus in practice do not represent large fragments. CORAL is quite versatile and can incorporate other features, provided by the user, which are not the classical chemical information. For example, the user can specify the size of the nanomaterial, the temperature, the duration of the experiment, etc.

8.8.6 SOM Tool

The National Institute of Chemistry in Slovenia developed this tool based on a counter-propagation neural network. Also in this case, some of the models developed in that institute with this methodology have been implemented in VEGA, but the user can develop his/her model using neural networks with this software.

8.8.7 OCHEM

Within the EC project CONCERT REACH (https://www.life-concertreach.eu/), a network of four systems has been established: VEGA, OCHEM, the Danish QSAR Database, and AMBIT. OCHEM is a software system offering many tools to develop in silico models, also applying some of the most recent approaches, such as deep learning.

8.8.8 AMBIT

AMBIT has been developed with the support of the CEFIC, the European Council of the Chemical Industries. This system, related to collections of chemicals derived from the registered substances in Europe, is quite powerful for read-across. It offers powerful systems for data exploration and takes into account the real substances, which means the substance with its components.

8.9 Conclusions

We need efficient, fast, and powerful tools to explore the potentially adverse effect of pharmaceuticals. The process of the development of new pharmaceuticals is very expensive and involves a set of sequential steps. We analysed above the in silico tools which can be applied to investigate adverse effects. We have seen that there are multiple tools and multiple purposes. The trend is towards a more and more systematic and pre-organized way to handle the information at the basis of the adverse properties. The sequential scheme, with steps to be done at successive times, is based on the fact that in the past these steps were done through laboratory experiments on real substances, involving multiple tests. Each of these tests has a cost, and it was convenient to find an optimal scheme. The use of the in silico model represents a novelty in this. Initially, some in silico models have been used to mimic some experimental tests. This is a possibility, but we have shown that the use of the in silico models may have a more advantageous impact, with dramatic influences on the scheme, which may be consistently modified. To run one or tens of models does not make a lot of difference for the computer. The number of endpoints, and the number of substances, would no more represent a barrier. This opens completely new perspectives. We can anticipate effects on a large palette of features.

Furthermore, the possibility to run tools in parallel implies a better possibility to identify links between properties and features, and the computer is able to cope with complexity in a better way than humans. We have seen that there are tools to integrate tens of models on different endpoints and platforms for exposure and hazard at the same time. New tools offer ways to identify safer substances, addressing both the adverse properties and the functional use. This is indeed the frontier. In silico models can help to organize these multiple, heterogeneous data. This process should be organized within the same architecture. However, part of the components have to be public, because necessary to evaluate the risk—and thus, both industry and regulators need access—and part of the components should be restricted. These are the tools to explore the beneficial, functional use of the substances.

Acknowledgements We thank the EC, LIFE programme for the LIFE CONCERT REACH project (LIFE 17 GIE/IT/000461).

References

1. Benfenati E (ed) (2022) In silico methods for predicting drug toxicity, 2nd edn. Springer, New York
2. EMA (2018) ICH M7 assessment and control of DNA reactive (mutagenic) impurities in pharmaceuticals to limit potential carcinogenic risk
3. ICH Harmonised Tripartite Guideline (2017) Assessment and control of DNA reactive (mutagenic) impurities in pharmaceuticals to limit potential carcinogenic risk—M7
4. European Parliament, Council of the European Union (2006) REGULATION (EC) No 1907/2006 OF THE EUROPEAN PARLIAMENT AND OF THE COUNCIL of 18 December 2006 concerning the Registration, Evaluation, Authorisation and restriction of Chemicals (REACH), establishing a European Chemicals Agency, amending Directive 1999/45/EC and repealing Council Regulation (EEC) No 793/93 and Commission Regulation (EC) No 1488/94 as well as Council Directive 76/769/EEC and Commission Directives 91/155/EEC, 93/67/EEC, 93/105/EC and 2000/21/EC
5. Gini G, Zanoli F, Gamba A et al (2019) Could deep learning in neural networks improve the QSAR models? SAR QSAR Environ Res 30:617–642. https://doi.org/10.1080/1062936X.2019.1650827
6. Gadaleta D, Porta N, Vrontaki E et al (2017) Integrating computational methods to predict mutagenicity of aromatic azo compounds. J Environ Sci Health C Environ Carcinog Ecotoxicol Rev 35:239–257. https://doi.org/10.1080/10590501.2017.1391521

7. Toropova AP, Toropov AA, Marzo M et al (2018) The application of new HARD-descriptor available from the CORAL software to building up NOAEL models. Food Chem Toxicol 112:544–550. https://doi.org/10.1016/j.fct.2017.03.060
8. Toropov AA, Toropova AP, Selvestrel G et al (2020) Prediction of no observed adverse effect concentration for inhalation toxicity using Monte Carlo approach. SAR QSAR Environ Res 31:1–12. https://doi.org/10.1080/1062936X.2020.1841827
9. Gini G, Benfenati E (2021) From data to models. In: Chemometrics and cheminformatics in aquatic toxicology. Wiley, pp 89–124
10. Gini G (2022) QSAR Methods. In: Benfenati E (ed) In silico methods for predicting drug toxicity. Springer, US, New York, NY, pp 1–26
11. Maertens A, Golden E, Luechtefeld TH et al (2022) Probabilistic risk assessment—the keystone for the future of toxicology. ALTEX 39:3–29. https://doi.org/10.14573/altex.2201081
12. Toropov AA, Toropova AP, Benfenati E (2009) QSAR modelling for mutagenic potency of heteroaromatic amines by optimal SMILES-based descriptors. Chem Biol Drug Des 73:301–312. https://doi.org/10.1111/j.1747-0285.2009.00778.x
13. Toma C, Manganaro A, Raitano G et al (2021) QSAR models for human carcinogenicity: an assessment based on oral and inhalation slope factors. Molecules 26:127. https://doi.org/10.3390/molecules26010127
14. Honma M, Kitazawa A, Cayley A et al (2019) Improvement of quantitative structure–activity relationship (QSAR) tools for predicting Ames mutagenicity: outcomes of the Ames/QSAR international challenge project. Mutagenesis 34:3–16. https://doi.org/10.1093/mutage/gey031
15. Mansouri K, Kleinstreuer N, Abdelaziz AM et al (2020) CoMPARA: collaborative modeling project for androgen receptor activity. Environ Health Perspect 128:027002. https://doi.org/10.1289/EHP5580
16. Mansouri K, Abdelaziz A, Rybacka A et al (2016) CERAPP: collaborative estrogen receptor activity prediction project. Environ Health Perspect 124:1023–1033. https://doi.org/10.1289/ehp.1510267
17. Mansouri K, Karmaus AL, Fitzpatrick J et al (2021) CATMoS: collaborative acute toxicity modeling suite. Environ Health Perspect 129:047013. https://doi.org/10.1289/EHP8495
18. Benfenati E, Golbamaki A, Raitano G et al (2018) A large comparison of integrated SAR/QSAR models of the Ames test for mutagenicity$. SAR QSAR Environ Res 29:591–611. https://doi.org/10.1080/1062936X.2018.1497702
19. Van Bossuyt M, Van Hoeck E, Raitano G et al (2018) Performance of In silico models for mutagenicity prediction of food contact materials. Toxicol Sci 163:632–638. https://doi.org/10.1093/toxsci/kfy057
20. Floris M, Manganaro A, Nicolotti O et al (2014) A generalizable definition of chemical similarity for read-across. J Cheminformatics 6:39. https://doi.org/10.1186/s13321-014-0039-1
21. Van der Stel W, Carta G, Eakins J et al (2021) New approach methods supporting read-across: two neurotoxicity AOP-based IATA case studies. Altern Anim Experimentation : ALTEX 38:615–635. https://doi.org/10.14573/altex.2103051
22. Gadaleta D, Bakhtyari AG, Lavado GJ et al (2020) Automated integration of structural, biological and metabolic similarities to improve read-across. ALTEX 37:469–481. https://doi.org/10.14573/altex.2002281
23. Helman G, Shah I, Williams AJ et al (2019) Generalised read-across (GenRA): a workflow implemented into the EPA CompTox Chemicals Dashboard. ALTEX 36:462–465. https://doi.org/10.14573/altex.1811292
24. Gini G, Franchi AM, Manganaro A et al (2014) ToxRead: a tool to assist in read across and its use to assess mutagenicity of chemicals. SAR QSAR Environ Res 25:999–1011. https://doi.org/10.1080/1062936X.2014.976267
25. Golbamaki A, Franchi AM, Manganelli S et al (2017) ToxDelta: a new program to assess how dissimilarity affects the effect of chemical substances. Drug Des 06. https://doi.org/10.4172/2169-0138.1000153
26. Ferrari T, Cattaneo D, Gini G et al (2013) Automatic knowledge extraction from chemical structures: the case of mutagenicity prediction. SAR QSAR Environ Res 24:365–383. https://doi.org/10.1080/1062936X.2013.773376

27. Committee ES, Hardy A, Benford D et al (2017) Guidance on the use of the weight of evidence approach in scientific assessments. EFSA J 15:e04971. https://doi.org/10.2903/j.efsa.2017.4971
28. Cassano A, Raitano G, Mombelli E et al (2014) Evaluation of QSAR models for the prediction of Ames genotoxicity: a retrospective exercise on the chemical substances registered under the EU REACH regulation. J Environ Sci Health C 32:273–298. https://doi.org/10.1080/10590501.2014.938955
29. Benfenati E, Chaudhry Q, Gini G, Dorne JL (2019) Integrating in silico models and read-across methods for predicting toxicity of chemicals: a step-wise strategy. Environ Int 131:105060. https://doi.org/10.1016/j.envint.2019.105060
30. COM (2020) Communication from the commission to the European parliament, the council, the European economic and social committee and the committee of the regions. https://ec.europa.eu/environment/pdf/chemicals/2020/10/Strategy.pdf
31. Moermond CTA, Janssen MPM, de Knecht JA et al (2012) PBT assessment using the revised annex XIII of REACH: a comparison with other regulatory frameworks. Integr Environ Assess Manag 8:359–371. https://doi.org/10.1002/ieam.1248
32. Böhnhardt A (2013) Identification of potential PBT/vPvB-Substances by QSAR methods. Federal Environment Agency (Germany)
33. Strempel S, Scheringer M, Ng CA, Hungerbühler K (2012) Screening for PBT chemicals among the "Existing" and "New" chemicals of the EU. Environ Sci Technol 46:5680–5687. https://doi.org/10.1021/es3002713
34. Wassenaar PNH, Rorije E, Janssen NMH et al (2019) Chemical similarity to identify potential substances of very high concern—an effective screening method. Comput Toxicol 12:100110. https://doi.org/10.1016/j.comtox.2019.100110
35. Wassenaar PNH, Rorije E, Vijver MG, Peijnenburg WJGM (2021) Evaluating chemical similarity as a measure to identify potential substances of very high concern. Regul Toxicol Pharmacol 119:104834. https://doi.org/10.1016/j.yrtph.2020.104834
36. Wassenaar PNH, Rorije E, Vijver MG, Peijnenburg WJGM (2022) ZZS similarity tool: the online tool for similarity screening to identify chemicals of potential concern. J Comput Chem 43:1042–1052. https://doi.org/10.1002/jcc.26859
37. Carlsen L, Walker J (2003) QSARs for prioritizing PBT substances to promote pollution prevention. QSAR Comb Sci 22:49–57
38. Shin S, Moon H-I, Lee KS et al (2014) A chemical risk ranking and scoring method for the selection of harmful substances to be specially controlled in occupational environments. Int J Environ Res Public Health 11:12001–12014. https://doi.org/10.3390/ijerph111112001
39. Papa E, Gramatica P (2010) QSPR as a support for the EU REACH regulation and rational design of environmentally safer chemicals: PBT identification from molecular structure. Green Chem 12:836. https://doi.org/10.1039/b923843c
40. Gramatica P, Cassani S, Sangion A (2015) PBT assessment and prioritization by PBT Index and consensus modeling: comparison of screening results from structural models. Environ Int 77:25–34. https://doi.org/10.1016/j.envint.2014.12.012
41. Davies M, Lane S, Shakir SF (2020) Principles of benefit-risk assessment: a focus on some practical applications. In: FPM. https://www.fpm.org.uk/blog/principles-of-benefit-risk-assessment-a-focus-on-some-practical-applications/
42. Dent MP, Vaillancourt E, Thomas RS et al (2021) Paving the way for application of next generation risk assessment to safety decision-making for cosmetic ingredients. Regul Toxicol Pharmacol 125:105026. https://doi.org/10.1016/j.yrtph.2021.105026
43. European Commission EC (2009) Regulation (EC) No.1223/2009 of the European parliament and of the council of 30 November 2009 on cosmetic products. Official J Eur Union L 342:59–209
44. Selvestrel G, Robino F, Baderna D et al (2021) SpheraCosmolife: a new tool for the risk assessment of cosmetic products. ALTEX 38:565–579. https://doi.org/10.14573/altex.2010221
45. SCCS—Scientific Committee on Consumer Safety (2021) SCCS notes of guidance for the testing of cosmetic ingredients and their safety evaluation—11th revision

46. Ciffroy P, Alfonso B, Altenpohl A et al (2016) Modelling the exposure to chemicals for risk assessment: a comprehensive library of multimedia and PBPK models for integration, prediction, uncertainty and sensitivity analysis—the MERLIN-Expo tool. Sci Total Environ 568:770–784. https://doi.org/10.1016/j.scitotenv.2016.03.191
47. EC—European Commission (2011) Commission Regulation (EU) No 10/2011 of 14 January 2011 on plastic materials and articles intended to come into contact with food Text with EEA relevance
48. EFSA Panel on Food Contact Materials, Enzymes, Flavourings and Processing Aids (CEF), Silano V, Bolognesi C et al (2008) Note for Guidance for the preparation of an application for the safety assessment of a substance to be used in plastic food contact materials. EFSA J 6:21r. https://doi.org/10.2903/j.efsa.2008.21r
49. EC—European Commission (2012) Regulation (EU) No 528/2012 of the European parliament and of the council of 22 May 2012 concerning the making available on the market and use of biocidal products. Official J Eur Union L 167, 1–123
50. European Chemicals Agency (2018) Revised emission scenario document for product type 14: rodenticides. Available on https://echa.europa.eu/documents/10162/16908203/esd_pt14_en.pdf/d27d3b7e-9aa6-8146-9228-f464901b526e. Publications Office, LU
51. Kar S, Sanderson H, Roy K et al (2022) Green chemistry in the synthesis of pharmaceuticals. Chem Rev 122:3637–3710. https://doi.org/10.1021/acs.chemrev.1c00631
52. Prat D, Hayler J, Wells A (2014) A survey of solvent selection guides. Green Chem 16:4546–4551. https://doi.org/10.1039/C4GC01149J
53. Grote M, van Bernem C, Böhme B et al (2018) The potential for dispersant use as a maritime oil spill response measure in German waters. Mar Pollut Bull 129:623–632. https://doi.org/10.1016/j.marpolbul.2017.10.050
54. United Nations (1992) 1992 Rio declaration on environment and development—Centre for international law
55. Cue BW, Zhang J (2009) Green process chemistry in the pharmaceutical industry. Green Chem Lett Rev 2:193–211. https://doi.org/10.1080/17518250903258150
56. Campos KR, Coleman PJ, Alvarez JC et al (2019) The importance of synthetic chemistry in the pharmaceutical industry. Science 363:eaat0805. https://doi.org/10.1126/science.aat0805
57. EPA Green Chemistry GCES Tool. In: American Chemical Society. https://www.acs.org/content/acs/en/greenchemistry/research-innovation/tools-for-green-chemistry.html. Accessed 1 Mar 2021

Chapter 9
Read-Across and RASAR Tools from the DTC Laboratory

Arkaprava Banerjee and Kunal Roy

Abstract In silico approaches for activity/toxicity predictions have gained attention recently, and these are accepted by various regulations like EU-REACH. Aspects like reproducibility, less ethical complications, no animal use and reduced time are some of the reasons why researchers nowadays are shifting toward the in silico approaches for prediction. Quantitative Structure–Activity Relationship (QSAR) is one of the most commonly used in silico approaches for the prediction of response, but the only drawback is that since it involves model-derived predictions, it is prone to erroneous results when the number of training data points is insufficient. In recent times, similarity-based algorithms like Read-Across are being adopted by researchers with the aim of data gap filling. The Read-Across approach does not involve model-derived predictions, rather it involves similarity-based predictions and thus can efficiently be used for data gap filling. The authors at the DTC Laboratory have developed a Java-based Read-Across tool (https://sites.google.com/jadavpuruniversity.in/dtc-lab-software/home) which utilizes three different similarity-based approaches (Euclidean Distance-based, Gaussian Kernel Similarity-based and Laplacian Kernel Similarity-based) for the prediction of responses of the query compounds along with the external validation metrics and the overall error measures. Moreover, the computation of certain compound-specific similarity and error-based metrics enables the user to identify the uncertainty in the Read-Across-based predictions, especially when the observed response values of the query compounds are unreported. The idea of clubbing the QSAR methodology and the Read-Across approach together has given rise to a novel chemometric prediction approach termed as Read-Across Structure–Activity Relationship (RASAR). The authors at the DTC Laboratory are the pioneers in reporting the quantitative predictions using the RASAR approach (q-RASAR). A Java-based RASAR descriptor calculator tool has also been developed which calculates the similarity and error-based descriptors based on the similarity-based approach selected by the user. The authors feel that these tools have a lot

A. Banerjee · K. Roy (✉)
Department of Pharmaceutical Technology, Drug Theoretics and Cheminformatics (DTC) Laboratory, Jadavpur University, Kolkata 700032, India
e-mail: kunal.roy@jadavpuruniversity.in
URL: http://sites.google.com/site/kunalroyindia/

S. Kar and J. Leszczynski (eds.), *Current Trends in Computational Modeling for Drug Discovery*, Challenges and Advances in Computational Chemistry and Physics 35,
https://doi.org/10.1007/978-3-031-33871-7_9

of potential in bridging data gaps and may prove to be very much essential for the predictions of various property/activity/toxicity endpoints in the future.

Keywords Read-across · RASAR · Tools · DTC laboratory

9.1 Introduction

In the context of risk assessment and environmental safety, chemical compounds are regulated by different legislations in the European Union (EU) like registration, evaluation, authorization and restriction of chemicals (REACH) (EC regulation No. 1907/2006) and classification, labeling and packaging of substances and mixtures (CLP) (EC regulation 1272/2008) in addition to different application-specific pieces of legislation for cosmetic, plant protection and biocidal products and legislation addressing food, novel food and food contact materials [1]. Although toxicity testing exercises have traditionally been performed in experimental animal-based studies, in the recent past, there has been increasing focus on the sustainability of these methodologies [2]. Reliable toxicity analysis methods to identify, assess and interpret the deleterious properties of any substance are urgently needed. To avoid the ethical complications and minimize animal use, the replacement, refinement and reduction in animal experimentations (3R principles) of Russell and Burch can be used [3]. There is strong support from regulatory bodies like US Environmental Protection Agency (US EPA), European Chemical Agency (ECHA), Organization for Economic Co-operation and Development (OECD), etc. for the development of New Approach Methodologies (NAMs) that meet regulatory preparedness [4]. NAMs include development novel omics, in vitro and computational methods including modeling and Read-Across, searchable databases that can be used for grouping and Read-Across purposes, computational modeling of quantitative structure–activity relationships, dose–response assessments and modeling, analyses of biological processes and toxicity pathways, etc. [5]. Computational toxicology helps to identify hazards of compounds even before synthesis, and thus, they help in predictions in very early stages of drug development. Computational methods can aid in gap-filling and guide risk minimization strategies in the chemical industries and also in regulatory settings. While there is a need to develop robust and reliable non-animal methods, no single alternative method is expected to provide a unique replacement for assays targeted for more complex toxicological endpoints. Hence, results from a combination of techniques including computational modeling, in vitro assays, high-throughput screening, omics and mathematical biology can provide complementary information to develop a comprehensive picture of the potential response of an organism to a chemical substance. Adverse outcome pathways (AOPs) of stressor chemicals and systems biology frameworks enable logical integration of relevant information from diverse sources [6]. Computational data include results obtained from quantitative structure–activity relationship (QSAR) models, chemical categories, grouping, Read-Across and physiologically-based pharmacokinetic

(PBPK) models and "big data" analysis [7]. There are general and more specific factors to be considered when using different computational methods as described in the multitude of guidance documents available to support their use for regulatory purposes. While computational methods are currently mainly used more for internal rather than regulatory decision-making, the situation may change as confidence grows in their applicability and predictivity. It is advisable to use computational methods within a weight of evidence (WoE) approach and with all available data. In one hand, computational models are valuable cheap alternatives to in vitro and in vivo experiments, and on the other, their use by non-experts can eventually be misleading [8].

Read-Across is a non-testing data gap-filling technique that provides information for toxicological hazard potential based on the known toxicity data of source compound(s) with a "similar" property or chemical profile [9]. Read-Across, i.e., the local similarity-based intrapolation of properties, is gaining importance with increasing data availability and guidelines on how to process and report it. It is mainly applied to in vivo test data as a gap-filling approach, but can as well be used for other incomplete datasets [10]. Molecular similarity provides a simple and popular method for virtual screening of chemical databases. Molecular diversity analysis explores the coverage of a given structural space and underlies many approaches for compound selection and design of combinatorial libraries. In chemoinformatics, molecular similarity and diversity measures are complementary. The measures of molecular similarity and/or diversity involve in general three main components: descriptors, their coefficients and the weighting scheme [11]. The increased usage of Read-Across was driven by the huge expenditure with respect to money, time and manpower apart from ethical issues associated with in vivo testing and also encouraged by regulatory frameworks (like EU-REACH) in order to minimize animal experimentation. ECHA and OECD have published several guidelines on the technicalities of a Read-Across study. Read-Across is an evolving method with several open issues as well as opportunities. As per the ECHA's Read-Across assessment framework, the starting point is chemical similarity [12]. There are several approaches and algorithms available for calculating chemical similarity based on molecular descriptors, fingerprints, distance/similarity measures and weighting scheme for specific endpoints. Toxicological endpoints are usually in the focus of Read-Across cases. In order to further enhance the quality of Read-Across cases, new approach methods can be very useful. While computational models offer major benefits to regulators and toxicologists, the absence of a guidance document for the execution of computational experiments and use of the results in an integrated framework may lead to uncertainty and contradictions across models and users, even for the same chemicals. Read-Across offers a strategy for deriving reference points or points of departure for risk assessment of untested chemicals, from the available experimental data for structurally similar compounds, mostly based on expert judgment [13]. While drug toxicity pathways can be extremely complex and difficult to fully understand [14], specific parts of the pathway may be simpler to understand. Every toxicity pathway starts with a molecular initiating event (MIE), which if well understood makes it possible to predict which compounds can be involved in that particular MIE with the help of computational techniques. Structural alerts can be used to identify chemicals which can form a covalent bond with a biological macromolecule. Prediction of the toxicity of a compound requires a comparison with similar compounds causing the same MIE and that are associated with known toxicological data. It is possible to form categories of compounds that are all thought to act via the same MIE and then use Read-Across within the category to make a toxicity prediction.

Enoch et al. [15] presented a mechanistic Read-Across for predicting the skin sensitization potential of alkenes acting via Michael addition using the electrophilicity index as a measure of similarity for sensitizing chemicals. The index was shown to offer a chemically interpretable qualitative ranking of the chemicals within the Michael acceptor domain. Schuurmann et al. [16] developed a Read-Across method based on atom-centered fragments (ACFs) for evaluating chemical similarity for predicting fish toxicity. The study showed that increasing the ACF minimum similarity increases the prediction quality while decreasing the application range. Kuhne et al. [17] presented a Read-Across approach that makes use of the atom-centered fragment (ACF) method as quantitative measure for structural similarity for quantitative prediction of the acute toxicity of organic compounds toward the water flea *Daphnia magna.* Hartung [10] presented a new web-based tool called REACH-across, which aims to support and automate structure-based Read-Across. Russo et al. [18] discussed identification and integration of biological data from various resources and used the in vitro bioassay data-driven profiling strategy for Read-Across modeling.

Although traditional Read-Across approaches are based on the chemical similarity principle to predict chemical toxicity, complexity in the mechanism of biological activity and/or toxicity makes the accuracy of such predictions often inadequate justifying the usage of biological similarity in addition to chemical similarity for Read-Across predictions. Low et al. [19] developed a hazard classification and visualization method using both chemical structural similarity and biological response similarity measured in multiple short-term assays. The Chemical–Biological Read-Across (CBRA) approach determines each compound's toxicity from both chemical and biological analogues whose similarities are determined by a similarity coefficient like Tanimoto coefficient. Ravenzwaay et al. [20] suggested that metabolomics can be used for chemical grouping and Read-Across from a biological perspective which can reduce animal testing and provide with mechanistic interpretation of the biological action. Przybylak et al. [21] stressed the importance of consideration of biotransformation to metabolites having the same mechanism of electrophilic reactivity, via the same metabolic pathway, with a rate of transformation sufficient to induce the same in vivo outcome for the rat oral repeated-dose toxicity of β-olefinic alcohols.

Schultz et al. [22] have identified a variety of uncertainties including the regulatory use of the prediction, the data for the endpoint being assessed, the Read-Across argumentation and the similarity justification that can potentially impact acceptance of a Read-Across argument. Alves et al. [23] introduced the multidescriptor Read-Across (MuDRA) method which is conceptually related to the well-known kNN approach using different types of chemical descriptors simultaneously for similarity assessment. They found that models derived from the MuDRA approach show high prediction Accuracy similar to that of conventional QSAR models. The authors claimed MuDRA to provide a powerful alternative to a much more complex consensus QSAR modeling.

Luechtefeld et al. [24] recently combined the chemical similarity concept (Read-Across) with supervised learning methods resulting in a new technique termed as Read-Across structure–activity relationship (RASAR). They used binary fingerprints and Jaccard distance to define chemical similarity. A large chemical similarity adjacency matrix was constructed from which feature vectors were derived for supervised learning. A "simple" RASAR trains a logistic regression model to predict chemical hazards based on the similarity to the closest chemical which has tested positive (maxPos) and similarity to the closest chemical tested negative (maxNeg). The "Data Fusion" (DF) RASAR extends this concept by expanding the feature vectors using all available property data rather than only the modeled endpoint. This version of RASAR trains random forest models from diverse chemical information of analogs.

Wu et al. [25] used standard properties of chemicals along with similarity measures in a DF-RASAR approach and showed efficient predictions of chemical hazards across taxa. They showed that DF-RASAR has several advantages in the integration of the data from different effects. AbdulHameed et al. [26] developed a chemical-similarity-based protocol for the prediction of the potential of a chemical to interact with different toxicity targets. They evaluated the performance of 2D and 3D similarity approaches in correctly ranking known interacting compounds using an external evaluation set from the ChEMBL database. They found that the 2D similarity-based predictions were superior to the 3D approaches.

This chapter summarizes the Read-Across and RASAR tools and different quality and evaluation metrics associated with this research developed in the Drug Theoretics and Cheminformatics (DTC) Laboratory and their applications in the prediction of different activity/toxicity endpoints.

9.2 The Theory Behind the Read-Across Approach

The concept of Read-Across utilizes the similarity between compounds to predict the response values of the query compounds. This technique has recently emerged as one of the most promising techniques for data gap filling, especially in cases where there is shortage of experimental data [27]. As supported by the Organization for Economic Co-operation and Development (OECD), the Read-Across approach can efficiently replace in vivo testing. If other techniques like High-Throughput Screening can be coupled with Read-Across, it can enhance the quality of predictions. The Registration, Evaluation, Authorization and Restriction of Chemicals (REACH) in the European countries have preferred experimentally unavailable toxicity data generated using in silico approaches [1]. Also, in the European Union (EU), there is a ban on animal experimentation for the evaluation of cosmetics, and such evaluation should be carried out using alternative and in silico approaches. To fulfill the requirements of REACH, the Read-Across approach needs to fulfill certain criteria:

- The results obtained should be sufficient to perform risk assessment, classification and labeling.
- There should be enough coverage of the key aspects in the testing methods.
- The duration of exposure should be comparably longer than the corresponding test method, if this parameter has sufficient relevance.
- There should be reliable and adequate documentation of the applied method.

Read-Across approach can be further classified based on the differences in the number of source and target compounds used for predictions. The four different strategies of Read-Across predictions are as follows:

- One-to-One: This approach makes use of the similarity of a single-source compound to predict the response value of a single-target compound.
- One-to-many: In this approach, there is utilization of a single-source compound to predict the response values of multiple-target compounds, based on the similarity levels.
- Many-to-one: This approach involves two or more source compounds to predict the response of a single-target compound, based on the similarity levels.
- Many-to-many: Involvement of two or more source compounds to predict the response values of multiple-target compounds based on similarity levels.

Although Read-Across is a very useful technique to fill data gaps, there are a couple of problems which can be encountered. First, due to the non-availability of sufficient proof supporting justification, it is difficult to ascertain the absence of toxicity from the Read-Across predictions. Secondly, it does not give us the idea about the uncertainty in predictions [27]. The first issue can be addressed by linking the Read-Across technique with Molecular Initiating Events and Adverse Outcome Pathways, thus giving rise to a concept called Biological Read-Across. The second issue has already been addressed in our Read-Across tool by computing similarity and error-based measures (vide infra) for each of the compounds, which enables the user to assess the uncertainty in predictions.

9.3 Read-Across Tool from the Drug Theoretics and Cheminformatics Laboratory

The Read-Across tool, developed by the DTC Lab, is a Java-based program which quickly computes Read-Across-based quantitative predictions of endpoints and their corresponding external validation metrics in terms of Q^2_{F1} and Q^2_{F2} (correlation-based metrics) [28]. It also computes overall error measures in terms of Root Mean Squares Error of Predictions (RMSEP) and Mean Absolute Error (MAE). Even without the availability of the observed endpoint values of the target compounds, this tool can efficiently predict the possible endpoint values, but the external validation metrics cannot be computed as they require the observed response values of the target compounds for their computation. This tool reaches one step ahead, as it also calculates various compound-specific error measures for the individual target compounds with respect to its nearest source compounds. With the help of this tool, it is also possible to perform classification-based Read-Across, and the input response does not necessarily need to be graded, i.e., it can handle quantitative response values even while performing classification-based Read-Across. The tool generates five output files—one of them displaying the close target compounds for each query compound with their response values and similarity levels in a sorted manner, while the second one is the main output file for Read-Across predictions and their quantitative validation metrics. The rest three files show classification-based validation metrics like Sensitivity, Specificity, Accuracy, Precision, Matthew's Correlation Coefficient (MCC), etc. The Receiver Operating Characteristic Curve (ROC Curve) is also generated by taking each response value of the target compound as the threshold and calculating the corresponding true-positive rate (Sensitivity) and false-positive rate (1-Specificity) along with computation of the Area Under the Curve (AUC). With the help of this tool, one can also proceed for quantitative Read-Across Structure–Activity Relationship (RASAR) (vide infra) [29] by utilizing the error measures for the individual target compounds as descriptors. This tool has been upgraded to incorporate such features, the current version of which is available as Read-Across-v4.1 from https://sites.google.com/jadavpuruniversity.in/dtc-lab-software/home. Figure 9.1 demonstrates the detailed workflow followed by the Read-Across tool.

This tool utilizes three distance/similarity-based approaches, namely Euclidean Distance-based, Gaussian Kernel Similarity-based and Laplacian Kernel Similarity-based, for computation of the predicted response values and validation with the external validation metrics. The descriptors are standardized at first to scale their range, which ultimately reduces noise during distance/similarity calculations. The Euclidean Distance approach computes the scaled Euclidean Distances between the query compound and the source compounds using the following equation:

$$d = \sqrt{\Sigma(q - p)^2}. \tag{9.1}$$

In Eq. (9.1), d stands for the Euclidean Distance, q and p are the descriptor vectors of the source and target compounds, respectively. It is important to know that the Euclidean Distance approach does not involve any hyperparameter, and thus, the optimization can only be performed with respect to the close source compounds and distance threshold values. The Gaussian Kernel Similarity method is derived from the Euclidean Distance method, and it computes the similarities between the query and the source compounds. Mathematically, the Gaussian Kernel Similarity can be represented as:

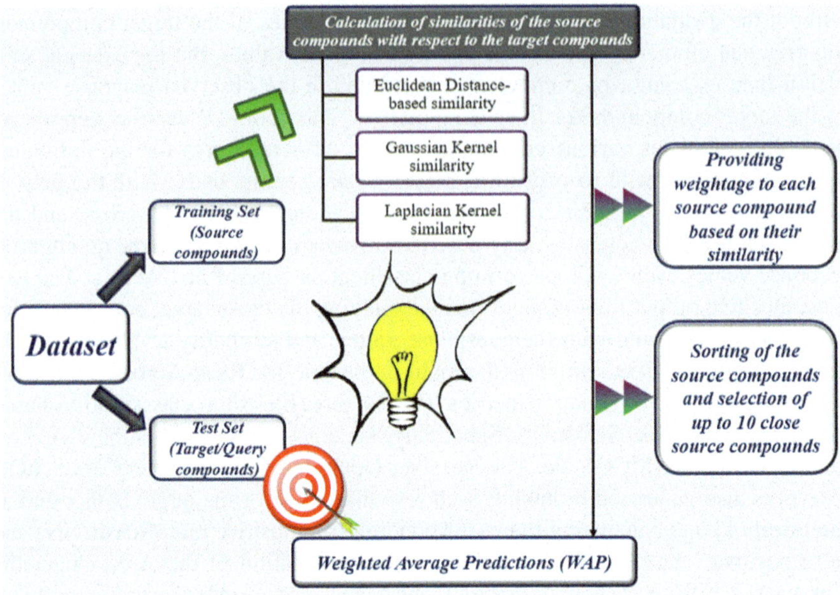

Fig. 9.1 Workflow of the Read-Across tool

$$f(\mathrm{GK}) = \mathrm{e}^{-\frac{||X_i - Y_i||^2}{2\sigma^2}}. \tag{9.2}$$

In Eq. (9.2), f(GK) stands for the Gaussian Kernel Similarity value, $||X - Y||^2$ is the L^2 norm or square of the Euclidean Distance, and σ is a hyperparameter which can be optimized. Thus, in the case of Gaussian Kernel Similarity-based predictions, the hyperparameter σ, the number of close source compounds and the similarity threshold are the optimizable entities, unlike Euclidean Distance-based approach, which allows only the optimization of the number of close source compounds and the distance threshold. The third similarity approach of predictions which this tool provides is the Laplacian Kernel Similarity method. Unlike Gaussian Kernel Similarity approach, this method utilizes the Manhattan Distance for the estimation of the similarity between a particular source compound and the target compounds. The mathematical representation for calculation of the Laplacian Kernel Similarity is as follows:

$$f(\mathrm{LK}) = \mathrm{e}^{(-\gamma ||X - Y||_1)}. \tag{9.3}$$

Equation (9.3) demonstrates the equation for calculation of the Laplacian Kernel Similarity. f(LK) is the Laplacian Kernel Similarity value, while $||X - Y||_1$ is the Manhattan Distance between the source and the target compounds. Here, γ is the hyperparameter, which allows the optimization of the Laplacian Kernel-based Read-Across predictions.

9.3.1 Pre-requisites for Using This Tool

9.3.1.1 System Specifications

Read-Across-v4.1 does not require exhaustive system resources, i.e., it can run on computers having standard memories for RAM and HDD/SSD. However, since it is a Java-based software tool, it is necessary that a particular system needs to have Java installed before running this tool. The Java Development Kit (JDK) can be downloaded from https://www.oracle.com/java/technologies/downloads/, and after successful installation, the Read-Across tool can be executed.

9.3.1.2 Input File Specifications

The program asks the user to enter two files, namely the training and test sets. The input files should be a Microsoft Excel workbook, having the extension **.xlsx**. The data tabulated in each of the training and test set files should have the following specific pattern:

1st column: compound number.

2nd to *n*th column: descriptors in subsequent columns.

($n + 1$)th column [last column]: biological activity/property/toxicity.

It is essential to note that the program can handle both quantitative and graded response values (i.e., 0 and 1) depending upon the requirement of the user. In case the input values are graded (0 and 1), the quantitative validation metrics should be ignored.

Figure 9.2 shows snapshots of the sample training and test set input files.

This tool provides another option for the user to calculate only the biological activity/property/toxicity of the target compounds without evaluating the quality of predictions. To implement this feature, the user only needs to put "**999**" in the first observed response value of the test set and any random entry for other compounds.

	A	B	C	D	E
1	S.No	X1	X2	Y	
2	1	2.2	5.7	2.32	
3	2	3.7	5.5	3.14	
4	3	6.1	6.1	3.2	
5	4	2.8	4.9	3.61	
6	5	4.2	5.9	2.79	
7					

Sample training set.xlsx

	A	B	C	D	E
1	S.No	X1	X2	Y	
2	6	3.8	4.11	2.11	
3	7	4.7	5.56	3.34	
4	8	5.9	6.1	3.1	
5	9	4.35	4.2	3.81	
6	10	3.98	5.78	2.79	
7					

Sample test set.xlsx

S.no = Compound numbers X1, X2 = Descriptors Y = Biological Activity

Fig. 9.2 Snapshots of the sample training and test set input files

In this case, the quantitative validation metrics and classification–based metrics are not computed, and the ROC Curves are not generated.

9.3.2 Downloading and Execution of the Software

- The software in the form of a **.zip** file has been made available at the DTC lab tools supplementary webpage (https://sites.google.com/jadavpuruniversity.in/dtc-lab-software/home).
- The.zip file needs to be downloaded and the contents need to be extracted. One can find that it consists of a folder (Read-Across-v4.1) inside which there is a **.jar** file, a library folder, and two sample input files (Fig. 9.3a).
- The training and test set files need to be placed inside this Read-Across-v4.1 folder, i.e., the same folder which contains the.jar file (Fig. 9.3b).
- By double clicking on the **Read-Across-v4.1.jar** file, the program will be executed (Fig. 9.3c).
- Certain dialog boxes which ask for appropriate data appear on the screen. The user needs to enter the file names for the training and test sets, some constant values (sigma and gamma, suggested value being 1), the number of close training set compounds (which can range from 2 to 10, provided that they lie inside the specified distance/similarity threshold), threshold values for distance (suggested value being 0.5; 1 in case of no known threshold) and similarity (suggested value in the range of 0–0.05; 0 in case of no known threshold). For classification, the user needs to enter the file name for the test set and the threshold value, if the input response values are quantitative (for example, the mean value of the responses of the source data set and 0.5 if there are graded response values as input) (Fig. 9.4).
- Sorting of the similarity measures will be automatically printed in a newly generated file, namely **TestSetFileName_Sort.xlsx,** and the biological activities along with the validation metrics will be automatically printed in a newly generated file, namely **TestSetFileName_Biological Activity.xlsx.** Additionally, three other files are also generated, namely **TestSetFileName_Euclidean.xlsx, TestSetFileName_Gaussian.xlsx** and **TestSetFileName_Laplacean.xlsx,** which contain the values of different classification metrics and the generated ROC Curves. The user can see all these files generated in the same folder (Read-Across-v4.1) (Fig. 9.5).

9.3.3 Analysis of the Output Files

This program generates a total of five different output files, each of which encodes certain chemometric information. The TEST_Biological Activity.xlsx file contains the predicted response values of the query compound(s), their external validation metrics in terms of Q^2_{F1} and Q^2_{F2} and the overall error measures in terms of Root Mean Squares Error of Predictions (RMSEP) and Mean Absolute Error (MAE). Apart from

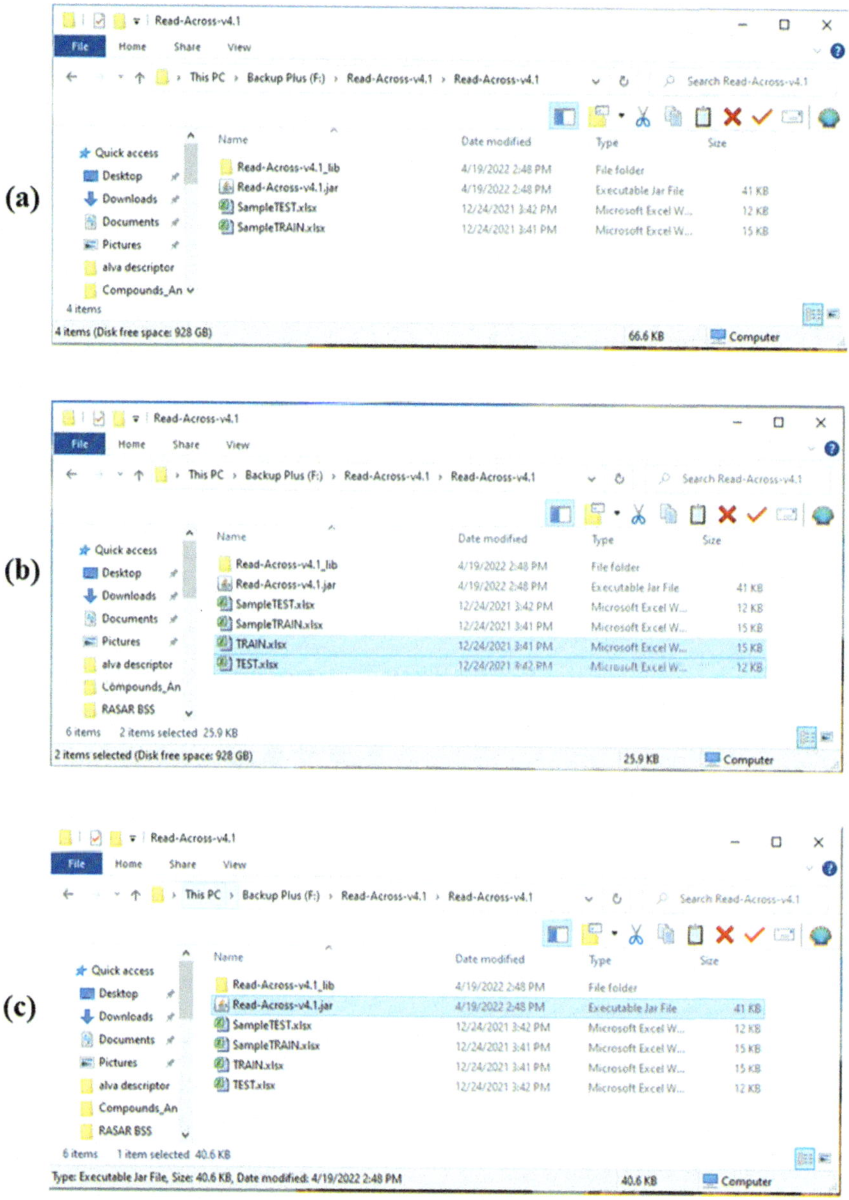

Fig. 9.3 Snapshots **a** after extraction of the .zip file, **b** after placing the training and test set files in the folder, **c** the executable.jar file

Fig. 9.4 Snapshots during execution of the tool

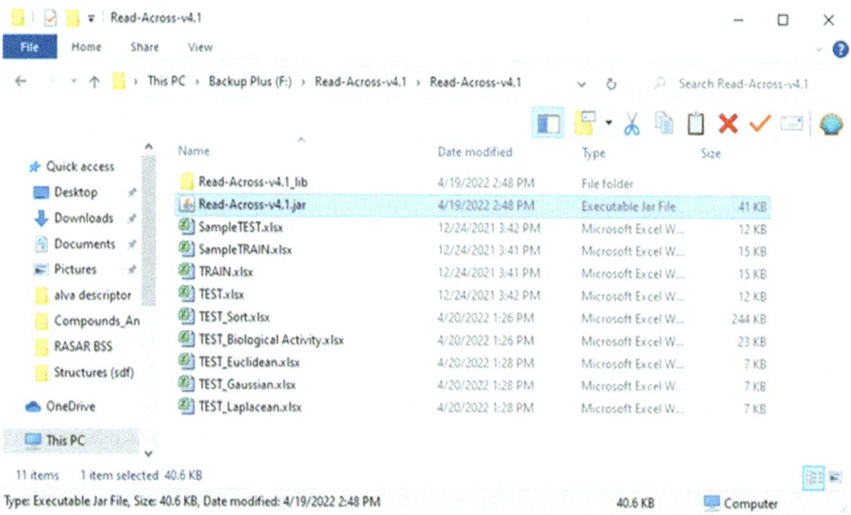

Fig. 9.5 Snapshot of the generated output files

that, various other compound-specific error measures are also generated which help the user to estimate the uncertainty of predictions for a compound with unreported response values. The metric SD_Activity computes the weighted standard deviation of the observed response values of the "*n*" close training compounds. SE stands for Standard Error of the activity values of the "*n*" close training compounds. CV_Activity represents the coefficient of variation of the observed response values of the close "*n*" source compounds, while CV_Similarity represents the coefficient of variation of their similarity values. The metric MaxPos represents the maximum similarity value of the target compound with respect to the closest source compounds having response values greater than the threshold (training set response mean). Similarly, MaxNeg denotes the maximum similarity value of the target compound with respect to the closest source compounds having a response value lower than the threshold (training set response mean). The metric g is a concordance measure, which takes into account the positive fraction (fraction of compounds among the close source compounds which have observed response values greater than the threshold), and uses it to estimate the uncertainty of Read-Across-based predictions. The mathematical representation of g is as follows:

$$g = 1 - 2|\text{Posfrac} - 0.5|. \tag{9.4}$$

The metric average similarity defines the average similarity values among the close "*n*" source compounds. SD_Similarity denotes the standard deviation of the similarity values of the close "*n*" source compounds. g_m [Banerjee-Roy coefficient] is the modified version of g, which takes into account both the MaxPos or MaxNeg and the Positive Fraction, and uses it to estimate the uncertainty in predictions. This was developed to establish a directionality to distinguish between the probable active and inactive compounds (with respect to the threshold observed response values). g_m can be mathematically represented as:

$$g_m = (-1)^n 2|\text{Posfrac} - 0.5|. \tag{9.5}$$

The value of n is 1, when the value of MaxPos < MaxNeg, and $n = 2$ when MaxPos > MaxNeg.

The TEST_Sort.xlsx file contains the sorted similarity values of the source compounds along with their response values for each query compound, calculated with the three different similarity-based algorithms (Euclidean Distance-based, Gaussian Kernel Similarity-based and Laplacian Kernel Similarity-based).

The TEST_Euclidean.xlsx, TEST_Gaussian.xlsx and TEST_Laplacian.xlsx files compute the true-positive and false-positive rates, i.e., Sensitivity and 1-Specificity values, taking each response value as the threshold and an ROC curve is computed for each of the three similarity-based measures. Apart from this, various other classification-based validation metrics are generated which determines the quality of the predictions while performing classification-based Read-Across.

9.3.4 *Application of the Read-Across Tool Developed in the DTC Laboratory*

The Read-Across tool, developed by the DTC Laboratory, has already been applied for the prediction of several different activity/toxicity endpoints. The advantage of this tool is that it provides a user-friendly GUI, which requires minimal system specifications and storage space. Several applications in the form of case studies have been mentioned below.

9.3.4.1 Case Study 1

Chatterjee et al. [28] utilized this tool for the quantitative predictions of nanotoxicity in three different datasets, aiming at data gap filling. Using the Euclidean Distance-based, Gaussian Kernel Similarity-based and Laplacian Kernel Similarity-based similarity calculations, the authors have obtained Sensitivity, Specificity, Accuracy, Precision and F-measure of up to 100%, thus demonstrating the data gap-filling ability of the tool. The hyperparameters were optimized for the computation of Gaussian Kernel Similarity-based and Laplacian Kernel Similarity-based predictions in addition to choosing the number of close source compounds along with distance and similarity thresholds. The predicted response values thus generated were calculated by the weighted average prediction of the individual observed response values, a technique somewhat similar to the Consensus Model 2, as proposed by Roy et al.

[30]. The Consensus Model 1 involves the average of the predictions derived from all the qualifying models (mean + 3 × SD) for a particular compound. The Consensus Model 2 is derived from the weighted average predictions of all the qualifying models. The Consensus Model 3 is obtained by selecting the model that provides the best prediction for a particular query compound. The Read-Across-based predictions in terms of Q^2_{F2} obtained in Dataset 1 were up to 0.96 using the Euclidean Distance-based approach. For Dataset 2, the Q^2_{F2} values were up to 0.91 in both the Gaussian Kernel-based and Laplacian Kernel-based predictions. The Q^2_{F2} values in Dataset 3 were up to 0.95, obtained by the Euclidean Distance-based approach. These values outperformed the previous QSAR and Read-Across studies on the same datasets.

9.3.4.2 Case Study 2

Chatterjee and Roy [31] have applied both QSAR and Read-Across techniques to predict the acute toxicities of mixtures of polar and non-polar narcotic substances present in the environment. They have calculated 2D descriptors and adopted a 2D-QSAR modeling technique, adhering strictly to the OECD principles, which provided sufficient robustness, predictivity and reproducibility. For the developed PLS model, the internal quality metrics are as follows: $r^2 = 0.82$ and $Q^2_{(LOO)} = 0.78$, suggesting that the model is robust, and the external validation metrics are $Q^2_{F1} = 0.87$ and $Q^2_{F2} = 0.87$, suggesting that the model is very predictive. The authors have also employed the Prediction Reliability Indicator tool to check the reliability of predictions of a true external set. Apart from the model-derived predictions, they have also employed a machine learning algorithm-derived similarity-based approach (Read-Across) and utilized our Read-Across tool for toxicity predictions. They have optimized the hyperparameters by dividing the training set into sub-training and sub-test sets, and the hyperparameters which provided the best prediction of the sub-test set, with respect to the sub-training set, were considered as the optimized hyperparameters. Using these optimized settings for the hyperparameters, the predictions were made for the original test set, with respect to the original training set. Interestingly, in their work, the authors have reported that the external validation metrics generated in case of Read-Across ($Q^2_{F1} = 0.94$, $Q^2_{F2} = 0.94$) were slightly higher than what they have obtained using QSAR technique, thus showing enhanced predictivity. This work demonstrates the importance and efficiency of similarity-based predictions over model-derived predictions.

9.3.4.3 Case Study 3

De et al. [32] worked on the identification of molecules which can potentially act as anti-SARS-CoV-2 drugs, using in silico approaches. They have utilized both QSAR and Read-Across algorithms to quantitatively predict the half maximal inhibitory concentration of the molecules. They have also used 2D descriptors and used them to develop 2D-QSAR models. They have developed four PLS models and also derived their consensus-based predictions. The internal quality metrics for the PLS models were as good as $r^2 = 0.672$ and $Q^2_{(\mathrm{LOO})} = 0.612$ which again suggest sufficient robustness, and the external validation metrics in terms of Q^2_{F1} and Q^2_{F2} were reported as up to 0.839 and 0.839, respectively. The quality of predictions was further enhanced by their consensus-based predictions, and the best predictions in terms of the Mean Absolute Error (MAE) were obtained in case of Consensus Model 3, where the reported values of Q^2_{F1} and Q^2_{F2} were 0.879 and 0.879, respectively. The authors have also employed the similarity-based Read-Across prediction technique. The optimization of the hyperparameters was done based on the source compounds, and the distance and similarity-based predictions were obtained using the optimized settings for the query chemicals. The source compounds (training set) was first divided into sub-training and sub-test sets, and the combination of hyperparameters which provided the best predictions for the sub-test set, with respect to the sub-training set, was considered as the optimized hyperparameters. This optimized setting was then used to predict the toxicity of the original query set compounds with respect to the original source compounds. In this work also, it has been reported that the external validation metrics obtained in case of Read-Across-based predictions were much better than the external validation metrics obtained from the Partial Least Squares models, as well as their consensus-based predictions. The external validation metrics for the Read-Across-based predictions were up to $Q^2_{F1} = 0.932$ and $Q^2_{F2} = 0.932$, while the best predictions for the PLS models were up to $Q^2_{F1} = 0.839$ and $Q^2_{F2} = 0.839$ and the consensus-based predictions $Q^2_{F1} = 0.879$ and $Q^2_{F2} = 0.879$. This work also demonstrates the increased Precision of the similarity-based predictions over model-derived predictions.

9.3.4.4 Case Study 4

Paul et al. [33] worked on the soil ecotoxicity predictions against *Folsomia candida* using computational techniques. In this work, the use of Read-Across on the ecotoxicity predictions of *Folsomia candida* was performed for the first time. Two of the most widely used in silico techniques—QSAR and Read-Across—were used to predict the half maximal effective concentration of a set of compounds on *Folsomia candida*. The authors have developed four individual PLS models which report an internal validation metrics in terms of r^2 and $Q^2_{(\mathrm{LOO})}$ as up to 0.762 and 0.633, respectively. The reported external validation metrics were up to $Q^2_{F1} = 0.714$ and $Q^2_{F2} = 0.642$, which suggest that the models have sufficient predictivity. The predictivity was further enhanced by the application of a consensus-based prediction algorithm which reported the Q^2_{F1} and Q^2_{F2} values of up to 0.726 and 0.656, respectively, based on consensus model 3, which reported the lowest MAE value. The authors have also performed Read-Across, an unsupervised machine learning approach, to check the effect on the quality of predictions. They have divided the dataset into sub-training and sub-test sets, and like the previous work, they have optimized the hyperparameters based on the prediction quality of the sub-test set. The optimized setting was then applied to the target compounds, while calculating the Read-Across-based predictions, with respect to the original source compounds. It is interesting to note that in this case also, the external validation results ($Q^2_{F1} = 0.775$ and $Q^2_{F2} = 0.717$) supersede the ones that were obtained from the QSAR approach, even after their consensus-based predictions. This work also proves that Read-Across-based predictions can potentially be a more effective tool in the quantitative predictions of toxicities than the conventional model-based QSAR approach.

9.3.4.5 Case Study 5

Banerjee et al. [34] reported in silico modeling of the androgen receptor binding affinity of various Endocrine Disruptor Compounds (EDCs) in rats. The authors have adopted the 2D-QSAR technique and Read-Across algorithm—two of the most commonly used in silico approaches for the prediction of response. The 2D-QSAR technique involved the steps like the collection of data, calculation of descriptors, division of the dataset into training and test sets, feature selection of the essential structural and physicochemical descriptors, development of initial MLR models and finally the development of a PLS model. The internal quality and validation metrics obtained in the QSAR approach were $R^2 = 0.737$ and $Q^2_{(LOO)} = 0.680$, which suggest that the developed model is robust. The external validation metric values in terms of Q^2_{F1} and Q^2_{F2} were acceptable ($Q^2_{F1} = 0.582$ and $Q^2_{F2} = 0.582$). The authors have also performed chemical Read-Across using the tool employing the features selected in the QSAR analysis. The training set data were further divided into sub-training and sub-test data sets, and a variety of combinations of hyperparameters were tried. The combination of the hyperparameters that provided the best results of the sub-test set in terms of its Q^2_{F1} and Q^2_{F2} values was considered as the optimized hyperparameters. These optimized settings of the hyperparameters were then employed to predict the original test set compounds with respect to the original training set compounds. The external validation metrics obtained after using the optimized hyperparameters were $Q^2_{F1} = 0.635$ and $Q^2_{F2} = 0.635$. From these results, it is evident that the predictions obtained in case of Read-Across are slightly better than the results obtained in case of QSAR. This work again shows that chemical Read-Across can be a potential approach for in silico predictions as an alternative to QSAR.

The fact that the Read-Across tool can potentially predict the response values of the query compounds, which do not have known observed response values, and we felt that it was essential to judge the confidence measure for the predictions of each query compound. To assess the quality of predictions of individual compounds, Banerjee et al. [35] have adopted certain error-based and similarity-based measures using which it is easier to identify the quality of predictions. In their work [35], the authors have elaborated these error and similarity measures, modeled them and identified the most important error and similarity measures using various techniques like the mean difference among the highest and the lowest residual compounds, Linear Discriminant Analysis of errors—a classification-based modeling technique and Sum of Ranking Differences (SRD) [36]. The measures which were used to assess the uncertainty in Read-Across-based predictions include SD_activity, which defines the weighted standard deviation of the observed response values of the close "n" source compounds to a particular query compound. The mathematical expression of the SD_activity is given in Eq. 9.6.

$$s_{\text{weighted}} = \sqrt{\frac{\sum_{i=1}^{n} w_i \left(x_i - \overline{x_{\text{wtd}}}\right)^2}{\sum_{i=1}^{n} w_i} \times \frac{n}{n-1}}, \tag{9.6}$$

$$\overline{x_{\text{wtd}}} = \frac{\sum_{i=1}^{n} w_i x_i}{\sum_{i=1}^{n} w_i}, \tag{9.7}$$

$$n = \frac{\left(\sum_{i=1}^{n} w_i\right)^2}{\sum_{i=1}^{n} \left(w_i^2\right)}. \tag{9.8}$$

The expression w_i denotes the similarity weightage, $\overline{x_{\text{wtd}}}$ signifies the weighted average prediction and n stands for the effective degree of freedom. CV_activity defines the coefficient of variation of the observed response values. It can be denoted mathematically as Eq. 9.9:

$$\text{CV}_{\text{activity}} = \frac{s_{\text{weighted}}}{\overline{x_{\text{wtd}}}}. \tag{9.9}$$

The Euclidean Distance-based similarity function can be described as the similarity value between two compounds that is obtained from their Euclidean Distance. It is expressed as in Eq. 9.10.

$$f(ED) = 1 - d(X, Y). \tag{9.10}$$

In this equation, $d(X, Y)$ denotes the Euclidean Distance between two compounds and $f(ED)$ signifies the Euclidean Distance-based similarity. Another similarity-based measure is the Gaussian Kernel-based Similarity function, which utilizes the L^2 norm of the Euclidean Distance, i.e., squared Euclidean Distance. The Gaussian Kernel Similarity and Laplacian Kernel Similarity were previously defined in Eqs. (9.2) and (9.3).

The term average similarity denotes the mean value of the similarities of all the close "n" source compounds, selected for each query compound. It can demonstrate the closeness of the source compounds to the target/query compound. The expression for the computation of average similarity is demonstrated in Eq. 9.11.

$$\text{Similarity}_{\text{average}} = \frac{\sum_{i=1}^{n} f_i}{n}. \tag{9.11}$$

The term f_i in this expression is the individual similarity values of the close "n" source compounds, with respect to the target compound. A dispersion measure, which is essential to estimate the uncertainty in Read-Across predictions, is the Standard Deviation of the Similarity values (SD_similarity) of the close "n" source compounds, with respect to a particular query compound. This is essential to check the dispersion of the similarity values among the selected close "n" source compounds. Mathematically, SD_similarity can be denoted as Eq. 9.12.

$$s_{\text{similarity}} = \sqrt{\frac{\sum_{i=1}^{n}(f-\overline{f})^2}{n-1}}. \tag{9.12}$$

In Eq. 9.12, $S_{\text{similarity}}$ denotes the SD_similarity, while $\overline{f}$ is the average similarity of the close "n" source compounds. Another similarity measure is MaxPos, which signifies the similarity value of the closest source compound, with respect to the target compound, and having an observed response value greater than the threshold (mean observed response values of all the source compounds). This measure is essential to estimate the closeness of a particular query compound toward positive or negative (with respect to the threshold observed response). Likewise, MaxNeg signifies the similarity value of the closest source compound, with respect to the target compound, and having an observed response value lower than the threshold. Again, MaxNeg provides the information on how close the query compound is to the negative source congeners. The metric Abs(MaxPos-MaxNeg) is the absolute differences in the MaxPos and MaxNeg similarity values, and a high value indicates that a particular query compound is significantly more similar to the positive or negative source compounds. A concordance measure g has been applied [25], which takes into account the fraction of close source compounds that have a higher observed response value than the threshold (Positive Fraction). The mathematical equation for the calculation of g has already been discussed in Eq. 9.4, and its value ranges from 0 to 1. The summary of these similarity and error measures has been provided in Table 9.1.

From the analysis of all the similarity and error-based measures, the authors have set the following criteria for the estimation of the uncertainty in Read-Across-based predictions in Table 9.2.

Reliability estimates: Very Good (All criteria met); Good (Criterion 1 and at least one of the rest, but not all); Moderate (Any one met); Bad (None of the criteria met).

9.4 Read-Across Structure–Activity Relationship—A Novel Concept

So far, in silico approaches for the assessment of activity/property/toxicity have centered around the use of QSAR and various Machine Learning (ML) approaches like Support Vector Machine (SVM) and Artificial Neural Networks (ANN).

Table 9.1 List of similarity and various error measures generated for each query compound during Read-Across predictions

Measures	Description	Comment
SD_activity (s_{weighted})	Weighted standard deviation of the observed response values of the close "*n*" source compounds for each query compound	Dispersion measure
CV_activity	Coefficient of variation of the response	Relative error measure
Euclidean distance-based similarity function	It determines the similarity between two compounds *X* and *Y* using the Euclidean distance approach	Similarity function (f)
Gaussian Kernel-based similarity function	It determines the similarity between two compounds *X* and *Y* using the Gaussian Kernel similarity approach	Similarity function (f)
Laplacian Kernel-based similarity function	It determines the similarity between two compounds *X* and *Y* using the Laplacian Kernel Similarity approach	Similarity function (f)
Average similarity	Mean similarity to the selected close source compounds for each query compound	Similarity measure
SD_similarity	Standard deviation of the similarity values of the selected close source compounds for each query compound	Dispersion measure
MaxPos	Maximum similarity level to the positive close source set compounds (based on the "training set" observed mean)	Similarity measure
MaxNeg	Maximum similarity level to the negative close source set compounds (based on the "training set" observed mean)	Similarity measure
AbsDiff *or* Abs(MaxPos-MaxNeg)	Absolute difference between MaxPos and MaxNeg	Similarity measure
g [25]	This is a concordance measure	Similarity measure

Table 9.2 Estimation of the reliability of Read-Across-based predictions by the levels of the similarity/dispersion measures [35]

Criteria	Dispersion/Similarity measures	Desired range
1	SD_activity (Euclidean)	≤0.75
2	*g* (Euclidean)	≤0.4[a]
3(a)	Average similarity (Euclidean)	≥0.85
3(b)	CV_similarity (Euclidean)	≥0.05

[a] Corresponds to PosFrac ≥0.8 or PosFrac ≤0.2

Recently, there has been a rise in the adoption of similarity-based approaches like Read-Across mainly for their simplicity and Accuracy of the predictions, as estimated from various external validation metrics. Read-Across has now become one of the most useful algorithms for data gap filling, especially in cases where there is a scarcity of experimental data, as this technique does not involve the development of a model, and thus, accurate predictions can be obtained based on similarity values. Luechtefeld et al. [24] in 2018 proposed the idea of Read-Across Structure–Activity Relationship which combines the Read-Across algorithm with the QSAR methodology. They adopted a machine learning approach and used the concept of MaxPos and MaxNeg (vide supra) to develop classification-based models. In a recent work [29], Banerjee and Roy have tried to club the advantages of QSAR modeling and Read-Across approach and derived a novel quantitative Read-Across Structure–Activity Relationship (q-RASAR) approach. The authors have also tried a variety of similarity and error-based descriptors in the generation of RASAR models unlike Luechtefeld et al., where they have only used the maximum similarity values with the positive and negative source compounds.

9.5 The RASAR Descriptor Calculator Tool from the DTC Laboratory

The RASAR Descriptor Calculator tool, developed by the DTC Laboratory, is a simple Java-based software application, which quickly computes the similarity and error-based measures for each particular compound that can be used for the development of RASAR models. We recommend to use this software only after successful Read-Across-based predictions with the optimized hyperparameters. These optimized hyperparameters are also taken as inputs in the RASAR descriptor calculator tool, based on which the RASAR descriptors are computed. During its execution, the tool asks the user for the similarity-based methods based on which it will compute the descriptors. The similarity-based measures which the user can select are the Euclidean Distance-based, the Gaussian Kernel-based and the Laplacian Kernel-based measures. If the user selects the Euclidean Distance-based approach, the tool asks for the number of similar training compounds the tool will consider and the threshold value of the distance. If the user selects the Gaussian Kernel Similarity-based approach, the tool asks for the σ value (an optimized hyperparameter), the number of similar training compounds and the threshold value for similarity. Selection of the Laplacian Kernel Similarity based approach prompts the tool to ask for the γ value (another optimizable hyperparameter), the number of similar training compounds and the threshold value for similarity. In each case, the program generates two output files, namely TESTsetfilename_Sort.xlsx and TESTsetfilename_RASAR_Descriptors.xlsx. The sort file contains the sorted values of the similarities for each query compound, with all the source compounds, according to the similarity measure specified by the user. The RASAR descriptor file contains the descriptors for the development of RASAR models. Apart from the previously discussed similarity and error-based uncertainty measures, this tool also computes five new measures. The first of these new measures is the product of the Banerjee-Roy coefficient and the average similarity of the close "*n*" source compounds (g_m × Avg.Sim), and the second is the product of the Banerjee-Roy coefficient and the Standard Deviation of the similarity values of the close "*n*" source compounds (g_m × SD_similarity). The other two measures include the average similarity of the compounds, constituting the close "*n*" neighbors having a response value greater than the threshold (Pos.Avg.Sim), and the average similarity of the close "*n*" source compounds having a response value lower than the threshold (Neg.Avg.Sim). Lastly, the RA_function is derived from Read-Across, which acts like a composite variable and contains all the information of the structural and physicochemical descriptors selected to perform Read-Across initially. The tool is freely available and can be downloaded from https://sites.google.com/jadavpuruniversity.in/dtc-lab-software/home.

9.5.1 Pre-Requisites for Using This Tool

9.5.1.1 System Specifications

Like the Read-Across tool, RASAR-Desc-Calc-v1.0 (*Note*: the current version is RASAR-Desc-Calc-v3.0.1) does not require exhaustive system resources, i.e., it can run on computers having standard memories for RAM and HDD/SSD. However, since it is a Java-based software tool, it is necessary that a particular system needs to have Java installed before running this tool. The link for downloading JDK has already been mentioned above.

9.5.1.2 Input File Specifications

The input file specifications for the RASAR Descriptor Calculator tool is the same as the Read-Across tool. The training and test set files must bear the extension of **.xlsx,** and in each of these files, the compound number constitutes the first column, the descriptors in subsequent columns and the observed response values at the last column.

9.5.2 Downloading and Execution of the Tool

- The software in the form of a **.zip** file has been made available at the DTC lab tools supplementary webpage (https://sites.google.com/jadavpuruniversity.in/dtc-lab-software/home)

- The user will need to download the.zip file and extract the contents. The folder consists of an executable **.jar** file, a library folder and sample training and test set files in.xlsx format.
- The training and test set files need to be placed inside this RASAR-Desc-Calc-v1.0 folder, i.e., the same folder which contains the.jar file.
- The user needs to double click on the executable.jar file and the program will be executed.
- The user needs to enter 1, 2, or 3 based on the similarity measure using which the user wants the descriptors to be calculated. If the user enters 1, the calculations will be based on the Euclidean Distance-based approach and the program will only take input of the number of close source compounds and the distance threshold. If the user enters 2, the calculations will be based on the Gaussian Kernel Similarity-based approach, which asks the user to enter the value for σ, the number of close source compounds and the similarity threshold. If the user enters 3, the calculation will be based on Laplacian Kernel-based Similarity, where the system asks for the value of γ, the number of close source compounds and the similarity threshold.
- The sorted similarity measures will be printed to TESTsetfilename_sort.xlsx, while the descriptors will be printed to TESTsetfilename_RASAR_Descriptors.xlsx.
- Using these descriptors, one can go for RASAR model development with or without involvement of the original structural and physicochemical descriptors.

9.5.3 *Analysis of the Output Files*

The above tool generates two different output files, namely TESTsetfilename_sort.xlsx, which prints the sorted similarity values of the source compounds according to the selected similarity-based measure, along with their observed response values, and TESTsetfilename_RASAR_Descriptors.xlsx, which prints the computed similarity and error measures which are used as RASAR descriptors for the generation of RASAR models.

9.5.4 *Application of the RASAR Descriptor Calculator Tool Developed by the DTC Laboratory*

Banerjee and Roy [29] have recently demonstrated the application of the q-RASAR approach taking a case study of androgen receptor binding affinity. With the novel idea of utilizing the similarity and error measures obtained from Read-Across-v4.1 (https://sites.google.com/jadavpuruniversity.in/dtc-lab-software/home) as descriptor values, the authors have clubbed the structural and physicochemical descriptors (obtained after feature selection) with the similarity and error measures. The total set of descriptors thus obtained was further subjected to feature selection, and various MLR models were generated based on the combination of these descriptors. Noise and intercorrelation among the descriptors were removed by generating Partial Least Squares models, and then, the prediction quality was further enhanced by consensus-based predictions.

The androgen receptor binding affinity data of various molecules on rats were collected from the Endocrine Disruptor Knowledge Base (EDKB) database (https://www.fda.gov/science-research/endocrine-disruptor-knowledge-base/accessing-ar-binding-dataset-androgen-receptor), and the descriptors were generated. The intercorrelated descriptors were removed by the technique of data pre-treatment. After removal of the intercorrelated descriptors, due to the absence of a true external set, the dataset was divided into training and test sets based on a certain pre-defined algorithm (Euclidean Distance-based division). Thereafter, feature selection algorithms like Genetic Algorithm-MLR and Best Subset Selection were employed, and consequently, the authors obtained a certain set of structural and physicochemical descriptors which were believed to contribute significantly toward the prediction of the androgen receptor binding affinity. Using these selected descriptors, the training set was further divided into sub-training and sub-test sets, and a series of Read-Across-based predictions were obtained, changing the hyperparameters in each case. The set of hyperparameters, which provided the best predictions for the sub-test set, in terms of the external validation metrics and the MAE, were used to perform Read-Across-based predictions for the original test set (query set), using the original training set compounds (source compounds). Although the initial concept of Read-Across demonstrates an unsupervised learning approach, the process of optimization of the hyperparameters in this work can be considered a supervised learning approach. The final output file generated from Read-Across-v4.1 had the predicted response values, the external validation metrics, the overall error measures (RMSEP and MAE) and the similarity and error measures for each query compound. The authors have utilized some of these similarity and error measures, namely SD_activity, CV_activity, average similarity, SD_similarity, MaxPos, MaxNeg and Abs(MaxPos-MaxNeg) (Table 9.1), and chosen them as the RASAR descriptors. These RASAR descriptors were then clubbed to the previously selected structural and physicochemical descriptors to obtain the descriptor pool. The complete descriptor pool was then subjected to feature selection using the Best Subset Selection technique—an approach which generates multiple MLR models from all possible combinations of descriptors. The best four MLR models were chosen based on their internal and external validation metrics, and four individual PLS models were developed. Also, three pooled PLS models were developed which consisted of the pooled descriptor combinations from the individual PLS models. To enhance the prediction quality, the authors have utilized an intelligent consensus-based prediction technique [30]. The consensus model 3, which utilizes the best prediction of a particular compound from a selected model out of all the available PLS models, showed the best prediction quality. Figure 9.6 demonstrates the workflow followed by Banerjee and Roy in their work.

The authors have then observed that the concordance measure *g*, proposed by Wu et al. [25], is not able to distinguish between the positive and the negative query

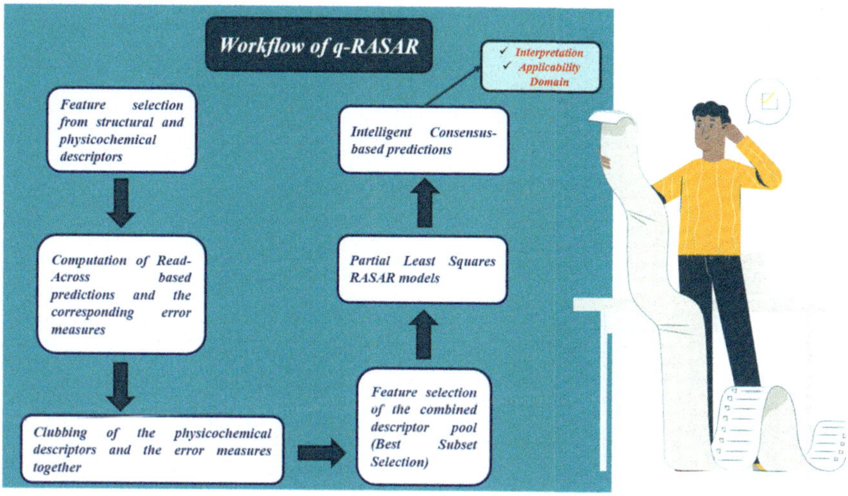

Fig. 9.6 Workflow of the q-RASAR methodology [29]

compounds, since it only takes the PosFrac into account (however, please note g was used in the original work in a different context). Therefore, according to the formula of g, a compound having PosFrac of 0.6 has the same value of g with another compound having a PosFrac of 0.4. Likewise, identical values of g can also be obtained in cases where PosFrac is 0.3 and 0.7, 0.2 and 0.8, 0.1 and 0.9, 0 and 1. Also, a compound may have a higher PosFrac value, but may possess a higher level of similarity to the negative source compounds and vice versa. To address all these aspects, the authors have developed novel Banerjee-Roy coefficient (g_m) which utilizes both the PosFrac and the MaxPos/MaxNeg values. Equation 9.5 denotes the mathematical form of this novel coefficient. The value of n in the equation is equal to 1, when MaxNeg > MaxPos, while $n = 2$ when MaxNeg < MaxPos. This incorporates a directionality among the query compounds as the compound which is more likely to belong to the negative class has a negative value of g, while a compound which has a tendency to be positive has a positive value of g. Using this modified g (g_m), the authors have re-developed one of the pooled PLS models, and it was observed that both the internal and external validation metrics obtained in this model were better than all the previous QSAR models and Read-Across (only external validation). The Mean Absolute Error obtained was even lower than what the authors have obtained in the previous consensus-based model. The summary of the internal and external validation metrics has been tabulated in Table 9.3.

Table 9.3 Summary of the internal and external validation metrics obtained in our work and their comparison with previous works (the best values of different metrics are shown in bold) [29]

PLS model(s)	LVs	R^2	$Q^2_{(LOO)}$	Q^2_{F1}	Q^2_{F2}	Q^2_{F3}	MAE -Fitted$_{(Train)}$	MAE -LOO$_{(Train)}$	$MAE_{(Test)}$
Pooled descriptor models									
P1 (M1 + M2)	3	0.718	0.666	**0.671**	**0.670**	**0.683**	0.479	0.517	0.478
P2 (M1 + M2 + M3)	2	**0.754**	**0.718**	0.630	0.629	0.644	**0.441**	**0.470**	0.504
P1m (Using g_m)	**4**	**0.753**	**0.698**	**0.674**	**0.674**	**0.686**	**0.434**	**0.477**	**0.461**
Intelligent consensus model									
ICP3 (M1 + M2 + M3 + M4) (CM3)	–	–	–	0.652	0.652	0.665	–	–	**0.463**
Previous 2D-QSAR model and Read-Across predictions (Banerjee et al. [34])									
2D-QSAR	3	0.737	0.680	0.582	0.582	0.606	0.456	0.497	0.539
Quantitative read-across (Gaussian Kernel Similarity-based)	–	–	–	0.635	0.635	0.656	–	–	0.468
Previous works done by other researchers									
3D-QSAR (CoMFA) by Hong et al. [37] *n*Training = 146 *n*Test = 8	–	0.902	0.571	–	–	–	–	–	–
Classification-based QSAR by Piir et al. [38] *n*Training = 1688 *n*Test = 5273	–	–	–	–	–	–	–	–	–

9.6 Conclusion

In compliance with the regulatory authorities like EU-REACH, in silico approaches of activity/toxicity estimation have had a surge in its applications in various fields. The most important aspect of this approach is that it avoids animal testing, experimental and instrumental errors and also requires minimal time to obtain accurate and reliable results. 2D-QSAR involves simple, transferable and interpretable models, while higher dimensional QSAR approaches deal with the spatial arrangement of atoms and molecules and involve various other steps like conformational analysis and alignment, which require exhaustive system resources, and reproducibility is compromised. The recent trend is to follow similarity-based prediction techniques that do not require a model to predict the response values of the compounds constituting the external set. In cases, where the number of training data points is very limited, model-derived predictions are more likely to produce biased or erroneous results due to an insufficient degree of freedom, but the similarity-based approaches like Read-Across can still be able to generate reliable predictions. Thus, the similarity-based approaches like Read-Across are very useful in data gap filling. In our Read-Across tool (https://sites.google.com/jadavpuruniversity.in/dtc-lab-software/home), we provide the predictions based on the Euclidean Distance approach, the Gaussian Kernel Similarity approach and the Laplacian Kernel Similarity approach and also the external validation metrics in terms of Q^2_{F1} and Q^2_{F2} along with the overall error measures in terms of RMSEP and MAE. This tool also generates certain measures using which one can estimate the uncertainty in the Read-Across predictions if the observed response values of the query compounds are not available. Our Read-Across approach deals with the local similarities with the close source compounds for each of the query compound, and thus, one may expect a slightly better prediction for the query compounds. The concept of RASAR, which was already introduced by Luechtefeld et al. [24], brings together the concept of Read-Across and QSAR. They have adopted machine learning approaches to develop classification-based RASAR models. We at DTC Laboratory performed quantitative Read-Across Structure–Activity Relationship (q-RASAR) and generated quantitative predictions by combining the advantages of Read-Across and QSAR. It was observed that the predictions obtained from q-RASAR were better in terms of both the internal and external validation metrics than majority of the work that have been done previously on the androgen receptor binding affinity of endocrine disruptors. Thus, we have developed a novel RASAR descriptor calculator tool (https://sites.google.com/jadavpuruniversity.in/dtc-lab-software/home) for the quick and efficient calculation of similarity and error-based descriptors to develop q-RASAR models. The development of data fusion RASAR models and linking them with multiple Molecular Initiating Events (MIEs) and Adverse Outcome Pathways (AOPs) can potentially be a useful algorithm for future drug discovery and development. We also believe that the Read-Across and q-RASAR models have a lot of potential for bridging data gaps, and probably, they may prove to be the essential prediction tools for the future.

Acknowledgements AB thanks Jadavpur University, Kolkata, for a scholarship. KR thanks the Science and Engineering Research Board (SERB), New Delhi, for financial assistance under the MATRICS scheme (MTR/2019/000008).

References

1. Mech A, Rasmussen K, Jantunen P, Aicher L, Alessandrelli M, Bernauer U, Bleeker EAJ, Bouillard J, Fanghella PDP, Draisci R, Dusinska M, Encheva G, Flament G, Haase A, Handzhiyski Y, Herzberg F, Huwyler J, Jacobsen NR, Jeliazkov V, Jeliazkova N, Nymark P, Grafström R, Oomen AG, Polci ML, Sandström CRJ, Shivachev B, Stateva S, Tanasescu S, Tsekovska R, Wallin H, Wilks MF, Zellmer S, Apostolova MD (2019) Insights into possibilities for grouping and read-across for nanomaterials in EU chemicals legislation. Nanotoxicology 13(1):119–141. https://doi.org/10.1080/17435390.2018.1513092
2. Fischer I, Milton C, Wallace H (2020) Toxicity testing is evolving! Toxicol Res 9(2):67–80. https://doi.org/10.1093/toxres/tfaa011
3. Hemmerich J, Ecker FG (2020) In silico toxicology: From structure–activity relationships towards deep learning and adverse outcome pathways. WIRES Comp Mol Sci 10:e1475. https://doi.org/10.1002/wcms.1475
4. Gomes SIL Scott-Fordsmand JJ Amorim MJB (2021) Alternative test methods for (nano) materials hazards assessment: Challenges and recommendations for regulatory preparedness. Nano Today 40:101242. https://doi.org/10.1016/j.nantod.2021.101242
5. Nymark P, Bakker M, Dekkers S, Franken R, Fransman W, García-Bilbao A, Greco D, Gulumian M, Hadrup N, Halappanavar S, Hongisto V, Hougaard KS, Jensen KA, Kohonen P, Koivisto AJ, Maso MD, Oosterwijk T, Poikkimäki M, Rodriguez-Llopis I, Stierum R, Sørli JB, Grafström R (2020) Toward rigorous materials production: new approach methodologies have extensive potential to improve current safety assessment practices. Small 16:1904749. https://doi.org/10.1002/smll.201904749
6. Madden JC, Enoch SJ, Paini A, Cronin MTD (2020) A review of in silico tools as alternatives to animal testing: principles, resources and applications. Alt Lab Ani 48(4):146–172. https://doi.org/10.1177/0261192920965977
7. Gellatly N, Sewell F (2019) Regulatory acceptance of in silico approaches for the safety assessment of cosmetic-related substances. Comp Toxicol 11:82–89. https://doi.org/10.1016/j.comtox.2019.03.003
8. Mangiatordi GF, Alberga D, Altomare CD, Carotti A, Catto M, Cellamare S, Gadaleta D, Lattanzi G, Leonetti F, Pisani L, Stefanachi A, Trisciuzzi D, Nicolotti O (2016) Mind the gap! a journey towards computational toxicology. Mol Inf 35:294–308. https://doi.org/10.1002/minf.201501017
9. Kovarich S, Ceriani L, Gatnik MF, Bassan A, Pavan M (2019) Filling data gaps by read-across: a mini review on its application, developments and challenges. Mol Inf 38:1800121. https://doi.org/10.1002/minf.201800121
10. Hartung T (2016) Making big sense from big data in toxicology by read-across. ALTEX 33(2). https://doi.org/10.14573/altex.1603091
11. Maldonado AG, Doucet JP, Petitjean M, Fan B (2006) Molecular similarity and diversity in chemoinformatics: from theory to applications. Mol Divers 10:39–79. https://doi.org/10.1007/s11030-006-8697-1
12. Ball N, Madden J, Paini A, Mathea M, Palmer AD, Sperber S, Hartung T, van Ravenzwaay B (2020) Key read across framework components and biology based improvements. Mutat Res Genet Toxicol Environ 853:503172. https://doi.org/10.1016/j.mrgentox.2020.503172
13. Benfenati E, Chaudhry Q, Gini G, Dorne JL (2019) Integrating in silico models and read-across methods for predicting toxicity of chemicals: a step-wise strategy. Environ Int 131:105060. https://doi.org/10.1016/j.envint.2019.105060

14. Ellison CM, Enoch SJ, Cronin MTD (2011) A review of the use of in silico methods to predict the chemistry of molecular initiating events related to drug toxicity. Expert Opin Drug Metabol Toxicol 7(12):1481–1495. https://doi.org/10.1517/17425255.2011.629186
15. Enoch SJ, Cronin MTD, Schultz TW, Madden JC (2008) Quantitative and mechanistic read across for predicting the skin sensitization potential of alkenes acting via michael addition. Chem Res Toxicol 21:513–520. https://doi.org/10.1021/tx700322g
16. Schuurmann G, Ebert RU, Kuhne R (2011) Quantitative read-across for predicting the acute fish toxicity of organic compounds. Environ Sci Technol 45:4616–4622. https://doi.org/10.1021/es200361r
17. Kühne R, Ebert RU, von der Ohe PC, Ulrich N, Brack W, Schüürmann G (2013) Read-Across prediction of the acute toxicity of organic compounds toward the water flea daphnia magna. Mol Inf 32:108–120. https://doi.org/10.1002/minf.201200085
18. Russo DP, Strickland J, Karmaus AL, Wang W, Shende S, Hartung T, Aleksunes LM, Zhu H (2019) Nonanimal models for acute toxicity evaluations: applying data-driven profiling and read-across. Environ Health Pers 127(4):047001. https://doi.org/10.1289/EHP3614
19. Low Y, Sedykh A, Fourches D, Golbraikh A, Whelan M, Rusyn I, Tropsha A (2013) Integrative chemical-biological read-across approach for chemical hazard classification. Chem Res Toxicol 26:1199–1208. https://doi.org/10.1021/tx400110f
20. van Ravenzwaay B, Sperber S, Lemke O, Fabian E, Faulhammer F, Kamp H, Mellert W, Strauss V, Strigun A, Peter E, Spitzer M, Walk T (2016) Metabolomics as read-across tool: a case study with phenoxy herbicides. Regulat Toxicol Pharmacol 81:288–304. https://doi.org/10.1016/j.yrtph.2016.09.013
21. Przybylak KR, Schultz TW, Richarz AN, Mellor CL, Escher SE, Cronin MTD (2017) Read-across of 90-day rat oral repeated-dose toxicity: a case study for selected β-olefinic alcohols. Comp Toxicol 1:22–32. https://doi.org/10.1016/j.comtox.2016.11.001
22. Schultz TW, Richarz AN, Cronin MTD (2019) Assessing uncertainty in read-across: questions to evaluate toxicity predictions based on knowledge gained from case studies. Comp Toxicol 9:1–11. https://doi.org/10.1016/j.comtox.2018.10.003
23. Alves VM, Golbraikh A, Capuzzi SJ, Liu K, Lam WI, Korn DR, Pozefsky D, Andrade CH, Muratov EN, Tropsha A (2018) Multi-descriptor read across (MuDRA): a simple and transparent approach for developing accurate quantitative structure–activity relationship models. J Chem Inf Model 58(6):1214–1223. https://doi.org/10.1021/acs.jcim.8b00124
24. Luechtefeld T, Marsh D, Rowlands C, Hartung T (2018) Machine learning of toxicological big data enables read-across structure activity relationships (RASAR) outperforming animal test reproducibility. Toxicol Sci 165(1):198–212. https://doi.org/10.1093/toxsci/kfy152
25. Wu J, D'Ambrosi S, Ammann L, Stadnicka-Michalak J, Schirmer K, Baity-Jesi M (2022) Predicting chemical hazard across taxa through machine learning. Environ Int 163:107184. https://doi.org/10.1016/j.envint.2022.107184
26. AbdulHameed MDM, Liu R, Schyman P, Sachs D, Xu Z, Desai V, Wallqvist A (2021) ToxProfiler: toxicity-target profiler based on chemical similarity. Comp Toxicol 18:100162. https://doi.org/10.1016/j.comtox.2021.100162
27. Manganelli S, Benfenati E (2016) Use of read-across tools. In: Benfenati E (ed) In silico methods for predicting drug toxicity. Humana Press, pp 305–322. https://doi.org/10.1007/978-1-4939-3609-0_13
28. Chatterjee M, Banerjee A, De P, Gajewicz A, Roy K (2022) A novel quantitative read-across tool designed purposefully to fill the existing gaps in nanosafety data. Env Sci: Nano. 9:189–203. https://doi.org/10.1039/D1EN00725D
29. Banerjee A, Roy K (2022) First report of q-RASAR modeling towards an approach of easy interpretability and efficient transferability. Mol Divers 26(5):2847–2862. https://doi.org/10.1007/s11030-022-10478-6
30. Roy K, Ambure P, Kar S, Ojha PK (2018) Is it possible to improve the quality of predictions from an "intelligent" use of multiple QSAR/QSPR/QSTR models? J Chemom 32:e2992. https://doi.org/10.1002/cem.2992

31. Chatterjee M, Roy K (2022) Application of cross-validation strategies to avoid overestimation of performance of 2D-QSAR models for the prediction of aquatic toxicity of chemical mixtures. SAR QSAR Env Res 33(6):463–484. https://doi.org/10.1080/1062936X.2022.2081255
32. De P, Kumar V, Kar S, Roy K, Leszczynski J (2022) Repurposing FDA approved drugs as possible anti-SARS-CoV-2 medications using ligand-based computational approaches: sum of ranking difference-based model selection. Struc Chem. https://doi.org/10.1007/s11224-022-01975-3
33. Paul R, Chatterjee M, Roy K (2022) First report on soil ecotoxicity prediction against *Folsomia candida* using intelligent consensus predictions and chemical read-across. Env Sci Pollut Res. https://doi.org/10.1007/s11356-022-21937-w
34. Banerjee A, De P, Kumar V, Kar S, Roy K (2022) Quick and efficient quantitative predictions of androgen receptor binding affinity for screening endocrine disruptor chemicals using 2D-QSAR and chemical read-across. Chemosphere 309:136579. https://doi.org/10.1016/j.chemosphere.2022.136579
35. Banerjee A, Chatterjee M, De P, Roy K (2022) Quantitative predictions from chemical read-across and their confidence measures. Chemom Intell Lab Syst 227:104613. https://doi.org/10.1016/j.chemolab.2022.104613
36. Heberger K (2010) Sum of ranking differences compares methods or models fairly. TrAC Trends Anal Chem 29(1):101–109. https://doi.org/10.1016/j.trac.2009.09.009
37. Hong H, Fang H, Xie Q, Perkins R, Sheehan DM, Tong W (2003) Comparative molecular field analysis (CoMFA) model using a large diverse set of natural, synthetic and environmental chemicals for binding to the androgen receptor. SAR QSAR Env Res 14(5–6):373–388. https://doi.org/10.1080/10629360310001623962
38. Piir G, Sild S, Maran U (2021) Binary and multi-class classification for androgen receptor agonists, antagonists and binders. Chemosphere 262:128313. https://doi.org/10.1016/j.chemosphere.2020.128313

Chapter 10
Databases for Drug Discovery and Development

Supratik Kar and Jerzy Leszczynski

Abstract Computational drug design and discovery have taken center stage attention during the time of COVID-19. The science community acknowledges the importance of ligand-based drug design (LBDD) and structure-based drug design (SBDD) to nullify the problem associated with a typical drug discovery process. In the modern era, a complement between experimental, theoretical, and computational approaches can make the drug discovery process rational, economical, and fast. Undoubtedly, computational power has increased manifold compared to the last few decades, making it possible to run many unthinkable calculations that cannot be imagined a few years ago. Along with the computational power, resources like open-access and commercial organic chemicals, phytochemicals, approved, experimental and investigational drugs, peptides, and metabolomic databases have increased enormously. Compared to designing a new drug, utilizing existing chemical and drug databases for virtual screening makes the process faster as the database chemicals are already synthesized (in most cases) and characterized. Even in a few instances, absorption, distribution, metabolism, excretion, and toxicity (ADMET) profiles are checked along with data for preclinical and clinical trials (primarily for investigational and/or in the process of approval drugs). A drug database is also a powerful resource for drug repurposing, where an old, approved drug for a specific disease can be used to treat another common/new/rare disease. The idea is increasingly becoming an attractive proposition as it comprises the use of already evaluated de-risked compounds which help lower the new drug development costs in a shorter time. Therefore, drug databases have an immense role to play as a repository of potential drugs for any common to a rare disease in the process of CADD and for the experimental scientists.

S. Kar (✉)
Department of Chemistry, Chemometrics and Molecular Modeling Laboratory, Kean University, 1000 Morris Avenue, Union, NJ 07083, USA
e-mail: skar@kean.edu

J. Leszczynski
Department of Chemistry, Physics and Atmospheric Sciences, Interdisciplinary Center for Nanotoxicity, Jackson State University, Jackson, MS 39217, USA

S. Kar and J. Leszczynski (eds.), *Current Trends in Computational Modeling for Drug Discovery*, Challenges and Advances in Computational Chemistry and Physics 35,
https://doi.org/10.1007/978-3-031-33871-7_10

Keywords Database · Drug design · Drug discovery · Biological activity · Targets

10.1 Introduction

Computer-aided drug designing (CADD) combines multiple computational approaches, decreasing the drug discovery process timeline manifold. Computational and literature resources are the backbone of the CADD [1]. With the advancement of computational resources, one can perform unimaginable calculations which could not be considered a few decades ago. Parallelly, storage capacity of webservers and cloud storage helped us to have a single platform of databases where researchers can have access to millions to billions of chemicals irrespective of organic chemicals, approved, investigational and experimental drugs, phytochemicals, peptides as well as databases for metabolomics and therapeutics targets too [2–4].

Standard drug design to discovery process [5] is a multistep process which is illustrated in Fig. 10.1. The most distinct steps are shown here where in each step, a different form of database can be utilized which shows the importance of databases in drug discovery [6, 7]. Based on the discovery step, researcher has to select the best possible database or combination of multiple databases for the study.

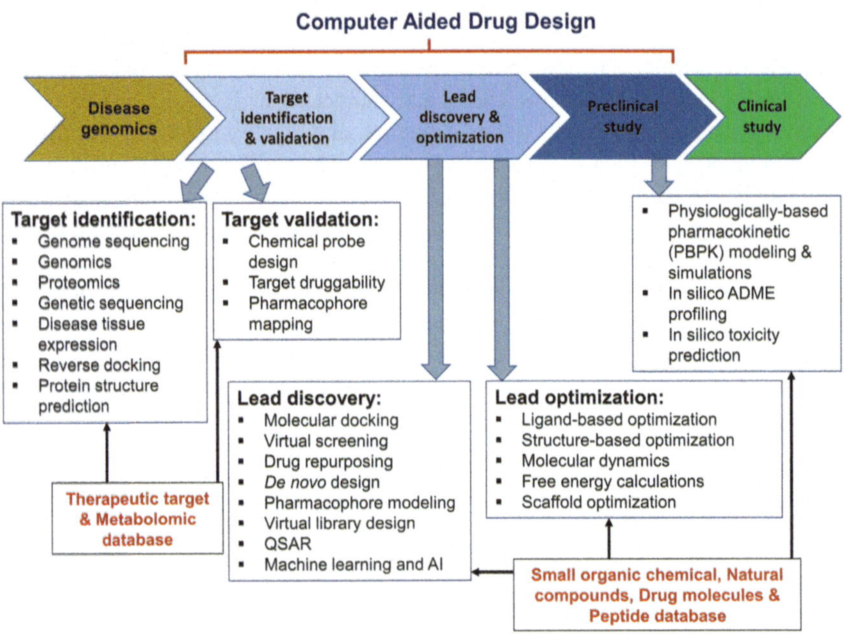

Fig. 10.1 Role of different databases in CADD

The databases are always not only limited to small organic molecules, but they can also be sometime peptides and biologicals molecules too. Databases are the key resources of any form off bioinformatics and cheminformatics project. These databases can be extremely big regarding the number of available molecules and result from text mining and automatic processes, whereas others can contain highly curated data. In recent time, a huge number of fairly large databases are freely available to use which has open more opportunists for open-access drug design and discovery process. The databases can be used entirely by downloading them or in many cases there is an option to search specific class and/or structural scaffolds from the big chunk of molecules. Therefore, depending on the requirements, a researcher can use the database for the study. A molecular cloud is represented in Fig. 10.2 using a small section of molecules (of SuperNatural II database only 150 from pool of 325,508 compounds) [8] to show the variety of structural scaffolds exist in a database.

Many times, these databases are directly connected with modeling and diverse analysis tools like docking, toxicity prediction, BLAST, etc. which help to utilize initial assessment of the database for the specific study. An ideal database should contain chemical structure information in form of SMILES/SDF, major physico-chemical properties, any form of experimental data related to activity and/or toxicity

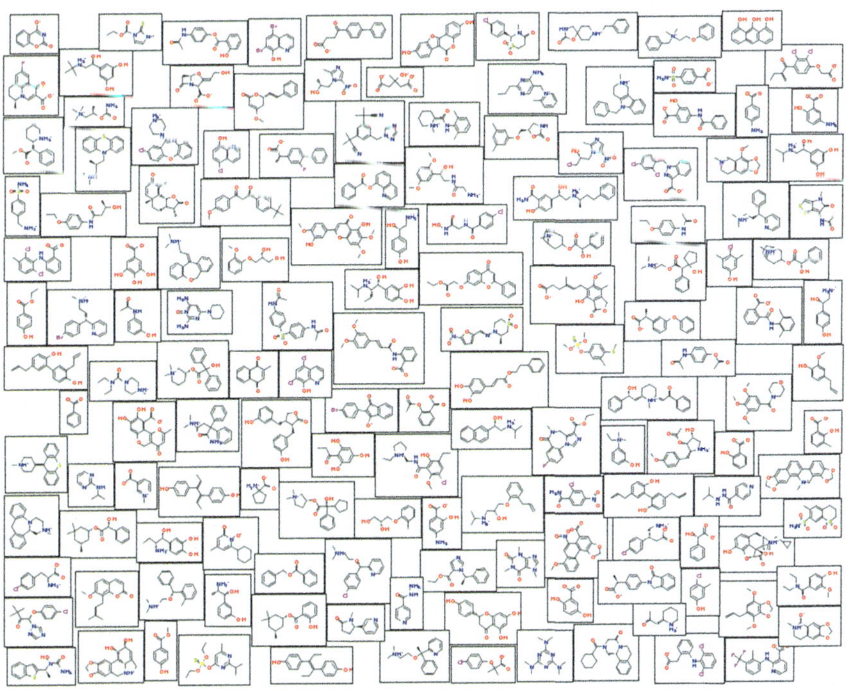

Fig. 10.2 Representation of a small section of chemicals from SuperNatural II database in form of molecular cloud

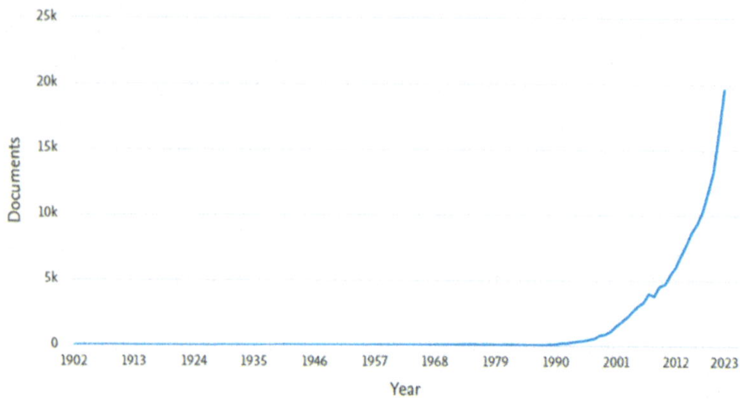

Fig. 10.3 Use of drug databases in published peer-reviewed research publications over the last 120 years according to Scopus

and potential vendors [9, 10]. If the molecules present in the database are already synthesized and characterized, it is good to have them in the study as if the molecule evolved as a potential drug, for experimental assay researcher can simply buy it from vendor for the further study which does not require any form of synthesis or initial characterization. A simple search of 'drug' and 'database' in Scopus can show immense growth of drug databases and their application in drug discovery where one can see sharp jump of the usage of databases from the year 2000 and it is reaching all time high at present time (Fig. 10.3).

10.2 Types of Databases for Drug Discovery

Drug databases can be classified according to the chemical nature of the drug molecule, disease-specific, target-oriented, and metabolomic pathways. Again, small chemical molecules can be classified as investigational and experimental drugs, while approved drugs are generally classified under drug molecules which are majorly used for drug repurposing. This chapter discusses databases for small chemical molecules, drug molecules, metabolomic, peptides, and therapeutic target information, as illustrated in Fig. 10.4. The researcher has to decide the types and requirement of the databases for the research. If researcher wants to do virtual screening, then going for multiple databases is a good option. Now, what types of databases one has to use thats completely depend on the requirement of the researcher [11–13]. On the other side, it is always better to take US FDA-approved drugs, investigational and experimental drugs under the radar of USFDA or any govt. approved agencies for drug repurposing purposes. Thus, selection of database is always important as this is the first step of CADD.

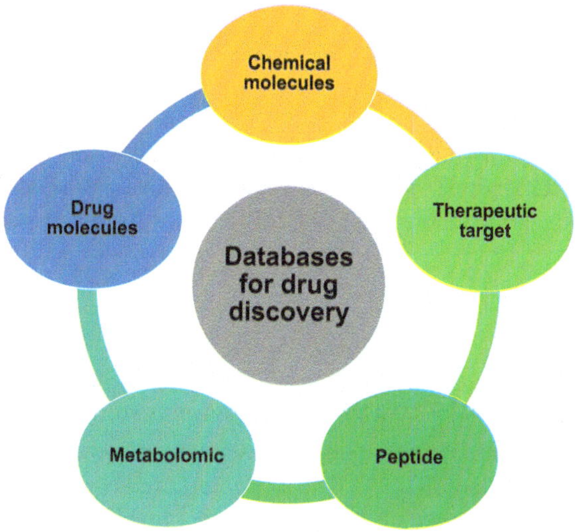

Fig. 10.4 Types of databases for drug discovery

10.3 Databases

10.3.1 Chemical Molecules Database

Chemical molecules databases are the essential and most powerful resources for virtual screening study. These databases can be used for ligand- and structure-based drug design and discovery, where quantitative structure–activity relationships (QSARs) [14] and machine learning models [15] can be strategically used for ligand-based drug discovery employing these databases. On the contrary, docking, pharmacophore, molecular dynamics [16–18] followed by ADMET profiling studies [19, 20] can be strategically used for virtual screening of chemicals databases. For better understanding, small chemical molecules databases are classified under two categories: a) Small organic chemicals databases and b) natural chemicals/compounds database.

10.3.1.1 Small Organic Chemicals

BindingDB

BindingDB is an open-access, web-accessible database of evaluated binding affinities, aiming principally on the interactions of protein believed to be drug targets with small, drug-like molecules [21, 22]. The database is accessible at https://www.bindingdb.org/bind/index.jsp. As of June 21, 2022, it consists of 41,296 entries containing 2,519,702 binding data for 1,080,101 small molecules and 8810 protein targets. There are 5988 protein–ligand crystal structures with BindingDB affinity measurements for proteins with 100% sequence identity, and 11,442 crystal structures allowing proteins to 85% sequence identity. BindingDB also presents the users a BLAST search page which allows input for an amino acid or nucleic acid sequence. Users can search the database based on target and compound. The user can refine their search for a target by giving ranges for experimental data collected for inhibitors like Ki, KD, EC50, ΔG, and pH. The database includes data extracted from the PubChemBioAssays, scientific reports, and ChEMBL entries for established targets.

ChEBI

ChEBI stands for 'Chemical Entities of Biological Interest,' a freely available database of 'small molecular entities' (any constitutionally or isotopically distinct molecule, atom, ion, ion pair, radical ion, radical, conformer, complex, etc.,), developed at the EBI [23]. ChEBI is referenced repository for molecular entities and their ontology based on small chemicals. ChEMBI is accessible at https://www.ebi.ac.uk/chebi/init.do. ChEBI is available from the EBI FTP site at https://ftp.ebi.ac.uk/pub/databases/chebi/. ChEBI can be downloaded in the following formats: SDF files, ontology files. Tools such as TopBraid, OWL-API, the NeOn toolkit and Protégé can be used with this ontology.

ChEMBL

ChEMBL is another prominent manually curated database of bioactive compounds with drug-like properties. The database contains chemical, bioactivity, and genomic data to aid the translation of genomic information into effective new drugs [24, 25]. ChEMBL is freely accessible at https://www.ebi.ac.uk/chembl/. The current release ChEMBL 30 contains 14,855 targets, 2,157,379 distinct compounds, around 14,000 drugs, 2000+ cells, 752 tissues, 6300+ mechanism, 19,286,751 activities, 84,092 publications as literature resources, and 194 deposited datasets. The database is available at https://www.ebi.ac.uk/chembl/ The ChEMBL offers multiple associated resources and tools which can be found in Table 10.1.

ChemDB

ChemDB is an open-source chemical database and consists of around 5 M commercially available small molecules [26]. The dataset is publicly accessible at http://cdb.ics.uci.edu/. These small molecules can be used as probes in systems biology, synthetic building blocks, and as leads for the drug discovery. Along with chemicals structure, it includes predicted or experimentally determined physicochemical properties, optimization of chemical structure. ChemDB primarily contains two types of datasets, namely chemical datasets and chemical reactivities datasets. The database includes a text-based search engine based on fuzzy text matching for searching of

Table 10.1 Resources and tools under ChEMBL

Resource/tools	Description
ADME SARfari	Tool for prediction and comparison of ADME targets
ChEBML-NTD	Primary screening and medicinal chemistry data of neglected tropical diseases. ChEMBL-NTD is maintained by EMBL-EBI at Hinxton in the United Kingdom
GPCR SARfari	Chemogenomics workbench for G protein-coupled receptor (GPCR)
Kinase SARfari	Chemogenomics workbench for kinases incorporating and linking kinase sequence
Malaria data	Compounds, targets, assays, and data for the malaria-related study
SureChEMBL	Publicly available database for patent information extracted from multiple patents documents and authorities
SARS-CoV-2 data	Contains 37 K activities, 8200+ compounds, 57 assays and 10 literature source
UniChem	UniChem is large-scale non-redundant database of pointers between chemical structures and EMBL-EBI chemistry resources. The cross-referencing between identifiers from different chemical databases

compounds based on over 65 M annotations from over 150 vendors. The built-in reaction models support searches through virtual chemical space, comprising of hypothetical compounds instantly synthesizable from the building blocks in ChemDB.

ChemSpider

ChemSpider is a publicly accessible chemical structure database https://www.chemspider.com/ offering structure search access to over 114 million chemical structures from 272 data sources [27]. Users can search chemical names based on systematic names, synonyms, trade names, and database identifiers. While chemical structure can be searched based on structure-based queries, drawing structure in the webserver, and using structure files from the computer, ChemSpider is maintained by the Royal Society of Chemistry.

Ligand Expo

Ligand Expo, previously known as Ligand Depot offers chemical and structural information of small molecules within the structure entries of the Protein Data Bank (PDB) [28]. The database provides tools to search the PDB for chemical components followed by identification of structure entries containing particular small molecules. Users can draw the new chemicals under sketch tool. The data is updated weekly. Access the data freely at http://ligand-depot.rutgers.edu/index.html. Data is available as per following formats: mmCIF, SDF/MOL, PDBML.

NCI Databases

National Cancer Institute databases [29] contain around 90 datasets. The database consists of 24 clinical, 23 epidemiological, 19 genomic, 14 imaging, 3 biological networks, and 3 patient registries databases. The databases are freely available under the following link: https://www.cancer.gov/research/resources/search?from=0&toolSubtypes=clinical_data&toolTypes=datasets_databases. Databases covering multitude information related to cancer treatment, biology, omics, screening and detection, cause and diagnosis, health disparities, prevention, public health, and overall statistics. Along with cancer, many datasets contain AIDS-related data too. For the information purpose, we have depicted clinical datasets and tools only, in Table 10.2 as this section is specific to chemical data only.

PubChem

PubChem is an open chemistry database at the National Institutes of Health (NIH) and maintained by National Center for Biotechnology Information (NCBI) [30]. One can freely access the data as well as deposit their scientific data which other users may use. PubChem is one of the most prominent resources for researchers, students, and scientists to avail the chemical information, structure of chemical compounds in diverse form like 2D/3D structure, 3D conformer, crystal structure information, IUPAC name/InChI/InChI key/Canonical SMILES along with multiple identifiers. Pubchem offers chemical and physical properties, spectral information, pharmacology and biochemistry, toxicity followed by multitude information associated with disorders, diseases, and biomolecular interactions and pathways. Users can access the PubChem site at https://pubchem.ncbi.nlm.nih.gov/. As per June 21, 2022, the PubChem data counts have been demonstrated in Table 10.3.

SuperLigands

SuperLigands is open-source database of ligand structures obtained from the PDB [31]. It combines knowledge about drug-likeness and binding properties of small molecules or ligands. The database provides ligands in the MDL Mol file format instead of the PDB format. Structural similarity can be estimated through the calculation of Tanimoto coefficients and by 3D superposition, while 2D similarity search can be performed by fingerprints. The database is an excellent source for drug discovery research.

Toxin and Toxin-Target Database (T3DB)

The T3DB (will be referred as the Toxic Exposome Database) is a distinctive source of toxin data with complete toxin target information [32]. The database consists of 3678 toxins including pollutants, pesticides, drugs, and food toxins, which are

Table 10.2 Clinical sub-databases and tools under NCI databases

Databases name	Description
AIDS Antiviral Screen Data	Checked tens of thousands of compounds for evidence of anti-HIV activity by the Developmental Therapeutics Program (DTP)
Cancer Data Access System (CDAS)	Submission and tracking system for data from the Prostate, the National Lung Screening Trial (NLST), Lung, Colorectal and Ovarian (PLCO) Cancer Screening Trial, and the Interactive Diet and Activity Tracking in AARP
CLAMP-Cancer	A tool used to build customized natural language processing pipelines for extracting cancer information from pathology reports through a user-friendly interface
Chemical Data	Compound sensitivity data for the NCI60 screen and similar screens run on small cell lung cancer cell lines and sarcoma cell lines, plus molecular target characterization data for the NCI60, sarcoma, and SCLC cell lines
Clinical Proteomic Tumor Analysis Consortium (CPTAC) Data Portal	CPTAC analyzes cancer biospecimens by mass spectrometry, characterizing and quantifying their constituent proteins, or proteome
Compound Inhibition Bulk Data	Contains compound sensitivity data for the NCI60 screen and similar screens run on sarcoma cell lines and small cell lung cancer cell lines
Extensible Neuroimaging Archive Toolkit (XNAT)	Imaging informatics platform designed to support institutional image repositories, image-based clinical trials, and translational imaging research
Genomic Data Commons (GDC)	A data sharing platform that offers harmonized genomic and de-identified clinical data from various large-scale cancer studies, along with tools for visualization and data analysis
Human Cancer Models Initiative (HCMI)	HCMI has two sections. Consent template and searchable catalog. Informed Consent Template is used for tissue accrual for cancer model development. A searchable catalog is for next-generation cancer models and associated clinical and molecular data
ivyGlimpse	A bioconductor-based interface to Ivy Glioblastoma Atlas Project (Ivy-GAP) data resources, allowing interactive selection of image features for scatter plotting, image sets for stratified survival distribution estimation, and gene sets for expression distribution comparison between strata
Molecular Target Data	The NCI60 lab will provide NCI60 cell line frozen cell pellets, DNA, or RNA for analysis in the external researchers' labs
NCI Brain Neoplasia Data	It integrates clinical and functional genomics data from clinical trials involving brain tumor patients and provides the ability to perform ad hoc querying, reporting, and analysis across multiple data domains, including gene expression, gene copy number, and clinical data

(continued)

Table 10.2 (continued)

Databases name	Description
Patient-Derived Xenograft (PDX) Finder	Tool for integrating, archiving, and disseminating information about PDX models and their associated data
TP53 Database	Contains TP53 variants, including functional/structural data, germline variants, somatic variants, cell lines, mouse models, and experimentally induced variants
Vizome	A data browser used to access molecular characterization and drug response studies of clinically annotated adult acute myeloid leukemia cases
Yeast Anticancer Drug Screen	Compound sensitivity data for the NCI60 screen and similar screens run on sarcoma cell lines and small cell lung cancer cell lines, plus molecular target characterization data for the NCI60, sarcoma, and SCLC cell lines

Table 10.3 PubChem data counts as per June 21, 2022

Data type	Counts	Description
Bioactivities	294,881,644	Biological activity data points reported in PubChem BioAssays
BioAssays	1,465,993	Biological experiments provided by PubChem contributors
Compounds	111,451,641	Unique chemical structures extracted from contributed PubChem Substance records
Data sources	862	Organizations contributing data to PubChem
Genes	103,628	Genes tested in PubChem BioAssays and those involved in PubChem Pathways and identified in PubChem Patents
Literature	34,208,642	Scientific publications with links in PubChem
Patents	41,796,860	Patents with links in PubChem
Pathways	238,908	Interactions between chemicals, genes, and proteins
Proteins	185,202	Proteins tested in PubChem BioAssays and those involved in PubChem Pathways and identified in PubChem Patents
Substances	279,279,532	Information about chemical entities provided by PubChem contributors
Taxonomy	112,603	Organisms of proteins/genes tested in PubChem BioAssays and those involved in PubChem Pathways and identified in PubChem Patents

linked to 2073 corresponding toxin targets. Totally, there are 42,374 toxin, toxin target associations. Each toxin record refers as ToxCard which comprises over 90 data fields and includes data like chemical properties and descriptors, cellular and molecular interactions, toxicity values, and medical information. The database is freely accessible at http://www.t3db.ca/ where records, data, structure and protein/gene sequences can be downloaded as XML, CSV/JASON, SDF and FASTA formats, respectively. The major aim of the T3DB is to offer precise mechanisms of toxicity and target proteins for every single toxin. T3DB is modelled after and carefully linked

Table 10.4 Resources available in the ZINC15 database

Resource	Contents	Approx. counts
Activities	Best ligand-gene affinity	220,000
Atc codes	Atc codes	1500
Catalogs	Vendor and annotated catalogs	400
Cat items	What vendors and annotated catalogs call the molecules in their source catalogs	1 billion
ecfp4s	Lala	100,000,000
Gene relations	Gene_relations	250,000
Genes	UniProt Gene Symbols	2800
Major classes	Major classes	15
Observations	Individual reports of ligand-gene associations	280,000
Organisms	asdf	5
Orthologs	UniProt accession codes, thus species specific	3800
Patterns	SMARTS patterns	535 patterns, 2.5 M entries
Predictions	SEA predictions	1 billion
Protomers	3D representations	6 million+
Rings	Ring systems	10,000
Subclasses	Subclasses	44
Substances	Molecules	200,000,000
Tool compounds	Tool compounds	3000

to the DrugBank and Human Metabolome Database (HMDB). Prospective purposes of T3DB consist of toxin/drug interaction prediction, toxin metabolism prediction, and general toxin hazard alertness by the people, creating it relevant to numerous fields.

Zinc

ZINC is an open-access database of commercially available compounds for virtual screening approach. ZINC is presently known as ZINC15 which contains over 230 million purchasable compounds in ready-to-dock accessible in 3D formats and also includes over 750 million purchasable compounds [33]. The database is accessible at https://zinc15.docking.org/. ZINC is created and maintained by the Irwin and Shoichet Laboratories in the Department of Pharmaceutical Chemistry at the University of California, San Francisco (UCSF). Users can freely download the data in different file formats, including mol2, SMILES, 3D SDF, and DOCK flexibase format. Searching, browsing, and molecular drawing interface facility are available on the ZINC database. ZINC15 is currently supporting multiple resources which are depicted in Table 10.4.

10.3.1.2 Natural Compounds Database

BIAdb

BIAdb is a compilation of benzylisoquinoline alkaloids (BIAs) database that stores information of around 846 unique BIAs where 196 entries from KEGG, 145 data from CTD, 171 entries from 171 and 334 data from other literature source [34]. The database can be downloaded from the following website at https://webs.iiitd.edu.in/raghava/biadb/. As BIAs have therapeutic properties, they can be a good resource for virtual screening to obtain potential lead molecules. Accessible natural alkaloids are produced by a range of organisms, like fungi, bacteria plants, and animals. The entire list of alkaloids underBIAdb is depicted in Table 10.5.

Dictionary of Natural Products Online

The Dictionary of Natural Products is an online resource for natural products [35]. It is resulting from a Dictionary of Organic Compounds (DOC), a repository of natural product and the data has been accumulated by a team at Chapman and Hall, UK. Comparable compounds are coordinated into a single entry streamlining the relationships of those strongly associated compounds. Compounds are indexed by their structural and biogenetic type. The dictionary is equipped with advanced search option with different properties like melting point and boiling point along with CAS, chemical and molecular formula.

Naturally Occurring Plant-Based Anticancer Compound-Activity-Target Database (NPACT)

NPACT is a compilation of plant-derived natural compounds showing anticancer activity under in vitro and in vivo experiments [36]. The present version contains around 1574 compound entries. The database offers chemical structure, data on in vitro and in vivo experiments along with inhibitory data like $ED_{50}/IC_{50}/GI_{50}/EC_{50}$ and physical, topological and elemental properties. User have also access to drug-likeness, target information, cancer types, references, and vendors information of respective compounds. NPACT can be a great starting point in the drug discovery of cancer.

SuperNatural II

SuperNatural II is an open-access database for natural products. It offers 325,508 natural compounds with information about the 2D structures, corresponding physicochemical properties, and predicted toxicity data [8]. Extreme chemically diverse natural products give enormous prospect for researchers to innovate new drug

Table 10.5 Types of alkaloids under BIAdb

Chemical class	Sub-class	Alkaloids name
Indole	Beta-carbolines	Harmine, harmaline, tetrahydroharmine
	Ergolines	Ergine, ergotamine, lysergic acid
	Mitragyna speciosa	Mitragynine, 7-hydroxymitragynine
	Strychnos nux-vomica	Strychnine, brucine
	Tabernanthe iboga	Ibogaine, voacangine, coronaridine
	Tryptamines	Serotonin, DMT, 5-MeO-DMT, bufotenine, psilocybin
	Vinca	Vinblastine, vincristine
	Yohimbans	Reserpine, yohimbine
Isoquinoline	Opium	Papaverine, narcotine, narceine
	–	Sanguinarine, hydrastine, berberine, emetine, berbamine, oxyacanthine
Phenanthrene	Opium	Morphine, codeine, thebaine
Phenethylamine	–	Mescaline, ephedrine, dopamine
Purine	Xanthines	Caffeine, theobromine, theophylline
Pyridine	–	Piperine, coniine
Pyrrolidine	–	Hygrine, cuscohygrine, nicotine
Quaternary ammonium		Muscarine, choline, neurine
Quinoline	–	Quinine, quinidine, dihydroquinine, dihydroquinidine, strychnine, brucine, veratrine, cevadine
Terpenoid	Aconite	Aconitine
	Steroid	Solanum alkaloids: solanidine, solanine, chaconine; Veratrum: veratramine, cyclopamine, cycloposine, jervine, muldamine; newt: samandarin
Tropane	–	Atropine, cocaine, ecgonine, scopolamine, catuabine
Miscellaneous	–	Capsaicin, cynarin, phytolaccine, phytolaccotoxin

discovery, nutritional products, cosmetics, and agrochemical research. The database is accessible at https://bioinf-applied.charite.de/supernatural_new/index.php?site=home. Users can search natural compounds based on properties, by name or by providing templates like amino acids, alpha sugars, D-sugars, aromatics, bases, bicycles, fused rings, heterocyclic rings.

10.3.2 Drug Molecules Database

10.3.2.1 DrugBank

DrugBank is a comprehensive, open-access, online database and consists of information on drugs and drug targets [11]. All drugs can be downloaded from https://go.drugbank.com/releases/latest. DrugBank is one of the most popular databases for drug design and discovery which contains detailed knowledge about drugs' chemical, pharmacological, and pharmaceutical data. The latest version of DrugBank (version 5.1.10, released 2023-01-04) consists of 15,321 drugs including 2734 approved small molecule drugs, 1572 approved biologics (peptides, proteins, vaccines, allergenics), 134 nutraceuticals, and over 6716 experimental (discovery-phase) drugs. Furthermore, 5294 non-redundant protein (i.e., drug target/enzyme/transporter/carrier) sequences are linked to these drug entries. Each item includes over 200 data fields with half of the information being devoted to drug/chemical data and the other half devoted to drug target or protein data. DrugBank also offers information related to pharmacological pathways, drug reactions, pharmacogenomics, metabolomics, transcriptomics, and proteomics. In recent time, DrugBank has created special dashboard for COVID-19-related drug information.

10.3.2.2 PharmGKB

Pharmacogenomics Knowledgebase (PharmGKB) offers data on the effect of human genetic variation on drug responses. PharmGKB is created by the NIH and managed at Stanford University. It is one of the partners of the NIH Pharmacogenomics Research Network (PGRN). The database contains clinical data, pharmacokinetics, and pharmacogenomics data in pulmonary, cancer, cardiovascular, and metabolic pathways domains. The database offers data on 832 drug label annotations, 201 curated pathways, 188 clinical guideline annotations, and 746 annotated drugs [37]. The drug labels comprising pharmacogenetic information approved by the US FDA, Swiss Agency of Therapeutic Products (Swissmedic), European Medicines Agency (EMA), Health Canada (Santé Canada) (HCSC), and Pharmaceuticals and Medical Devices Agency Japan (PMDA). All the available resources can be downloaded at free of charge from https://www.pharmgkb.org/downloads.

10.3.2.3 SuperDRUG2

SuperDRUG2 is a comprehensive knowledgebase of approved and marketed drugs [12]. The database offers around 4600 active pharmaceutical ingredients as per last release 2018.2.7. SuperDRUG2 annotated drugs with regulatory information, chemical structures, physicochemical properties, dosage, biological targets, side effects, and pharmacokinetic data. Users can search the chemical space of approved drugs through a different mechanism. It also offers a 2D chemical structure search on top of a 3D superposition feature that superposes a drug with ligands found in the protein–ligand complexes. The interaction check feature can detect possible drug–drug interactions, which includes alternate suggestions for geriatric patients. SuperDRUG2 can be accessed freely for academia and requires a free browser plugin called "Chime" for visualization.

10.3.3 Therapeutic Target Database

Proteins, enzymes, and nucleic acids are potential therapeutic targets for diseases. Therefore, binding interaction of small drug molecules to macromolecules like protein and/or protein–protein interactions is significant to understand developing new drug candidates for a specific disease. Understanding protein's structure and functions is essential to understand the pharmacological mechanism of small molecules binding to a specific protein. A therapeutic target database can offer more detailed information about different drug design and discovery targets. The most commonly employed therapeutic target databases are discussed below.

10.3.3.1 Herbal Ingredients Targets Database (HIT) 2.0

HIT 2.0 is developed based on the most updated curated database focusing on Herbal Ingredients' Protein Targets covering PubMed literature 2000–2020 and precursors for FDA-approved drugs [38]. HIT 2.0 hosts 10,031 ingredient-target activity pairs with quality indicators between 2208 biological targets and 1237 herbal ingredients from 1250 source herbs. The database has also consisted of 1231 therapeutic targets and 56 micro RNA targets. The molecular targets cover those genes/proteins that are directly/indirectly activated/inhibited, protein binders, and enzyme substrates or products. Also included are those genes regulated under the treatment of individual ingredients. HIT can be freely reachable at http://hit2.badd-cao.net.

HIT facilitates automated target-mining and My-target curation, where users can retrieve and download the latest abstracts containing potential targets for concerning herbs. The database contains molecular target information, which encompasses those proteins being activated or inhibited, protein binders, and enzymes whose substrates or products are interesting compounds. On the other hand, users can enter the 'My-target' curation system to curate the comprehensive ingredient-target relationship and create the latest individual targeting profiles.

10.3.3.2 Molecular Modeling Database (MMDB)

The MMDB is an open-access database that contains experimentally determined 3D biological macromolecular structures, including proteins and polynucleotides.

The database is maintained by the National Center for Biotechnology Information (NCBI), USA. MMDB can be freely accessed at http://www.ncbi.nlm.nih.gov/structure. It is linked to NCBI's Entrez search and retrieval systems, including contents of protein structures of PDB, PubMed, nucleotide and protein sequences, taxonomy, complete genomes, etc. [39]. MMDB offers accurate and pre-computed structural alignments obtained by the Vector Alignment Search Tool (VAST) (accessed at: http://www.ncbi.nlm.nih.gov/Structure/VAST/vastsearch.html) and also provides visualization tools for 3D structure and sequence alignment with molecular graphics tool Cn3D (available at: http://www.ncbi.nlm.nih.gov/Structure/CN3D/cn3d.shtml. CBLAST (accessed at: http://www.ncbi.nlm.nih.gov/Structure/cblast/cblast.cgi) is another web service that visualizes similarities between proteins in NCBI's Entrez database and those with known 3D structures tracked in MMDB.

10.3.3.3 Protein Data Bank (PDB)

The PDB is a freely accessible (https://www.rcsb.org/) archive for the three-dimensional (3D) structures of biological macromolecules like proteins, nucleic acids, and complex assemblies [13]. The obtained data is created through 3D structure elucidation techniques such as X-ray crystallography, NMR spectroscopy, or cryo-electron microscopy. Users can get the 3D structure of proteins and their complexes with other molecules with different resolutions, organisms, and expression systems at PDB. For docking and molecular dynamics study, PDB is the most commonly accessible site to obtain the target for structural biology and drug designing. Users cannot only access the X-ray crustal structure of PBD but also deposit their analyzed X-ray crystal structure. The PDB database consists of advanced 'Search,' 'Visualize,' and 'Analyze' tabs with many scientific analysis options. One can avail of the PDB site from the multiple websites of its member organizations like PDBe, RCSB, and PDBj. The PDB archive is maintained and organized by Worldwide Protein Data Bank (wwPDB) [40].

10.3.3.4 Therapeutic Target Database (TTD)

TTD is a database about the known therapeutic protein, nucleic acid targets, the targeted disease, pathway information, and the corresponding drug targets [41]. The database is open-access at http://db.idrblab.net/ttd/. The database is created and maintained by constructed by the Innovative Drug Research and Bioinformatics Group (IDRB) (Zhejiang University, China) and the Bioinformatics and Drug Design Group (BIDD) (National University of Singapore). Users can search the database based on 'Search for drugs,' 'Search Drugs and Targets by Disease or ICD Identifier,' 'Search for Biomarkers,' and 'Search for Drug Scaffolds.' TTD is well referenced to related databases accumulating information on sequence, target function, 3D structures, drug structure, therapeutic class, enzyme nomenclature, ligand properties, and category of clinical development. Multi-target agents have been studied to improve the safety profiles, therapeutic activity, and resistance by modulating the activity of a primary target. The recently updated database includes 1308 targets with 12,683 non-binders and 34,861 poor binders; (1) 1127 co-targets of 672 targets regulated by 642 approved and 624 clinical trial drugs; (2) 534 prodrug-drug pairs for 121 targets; (3) the profiles of drug-like properties of 33,598 agents' of 1102 targets; and (4) structure–activity landscapes of 427,262 active agents of 1565 targets [41]. The database offers further data and function, including cross-links to the target structure in PDB and AlphaFold, 159 and 1658 newly emerged targets and drugs.

10.3.3.5 Universal Protein Resource (UniProt)

The UniProt consortium is created by the European Bioinformatics Institute (EBI), the Swiss Institute of Bioinformatics (SIB), and the Protein Information Resource (PIR) [42]. It offers data on protein sequences and their functions where entries originated from genome sequencing projects. UniProt is freely accessible at https://www.uniprot.org/. The database for protein sequences is specifically known as UniProtKnowledgeBase (UniProtKB), which has two sections named UniProtKB/Swiss-Prot and UniProtKB/TrEMBL. The first consists of manually annotated entries curated from the literature and curator-evaluated computational analysis. In contrast, the latter one stores computer-annotated entries, which await complete manual annotation. In the UniProt database, sequence clusters, sequence archives, and proteome sets are available under UniRef, UniParc, and Proteomes tabs.

1. **UniRef**: UniRef offer clustered sets of sequences from the UniProtKnowledgeBase and selected UniParc records. UniRef hides superfluous sequences and gets comprehensive coverage of the sequence space at three different resolutions. UniRef100 contains identical sequences and sub-fragments with 11 or more residues from any organism into a single UniRef entry. UniRef90 is prepared by clustering UniRef100 sequences that have at least 90% sequence identity and 80% overlap with the longest sequence. UniRef50 is created by clustering UniRef90 seed sequences that have at least 50% sequence identity and 80% overlap with the longest sequence in the cluster.
2. **UniParc**: UniParc is a complete and non-redundant database that contains publicly available protein sequences. UniParc avoids duplicate copies of protein by storing each unique sequence only once and providing it a unique identifier (UPI). UniParc contains only protein sequences, while cross-referencing retrieves other information about the protein from the source databases.
3. **Proteomes**: A proteome is a set of proteins considered to be expressed by an organism. UniProt proteomes provide proteomes for species with completely sequenced genomes.

10.3.4 Peptide Database

10.3.4.1 Antimicrobial Peptide Database (APD)

The APD contains 3425 antimicrobial peptides (AMP) from six life kingdoms of natural sources [43]. The antimicrobial activities are demonstrated for the included peptides, and the activity range is either MIC < 100 μM or 100 μg/ml. The natural sources are the following: 2489 from animals, including some synthetic peptides, 385 isolated/predicted bacteriocins/peptide antibiotics from bacteria, 368 from plants, 25 from fungi, 8 from protists, and 5 from archaea). A complete list of 3425 AMPs under APD is listed in Table 10.6.

To avail of the database, one should visit https://aps.unmc.edu/. The database comprises a pipeline of search functions for innate immune peptides. One can search for peptide information utilizing APD ID, amino acid sequence, chemical modification, peptide name, peptide motif, length, hydrophobic content, charge, 3D structure, PDB ID, peptide source organism, methods for structural determination, peptide family name, life domain/kingdom, biological activity, target microbes, synergistic effects, molecular targets, mechanism of action, and publications details.

10.3.4.2 CAMPR3

The collection of antimicrobial peptides (CAMPR3) is a rich resource for antimicrobial peptide family-based information. Antimicrobial peptides are family-specific sequence compositions that can be used to design and discover novel AMPs [44]. The CAMPR3 comprises data on the preserved sequence signatures captured as patterns and hidden Markov models (HMMs) in 1386 AMPs characterized by 45 families. Information connected to protein definition, sequence, activity, accession numbers, target organisms, source organism, protein family descriptions is freely available. The database also offers pattern creation, sequence alignment, AMP prediction, pattern and HMM-based search tools. The database is also linked to databases like PubMed, UniProt, and other antimicrobial peptide databases. The site is accessible at: http://www.camp.bicnirrh.res.in/. A detailed, accessible data statistic is shown in Table 10.7.

10.3.4.3 CancerPPD

CancerPPD is a database that offers experimentally verified anticancer peptides (ACPs) and proteins. All the available data was mined manually from peer-reviewed literature and patents [45]. The database predicted tertiary structures of anticancer peptides employing the PEPstr method and secondary structure states are assigned using DSSP. One can browse the database based on protein, peptide, tissue, cell line,

Table 10.6 Antimicrobial activity, resource statistics of peptides under APD

Function information	Count
Antibacterial peptides	2891
Antiviral peptides	200
Antifungal peptides	1252
Anti-candida peptides	721
Antibiofilm peptides	75
Antiparasital peptides	140
Insecticidal peptides	41
Spermicidal peptides	14
Anti-HIV peptides	109
Anticancer (antitumor) peptides	264
Chemotactic peptides	67
Wound healing peptides	25
Antioxidant peptides	30
Enzyme/protease inhibitory peptides	33
Immobilized peptides	31
Anti-MRSA peptides	208
Antitoxin peptides	15
Channel inhibitors	7
Antiinflammatory peptides	32
Antidiabetic peptides	16
Anti-TB peptides	14
Antiendotoxin peptides	88
Two-chain peptides	35
Synergistic peptides	47
TOTAL	3425
Resources	Counts
Human host defense peptides	147
Active peptides from amphibians (frogs/toads)	1148 (1070/74)
Fish peptides	140
Reptile peptides	45
Mammals annotated	352
Birds	43
Molluscs	47
Protozoa	6
Insects	339
Crustaceans	73

(continued)

Table 10.6 (continued)

Resources	Counts
Myriapods	8
Spiders	44
Scorpions	93

Table 10.7 Statistics of the CAMPR3 database

Database types	Information
Sequences	Contains 8164 AMP sequences covering taxonomy of algae, amoebozoa, animalia, archaea, bacteria, fungi, heterolobosea, viridiplantae, virus, synthetic construct
Structures	757 AMP structures covering activity for antibacterial, antifungal, antiviral, unclassified
Patents	2083 patented AMPs are available
Signatures	36 patterns and 78 HMMs

assay, etc. The database is available at: http://crdd.osdd.net/raghava/cancerppd/index.php. The available total peptides, cell lines, and tissue types under CancerPPD are 3491, 249, and 21, respectively. Major features under this database are (1) data retrieval where data fetching and advanced search are possible along with search of peptides; (2) data analysis through multiple tools like BLAST, Smith-waterman, sequence and structure mapping followed by similarity-based search, (3) availability of ACPs SMILES and structures, (4) prediction of tertiary structures of all ACPs are also accessible. The CancerPPD also offers data connected to diverse chemical modifications like D-amino acids, non-natural, modified-amino acid like ornithine.

10.3.4.4 StraPep

Structure database of bioactive peptides (StraPep) is a dedicated database of bioactive peptides with known structures [46]. The present version of the database contains 3791 bioactive peptide structures and comprises 1312 unique bioactive peptide sequences. The StraPep is categorized into six functional groups counting antimicrobial peptide (404/833) with the distribution of 30.79%, toxin and venom peptide (464(unique sequences)/885 (structures)) covering 35.37%, cytokine and growth factor (217/901) having 16.54%, hormone (141/860) having 10.75%, neuropeptide (39/60) of 2.97%, and others (47/252) sharing 3.58%. The database is accessible at: http://isyslab.info/StraPep/. The peptides can be browsed based on classification, organism, disulfide bond, and cystine knot. The StraPep is also connected with tools like Blastp, Map, and Secondary Structure Composition search.

10.3.5 Metabolomic Database

10.3.5.1 Biochemical Genetic and Genomic (BiGG)

BiGG is a knowledgebase of large-scale biochemically, genetically and genomically structured genome-scale metabolic reconstructions under the constraint-based reconstruction and analysis (COBRA) framework which are valuable tools for evaluating the metabolic capacities of organisms and interpretation of experimental data [47]. BiGG is freely available for academic users at http://bigg.ucsd.edu/. BiGG can be utilized to browse model content, visualize metabolic pathway maps, and export SBML files of the models for additional assessment. Users may follow links from BiGG to several external databases to obtain additional information on proteins, genes, metabolites, reactions, and citations of interest. BiGG contains 9088 metabolites with specific ID and names. This database focuses on the need for systems biology scientists by delivering 75 genome-scale high-quality metabolic models under BiGG models [48].

10.3.5.2 BioCyc

BioCyc is a compilation of 20,005 pathway/genome databases (PGDBs) for model eukaryotes and thousands of microbes and tools for exploring them [49]. BioCyc is encyclopedic which contains curated data from 130,000 literatures. The BioCyc is available at https://biocyc.org/ but requires a subscription to use. Under BioCyc, 470 databases are for archaea, 19,416 for bacteria, 37 for eucaryota, and the remaining are for metabolic databases (named MataCyc). BioCyc incorporates knowledge from other bioinformatics databases, for instance, protein feature and Gene Ontology information from UniProt, gene-essentiality datasets from OGEE, and regulatory information from RegTransBase. BioCyc offers a suite of bioinformatics tools like Search across organisms and databases, visualization, genome browser, omics data analysis, SmartTable, metabolic route search, comparative analysis, sequence analysis, and pathway tools software. The database has three tier PGDBs where Tier 1 can be manually curated and frequently updated and include EcoCyc, HumanCyc, MetaCyc, AraCyc, YeastCyc, and the BioCyc Open Compounds Database (BOCD). Tier 2 is generated computationally by PathoLogic program used to predict their metabolic pathways, persons, and pathways hole filers. It has moderate manual updating. It contains 64 databases as per the present version. However, Tier 3 has 19,936 databases which were computationally generated and receive no manual updates.

10.3.5.3 Human Metabolome Database (HMDB)

The HMDB database contains information about small molecule metabolites in the human body [50]. The database is anticipated to be applied in clinical chemistry, metabolomics, and biomarker discovery. The database has three forms of data: (1) clinical data, (2) chemical data, and c) molecular biology/biochemistry data. The database is freely accessible at https://hmdb.ca/. The HMDB database is released every two years with monthly corrections and updates. The current Version (5.0) contains 220,945 metabolites and 8610 protein sequences, including enzymes and transporters linked to these metabolite entries. Metabolite structures are available in SDF format, protein and gene sequences are in FASTA format, and metabolite and protein data are in XML format. Individual MetaboCard entry contains 130 data fields, with two-thirds of the material being dedicated to chemical and clinical data and the other one-third committed to enzymatic or biochemical data. The HMDB endorses extensive text, chemical structure, sequence, NMR, and MS spectral query searches. Databases like T3DB, DrugBank, FooDB, and SMPDB are also part of the HMDB suite. Metabolites can browse by multiple filter options like metabolite status, biospecimen (saliva, blood, urine, feces, cerebrospinal fluids, breast milk, bile, sweat, amniotic fluids), origin (endogenous, exogenous, plant, food, microbials, toxins, cosmetics, drugs) and cellular location (cell membrane, cytoplasm, mitochondria, nucleus).

10.3.5.4 Kyoto Encyclopedia of Genes and Genomes (KEGG)

KEGG is an excellent resource for genomes, diseases, biological pathways, drugs, and chemicals, offering an understanding of high-level functions and utilities of biological systems [51]. It is a large-scale molecular database created by genome sequencing and other high-throughput experimental technologies. The current release of the database can be found at https://www.genome.jp/kegg/. KEGG offers many information on chemicals, genes, and genomes, followed by health information. It consists of a set of tools for diverse analysis, which can be found in Table 10.8.

10.3.5.5 MetaboLights Database

MetaboLights is a database for metabolomics experiments and derived information [52, 53]. The database is cross-platform, cross-species, cross-technique metabolomic research performed at the European Bioinformatics Institute (EMBL-EBI). The database offers experimental data from metabolomics experiments compliant with Metabolomics Standards Initiative (MSI). The database also offers metabolite structures, reference spectra, biological roles, locations, and concentrations. It has robust reporting capabilities and provides user-friendly submission tools. The database is accessible at https://www.ebi.ac.uk/metabolights/. Users can search data based on studies, compounds, and species.

Table 10.8 Information available under KEGG database

Database	Narration
Chemical information	
COMPOUND	A repository of small molecules, biopolymers, and other chemicals relevant to biological systems
ENZYME	Collection of information about enzyme nomenclature (EC number system) based on ExplorEnz database
GLYCAN	A compilation of experimentally determined glycan structures
REACTION	A depository of chemical reactions from KEGG metabolic pathway maps enzyme nomenclature
Genomic information	
GENES	A collection of gene catalogs from NCBI RefSeq and GenBank. The catalog contains 41,718,457 genes in KEGG organisms, 595,312 viral genes, 284 viral mature peptides, and 4106 addendum proteins
GENOME	A collection of organisms with complete genome sequences and selected viruses with relevance to diseases
ORTHOLOGY	A database of molecular functions as functional orthologs with the present tally of 25,221
Health information	
DRUG	Collection of approved drugs in the USA, Europe, and Japan
DISEASE	A collection of human disease entries aiming only on the perturbation basis
ENVIRON	A collection of health-promoting natural products of plants such as crude drugs, essential oils, etc
MEDICUS	An integrated resource of diseases, drugs, and health-related substances
NETWORK	To capture knowledge on diseases and drugs in terms of perturbed molecular networks
Organism information	
Organisms	Collection of complete genomes of 762 eukaryotes, 7043 bacteria, and 389 archaea
Systems information	
BRITE	A collection of hierarchical text (htext) files storing functional hierarchies of biological objects (KEGG objects). The functional hierarchies and reference (total) are 185 and 318,231, respectively
MODULE KEGG	Modules are manually defined functional units identified by the M numbers. These are used for annotation of sequenced genomes. The number of KEGG and reaction modules are 459 and 46, respectively
PATHWAY	Contains pathway maps with molecular interaction and reaction. The number of pathway maps and reference (total) is 551 and 937,937, respectively
Tools for analysis	
KEGG mapper	KEGG PATHWAY/BRITE/MODULE mapping tools
BlastKOALA	BLAST-based KO annotation and KEGG mapping
GhostKOALA	GHOSTX-based KO annotation and KEGG mapping

(continued)

Table 10.8 (continued)

Database	Narration
KofamKOALA	HMM profile-based KO annotation and KEGG mapping
BLAST/FASTA	Sequence similarity search
SIMCOMP	Chemical structure similarity search

10.3.5.6 Small Molecule Pathway Database (SMPDB)

SMPDB is an interactive and visual database comprising more than 30,000 small molecule pathways found in humans only [54, 55]. The database is unique as most of the included pathways are not accessible in any other pathway database. The main aim of this database is to support pathway discovery and its elucidation in proteomics, metabolomics, transcriptomics, and systems biology. It offers detailed information about each pathway, fully searchable, hyperlinked diagrams of human metabolic pathways, metabolite signaling pathways, metabolic disease pathways, and drug-action pathways. The database is accessible at https://www.smpdb.ca/. The most recent version is SMPDB v2.75. SMPDB pathways include knowledge of related organs, subcellular compartments, protein_complex locations, protein_complex cofactors, protein_complex quaternary structures, chemical structures, and metabolite locations. Gene, metabolite, and protein_complex concentration data can also be visualized through SMPDB's mapping interface. SMPDB's images, image maps, descriptions, and tables are downloadable.

10.3.5.7 WikiPathways

WikiPathways is an open-accessed collective platform for acquiring and distributing models of biological pathways for data visualization and analysis [56]. WikiPathways is accessible at http://www.wikipathways.org. It offers services to support pathway analysis and visualization via popular standalone tools, like PathVisio and Cytoscape, web applications, and standard programming environments. WikiPathways platform is also open to community participation (https://github.com/wikipathways). WikiPathways comprises over 2300 pathways across over 25 different species. The human pathway compilation is the biggest and most active collection by species, having expanded sixfold to include 640 pathways. In terms of coverage of unique human genes, WikiPathways is comparable to KEGG. Additionally, it consists of more than 640 pathways from humans encompassing more than 7500 genes and stores pathways with more than 1000 metabolites.

10.4 How to Select the Database for the Research?

Databases are critical resources for drug discovery using virtual screening and drug repurposing approaches. Most of the time, the researcher faces difficulty choosing the correct database for their research work. Therefore, a thorough understanding of the database's nature, type of information and data under the database, source of the data, accessibility of data in the form of download, as well as open-source or commercial nature needs to be understood. All drug databases have some pros and cons. One cannot mention that any specific database is perfect; each is unique on its own. We have identified five significant characteristics or features summarized in the user-friendliness score. The highest score is five (all five characteristics are present for an ideal database), and the lowest score is zero (no characteristic is present). The characteristics are discussed below:

1. **Updated**: The databases should have a registered website with regular updates. In many cases, we have seen after the first release of the database, they are not updated or maintained for several years. In many cases, even they have stopped the update. As data is changing daily, data updates and all respective information related to data need to be available and updated.
2. **Advanced search option**: Each researcher's requirement differs from one another. Therefore, an advanced search option with filters is one of the significant criteria. DrugBank is one of the ideal databases for the advanced search option where users can perform a multidimensional search based on chemical information, pharmacological aspects, metabolism, etc.
3. **Downloadable**: This is one of the significant characteristics as users need to use the database for further analysis, for example, virtual screening. Now, if a database is not downloadable and exportable in major acceptable formats, then the database is useless. Instead, the database is merely information provided without further user analysis.
4. **Classified**: The drug databases are large and contain multitude of information. Thus, classifying them into multiple categories can assist the researcher in obtaining the necessary data with a proper approach. Classification not only categorizes the data but also helps to narrow down the requirement of the researchers.
5. **Accessibility**: Drug discovery is a time-consuming and expensive process. Thus, in many cases, academics and independent researchers cannot have access to many commercial databases. They must rely on open-access/freely available drug databases for their research. Therefore, open-access databases not only help the researcher to gather enormous resources in no time without spending money. Once they know which drug they have to look for experimental analysis, they can only buy the small molecule/peptide directly from the supplier, either connected with the databases or from external resources.

Considering the discussed characteristics, major databases are plotted using a user-friendliness score in Fig. 10.5, which shows that CheMBL, DrugR+, DTC,

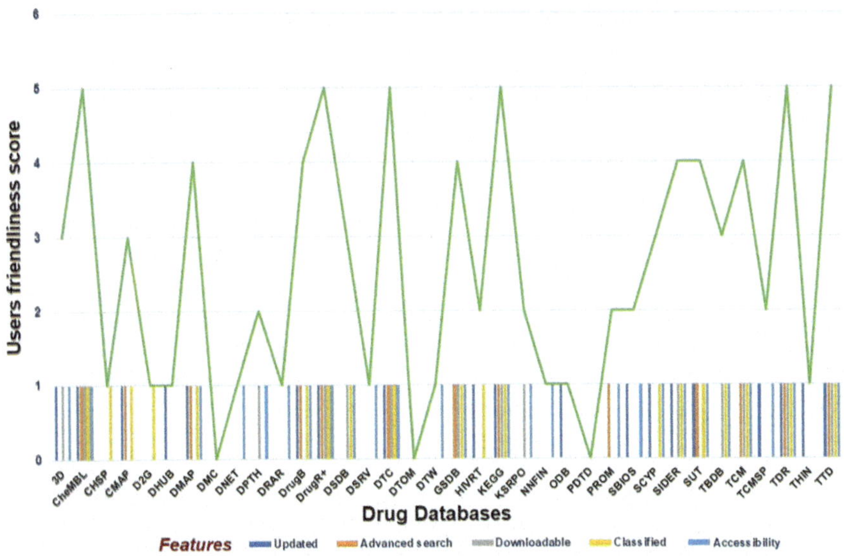

Fig. 10.5 User-friendliness of different drug databases. *Full form of the databases can be found in the abbreviation section

KEGG, TDR, and TTD scored 5. Here, it is essential to mention that we have only classified and scored based on our denoted features. Other databases also can be a better choice depending on the researcher's requirements. The main idea is to show how to choose databases for drug discovery only, not to identify the best database.

10.5 Overview and Conclusion

Computational drug discovery research is growing daily, with high-performing computational resources available. Therefore, with minimal time, the researcher can screen million to billion of compounds based on their studied targets and diseases. Drug repurposing is another essential method where an existing USFDA drug for a specific disease can be used to cure another disease with the help of ligand- and structure-based drug design and screening approaches. We have seen the use of remdesivir, a nucleotide analog prodrug initially developed for the treatment of the Ebola virus, was observed to inhibit the replication of coronaviruses in vitro and in preclinical studies. This classic example helps us understand that proper screening of existing approved drugs, drugs under clinical trials, drugs under investigation, drugs under experimental state, etc., can be beneficial to finding the cure for many diseases. In this perspective, this enormous number of databases consisting of small drug molecules, peptides, metabolites, and natural products can be used for the future mining of potential drug candidates for rare and neglected diseases with minimal time and money. Futuristically, integrating computational tools and databases can make the computer-aided drug design process more realistic as all the resources can be found in a single place and are more accessible. Another essential aspect that must be addressed for existing and future databases is the transparency of the data so that users can have high confidence to employ the data in their research.

Declaration of Competing Interest
The authors declare that they have no known competing financial interests or personal relationships that could have appeared to influence the work reported in this chapter.

Acknowledgements SK wants to thank the administration of Dorothy and George Hennings College of Science, Mathematics and Technology (HCSMT) of Kean University for providing research opportunities and resources.

Abbreviations

CHM	CheMBL
CHSP	Anticancer Herbs database for System Pharmacology;
CMAP	Complement Map Database
D2G	Drug to Gene
DHUB	Drug Repurposing Hub
DMAP	Drug-Protein Connectivity MAP
DMC	Drug Map Central
DNET	Drug-Disease Network Database
DPTH	Drug Pathway Database
DRAR	Drug Repurposing Adverse Reaction
DrugB	DrugBank
DSDB	Drug Signatures Database
DSRV	Drug Survival Database
DTC	Drug Target Commons
DTOM	Drug Target Interactome Database
DTW	Drug Target Web
GSDB	Gene Set Database
HIVRT	HIV Drug Resistance Database
KEGG	Kyoto Encyclopedia of Genes and Genomes
KSRPO	A Platform for Drug Repositioning
NNFIN	Network-based Similarity Finder
ODB	Ontario Database
PDTD	Potential Drug Target Database
PROM	Promiscuous
SBIOS	Swiss BIOisostere
SCYP	Super Cytochrome P450
SIDER	Side Effect Resource

SUT	SuperTarget database
TBDB	Tuberculosis Database
TCM	Traditional Chicness Medicine
TCMSP	Traditional Chinese Medicine Platform
TDR	Tropical Diseases Research
THIN	The Health Improvement Network
TTD	Therapeutic Target Database

References

1. Medina-Franco JL (2021) Grand challenges of computer-aided drug design: the road ahead. Front Drug Discov 1:728551
2. Mohs RC, Greig NH (2017) Drug discovery and development: Role of basic biological research. Alzheimers Dement 3:651–657
3. Wouters OJ, McKee M, Luyten J (2020) Estimated research and development investment needed to bring a new medicine to market, 2009–2018. JAMA 323:844–853
4. Jumper J, Evans R, Pritzel A et al (2021) Highly accurate protein structure prediction with AlphaFold. Nature 596:583–589
5. Tang Y, Zhu W, Chen K, Jiang H (2006) New technologies in computer-aided drug design: toward target identification and new chemical entity discovery. Drug Discov Today: Technol 3:307–313
6. Potemkin V, Potemkin A, Grishina M (2018) Internet resources for drug discovery and design. Curr Top Med Chem 18:1955–1975
7. Miller M (2002) Chemical database techniques in drug discovery. Nat Rev Drug Discov 1:220–227
8. Banerjee P, Erehman J, Gohlke BO, et al. (2015) Super natural II—A database of natural products. Nucleic Acids Res 43(Database):D935–D939
9. Gosh S, Kar S, Leszczynski J (2020) Ecotoxicity databases for QSAR modeling. In: Roy K (ed) Ecotoxicological QSARs. Humana, New York, pp 709–758
10. Kumar V, Roy K (2020) Development of a simple, interpretable and easily transferable QSAR model for quick screening antiviral databases in search of novel 3Clike protease (3CLpro) enzyme inhibitors against SARS-CoV diseases. SAR QSAR Env Res 31:511–526
11. Wishart DS, Feunang YD, Guo AC et al (2018) DrugBank 5.0: a major update to the DrugBank database for 2018. Nucleic Acids Res 46:D1074–D1082
12. Siramshetty VB, Eckert OA, Gohlke BO et al (2018) SuperDRUG2: a one stop resource for approved/marketed drugs. Nucleic Acids Res 46:D1137–D1143
13. Berman HM, Westbrook J, Feng Z et al (2000) The protein data bank. Nucleic Acids Res 28:235–242
14. Roy K, Kar S, Das RN (2015) Understanding the basics of QSAR for applications in pharmaceutical sciences and risk assessment. Academic Press
15. Kar S, Leszczynski L (2021) QSAR and machine learning modeling of toxicity of nanomaterials: a risk assessment approach. In: Njuguna J, Pielichowski K, Zhu H (eds) Health and environmental safety of nanomaterials. Woodhead Publishing, pp 417–441
16. Ojha PK, Mitra I, Kar S, Das RN, Roy K (2012) Lead hopping for PfDHODH inhibitors as antimalarials based on pharmacophore mapping, molecular docking and comparative binding energy analysis (COMBINE): a three-layered virtual screening approach. Mol Inform 31:711–718
17. Kumar V, Kar S, De P, Roy K, Leszczynski J (2022) Identification of potential antivirals against 3CLpro enzyme for the treatment of SARS-CoV-2: a multistep virtual screening study. SAR QSAR Env Res 33:357–386

18. Kar S, Roy K (2013) Prediction of milk/plasma concentration ratios of drugs and environmental pollutants using in silico tools: classification and regression based QSARs and pharmacophore mapping. Mol Inform 32:693–705
19. Kar S, Leszczynski L (2020) Open access in silico tools to predict the ADMET profiling of drug candidates. Expert Opin Drug Discov 15:1473–1487
20. Kar S, Roy K, Leszczynski L (2020) In silico tools and software to predict ADMET of new drug candidates. In: Benfenati E (ed) In Silico Methods for Predicting Drug Toxicity. Humana, New York, pp 85–115
21. Gilson MK, Liu T, Baitaluk M, Nicola G, Hwang L, Chong J (2016) BindingDB in 2015: A public database for medicinal chemistry, computational chemistry and systems pharmacology. Nucleic Acids Res 44:D1045–D1063
22. Chen X, Liu M, Gilson MK (2001) Binding DB: a web-accessible molecular recognition database. Combi Chem High-Throughput Screen 4:719–725
23. de Matos P, Alcántara R, Dekker A, et al (2010) Chemical entities of biological interest: an update. Nucleic Acids Res 38(Database):D249–D254
24. Mendez D, Gaulton A, Bento AP et al (2019) ChEMBL: towards direct deposition of bioassay data. Nucleic Acids Res 47:D930–D940
25. Davies M, Nowotka M, Papadatos G et al (2015) ChEMBL web services: streamlining access to drug discovery data and utilities. Nucleic Acids Res 43:W612–W620
26. Chen JH, Linstead E, Swamidass SJ, Wang D, Baldi P (2007) ChemDB update-full-text search and virtual chemical space. Bioinformatics 23:2348–2351
27. Pence HE, Williams A (2010) ChemSpider: an online chemical information resource. J Chem Educ 87:1123–1124
28. Feng Z, Chen L, Maddula H, Akcan O, Oughtred R et al (2004) Depot: a data warehouse for ligands bound to macromolecules. Bioinformatics 20(13):2153–2155
29. National Cancer Institute, Washington, DC (1997). http://rex.nci.nih.gov. Accessed on 15 Oct 2022
30. Kaiser J (2005) Science resources. Chemists want NIH to curtail database. Science 308(5723):774
31. Michalsky E, Dunkel M, Goede A, Preissner R (2005) SuperLigands—A database of ligand structures derived from the Protein Data Bank. BMC Bioinformatics 6:122
32. Wishart D, Arndt D, Pon A, (2015) T3DB: the toxic exposome database. Nucleic Acids Res 43(Database issue):D928–D934.
33. Sterling T, Irwin JI (2015) ZINC 15—Ligand discovery for everyone. J Chem Inf Model 55:2324–2337
34. Singla D, Sharma A, Kaur J, Panwar B, Raghava GP (2010) BIAdb: a curated database of benzylisoquinoline alkaloids. BMC Pharmacol 10:4
35. Dictionary of natural products online. http://dnp.chemnetbase.com. Accessed on 15 Oct 2022
36. Mangal M, Sagar P, Singh H, Raghava GP, Agarwal SM (2013) NPACT: naturally occurring plant based anti-cancer compound-activity-target database. Nucleic Acids Res 41(Database):D1124–D1129
37. McDonagh EM, Whirl-Carrillo M, Garten Y, Altman RB, Klein TE (2011) From pharmacogenomic knowledge acquisition to clinical applications: the PharmGKB as a clinical pharmacogenomic biomarker resource. Biomark Med 5(6):795–806
38. Yan D, Zheng G, Wang C et al (2020) HIT 2.0: an enhanced platform for Herbal Ingredients' targets. Nucleic Acids Res 50:D1238–D1243
39. Madej T, Addess KJ, Fong JH et al (2012) MMDB: 3D structures and macromolecular interactions. Nucleic Acids Res 40(Database):D461–D464
40. Berman H, Henrick K, Nakamura H, Markley JL (2007) The worldwide Protein Data Bank (wwPDB): ensuring a single, uniform archive of PDB data. Nucleic Acids Res 35(Database):D301–D303.
41. Zhou Y, Zhang YT, Lian XC et al (2022) Therapeutic target database update 2022: facilitating drug discovery with enriched comparative data of targeted agents. Nucleic Acids Res 50(D1):1398–1407

42. The, UniProt, Consortium (2021) UniProt: the universal protein knowledgebase in 2021. Nucleic Acids Res 49:D480–D489
43. Wang G, Li X, Wang Z (2016) APD3: the antimicrobial peptide database as a tool for research and education. Nucleic Acids Res 44:D1087–D1093
44. Waghu FH, Idicula-Thomas S (2020) Collection of antimicrobial peptides database and its derivatives: Applications and beyond. Protein Sci 29(1):36–42
45. Tyagi A, Tuknait A, Anand P et al (2015) CancerPPD: a database of anticancer peptides and proteins. Nucleic Acids Res 43(Database issue):D837–D843
46. Wang J, Yin T, Xiao X, He D, Xue Z, Jiang X, Wang Y (2018) StraPep: a structure database of bioactive peptides. Database 2018:bay038
47. Schellenberger J, Park JO, Conrad TM et al (2010) BiGG: a biochemical genetic and genomic knowledgebase of large scale metabolic reconstructions. BMC Bioinf 11:213
48. King ZA, Lu JS, Dräger A et al (2016) BiGG models: a platform for integrating, standardizing, and sharing genome-scale models. Nucleic Acids Res 44(D1):D515–D522
49. Karp PD, Billington R, Caspi R et al (2019) The BioCyc collection of microbial genomes and metabolic pathways. Brief Bioinform 20:1085–1093
50. Wishart DS, Guo AC, Oler E, et al. (2022) HMDB 5.0: the Human Metabolome Database for 2022. Nucleic Acids Res 50(D1):D622–D631
51. Kanehisa M, Goto S (2000) KEGG: Kyoto encyclopedia of genes and genomes. Nucleic Acids Res 28:27–30
52. Kenneth Haug K, Keeva Cochrane K, Venkata Chandrasekhar Nainala VC et al (2020) MetaboLights: a resource evolving in response to the needs of its scientific community. Nucleic Acids Res 48(D1):D440–D444
53. Kale NS, Haug K, Conesa P et al. (2016) MetaboLights: an open-access database repository for metabolomics data. Curr Protoc Bioinf 53:14.13.1–14.13.18
54. Wishart DS, Frolkis A, Knox C et al. (2010) SMPDB: the small molecule pathway database. Nucleic Acids Res 38(Database issue):D480–D487
55. Jewison T, Su Y, Disfany FM, et al. (2014) SMPDB 2.0: Big improvements to the small molecule pathway database. Nucleic Acids Res 42(Database issue):D478–D484
56. Kutmon M, Riutta A, Nunes N et al (2016) WikiPathways: capturing the full diversity of pathway knowledge. Nucleic Acids Res 44(D1):D488–D494

Index

S. Kar and J. Leszczynski (eds.), *Current Trends in Computational Modeling for Drug Discovery*, Challenges and Advances in Computational Chemistry and Physics 35,
https://doi.org/10.1007/978-3-031-33871-7

The manufacturer's authorised representative in the EU is Springer Nature Customer Service Centre GmbH, Europaplatz 3, 69115 Heidelberg, Germany. If you have any concerns regarding our products, please contact ProductSafety@springernature.com

Printed and bound by CPI Group (UK) Ltd, Croydon, CR0 4YY
28/09/2026
02224078-0004